GRAPH THEORY
AN ALGORITHMIC APPROACH

Computer Science and Applied Mathematics

A SERIES OF MONOGRAPHS AND TEXTBOOKS

Editor
Werner Rheinboldt
University of Maryland

HANS P. KÜNZI, H. G. TZSCHACH, and C. A. ZEHNDER. Numerical Methods of Mathematical Optimization: With ALGOL and FORTRAN Programs, Corrected and Augmented Edition

AZRIEL ROSENFELD. Picture Processing by Computer

JAMES ORTEGA AND WERNER RHEINBOLDT. Iterative Solution of Nonlinear Equations in Several Variables

AZARIA PAZ. Introduction to Probabilistic Automata

DAVID YOUNG. Iterative Solution of Large Linear Systems

ANN YASUHARA. Recursive Function Theory and Logic

JAMES M. ORTEGA. Numerical Analysis: A Second Course

G. W. STEWART. Introduction to Matrix Computations

CHIN-LIANG CHANG AND RICHARD CHAR-TUNG LEE. Symbolic Logic and Mechanical Theorem Proving

C. C. GOTLIEB AND A. BORODIN. Social Issues in Computing

ERWIN ENGELER. Introduction to the Theory of Computation

F. W. J. OLVER. Asymptotics and Special Functions

DIONYSIOS C. TSICHRITZIS AND PHILIP A. BERNSTEIN. Operating Systems

ROBERT R. KORFHAGE. Discrete Computational Structures

PHILIP J. DAVIS AND PHILIP RABINOWITZ. Methods of Numerical Integration

A. T. BERZTISS. Data Structures: Theory and Practice, Second Edition (in preparation)

N. CHRISTOPHIDES. Graph Theory: An Algorithmic Approach

In preparation
ALBERT NIJENHUIS AND H. S. WILF. Combinatorial Algorithms

GRAPH THEORY

An Algorithmic Approach

Nicos Christofides

Management Science
Imperial College
London

1975

ACADEMIC PRESS
New York   London
San Francisco

A Subsidiary of Harcourt Brace Jovanovich, Publishers

ACADEMIC PRESS INC. (LONDON) LTD.
24/28 Oval Road.
London NW1

United States Edition published by
ACADEMIC PRESS INC.
111 Fifth Avenue
New York, New York 10003

Library of Congress Catalog Card Number: 74–5664
ISBN—0 12 174350 0

Set in 'Monophoto' Times and printed offset litho in Great Britain by
Page Bros (Norwich) Ltd. Norwich

Preface

It is often helpful and visually appealing, to depict some situation which is of interest by a graphical figure consisting of points (vertices)—representing entities—and lines (links) joining certain pairs of these vertices and representing relationships between them. Such figures are known by the general name *graphs* and this book is devoted to their study. Graphs are met with everywhere under different names: "structures" in civil engineering, "networks" in electrical engineering, "sociograms", "communication structures" and "organizational structures" in sociology and economics, "molecular structure" in chemistry, "road maps", gas or electricity "distribution networks" and so on.

Because of its wide applicability, the study of graph theory has been expanding at a very rapid rate during recent years; a major factor in this growth being the development of large and fast computing machines. The direct and detailed representation of practical systems, such as distribution or telecommunication networks, leads to graphs of large size whose successful analysis depends as much on the existence of "good" algorithms as on the availability of fast computers. In view of this, the present book concentrates on the development and exposition of algorithms for the analysis of graphs, although frequent mention of application areas is made in order to keep the text as closely related to practical problem-solving as possible. By so doing, it is hoped that the reader will be left in a position to relate and adapt the basic concepts to his own field of application, and, indeed, be able to derive new methods of solution to his specific problem.

Although, in general, algorithmic efficiency is considered of prime importance, the present text is not meant to be a handbook of efficient algorithms. Often a method is discussed because of its close relation to (or derivation from) previously introduced concepts, in preference to another algorithm which may be equally—and in some cases slightly more—efficient. The overriding consideration is to leave the reader with as coherent a body of knowledge with regard to graph analysis algorithms, as possible.

The title *Graph Theory* must, to some extent, be a misnomer on any single volume, since it is quite impossible to cover even remotely the subject in such a short space. The present book is no exception and its contents reflect, as they must, the author's interest and background in Operations Research, Computer and Management Science.

Chapter 2 discusses basic reachability and connectivity properties of graphs, the computation of strong components and bases and their application to the formation of power-groups and coalitions in organizations.

Chapter 3 considers two problems in connection with choosing extremal subsets of vertices with prescribed properties. The problems of computing maximal independent or minimal dominating sets are discussed, the latter problem generalizing to the set covering problem. Applications of the set covering problem to airline crew scheduling, vehicle scheduling and information retrieval are given and the transformation of other graph theory problems to the set covering problem discussed.

Chapter 4 considers the vertex colouring problem, a specific case of the set covering problem discussed in the previous Chapter. Applications of the colouring problem to the scheduling of timetables and resource allocation are given, together with generalizations to the loading problem.

The next two Chapters are concerned with location problems on graphs. Chapter 5 considers the problem of locating multi-centres on a graph and is applicable to cases of locating emergency facilities such as fire, police or ambulance stations on a road network. Chapter 6 considers the problem of locating multi-medians on a graph and is applicable to cases of locating facilities such as depots or switching centres in goods-distribution or telecommunication networks. Chapter 5 deals with a minimax and Chapter 6 with a minisum location problem.

Chapter 7 is concerned with trees, shortest spanning trees and the Steiner problem, and their application to the construction of electric power or gas pipeline networks.

Chapter 8 deals with the problem of finding shortest paths between pairs of vertices and its applications to routing problems for maximizing capacity or reliability, and to the special case of PERT networks and critical path analysis. The Chapter also deals with the determination of negative cost circuits in graphs and its applications to other graph theoretic problems. The calculation of second, third, etc. shortest paths is also discussed.

The next two Chapters deal with problems of finding circuits in graphs. Chapter 9 discusses general circuits and cut-sets. The Chapter also deals with Eulerian circuits and the Chinese postman's problem with its applications to refuse collection, delivery of milk or post and inspection of distributed parameter systems. Chapter 10 is in two parts. The first part deals with the problem of finding a Hamiltonian circuit in a graph that is not complete and its application to machine scheduling. The second part of this Chapter deals with the problem of finding the shortest Hamiltonian circuit i.e. with the well known "travelling salesman problem" and its applications to vehicle routing.

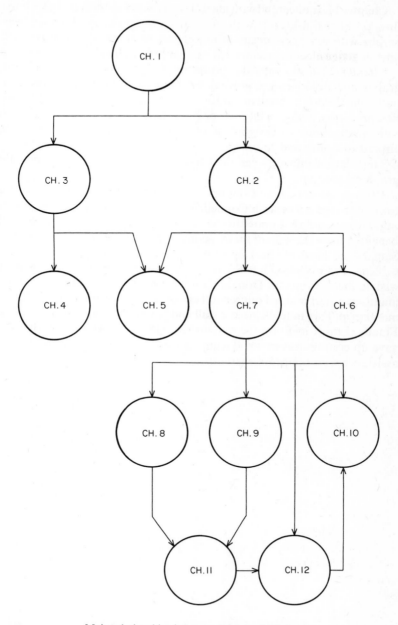

Main relationships between chapters of this book

Chapter 11 is concerned with maximum flow and minimum cost—maximum flow problems in graphs with arc capacities and costs. Flow problems in graphs with arc gains occur in mathematical models of arbitrage, current flow in active electric circuits, etc., and are also considered.

Chapter 12 deals with the problem of finding maximum matchings in graphs and describes the generalized hungarian algorithm. The algorithm particularizes to the bipartite graphs case for the assignment and transportation problems, both of which are of great significance in Operations Research with applications to assignment of people to jobs, facilities to locations, aircraft to routes and so on.

The relationships between the Chapters of this book are portrayed by the graph on page vii.

Parts of the contents of the book were developed with the help of grants from the Science Research Council, for research in mathematical programming. This assistance is gratefully acknowledged. In preparing this book I have benefited from the help of many people. I want especially to thank Professor Sam Eilon, head of the Department of Management Science at Imperial College, Peter Viola, Geff Selby, Peter Brooker and Sam Korman. I also wish to thank Professor Donald Knuth who has read an earlier version of the manuscript and pointed out several errors. For the arduous task of typing the manuscript I am indebted to the skill and patience of Miss Margaret Hudgell. Finally, I must mention the invaluable help of my wife Ann, who not only gave up countless evenings during the writing of this book but also for her assistance with the proof reading.

Nicos Christofides
October 1973
London.

List of Symbols

$A = [a_{ij}]$	Adjacency matrix
A	Set of arcs
a_j	j^{th} arc
$B = [b_{ij}]$	Incidence matrix
$C = [c_{ij}]$	Arc cost matrix
$c_{ij} = c(x_i, x_j)$	Cost of arc (x_i, x_j)
c_j	Cost of arc a_j
$D = [d_{ij}]$	Shortest distance matrix
$d_{ij} = d(x_i, x_j)$	Shortest distance (cost of least cost path) from x_i to x_j
$d_i = d(x_i)$	Degree of vertex x_i
$d_i^H = d^H(x_i)$	Degree of vertex x_i with respect to graph H
$d_0(x_i), d_t(x_i)$	Outdegree and indegree of vertex x_i respectively
E	Covering
f_{ij}	Value of the maximum flow from x_i to x_j
$G = (X, A)$	Graph with set of vertices X and set of arcs A
$G(\xi)$	Graph G with flow pattern ξ flowing in it
$G^\mu(\xi)$	Incremental graph of $G(\xi)$
$\bar{G}$	Graph G with all arc directions ignored
$\tilde{G}$	Graph complementary to G
g_{ij}	Gain of arc (x_i, x_j)
$K = [k_{ij}]$	Cut-set matrix
K	Cut-set
m	Number of arcs in graph
M	Matching
n	Number of vertices in graph
$p(x_i)$	Predecessor to vertex x_i
Q	Reaching matrix
$Q(x_i)$	Reaching set of vertex x_i
$q_{ij} = q(x_i, x_j)$	Upper limit (capacity) on the flow in arc (x_i, x_j)
R	Reachability matrix
$R(x_i)$	Reachability set of vertex x_i
r_{ij}	Lower limit on the flow in arc (x_i, x_j)
s	Initial vertex in shortest path or flow calculation
t	Final vertex in shortest path or flow calculation
v_i	"Weight" of vertex x_i
x_i	i^{th} vertex
X	Set of vertices of graph G
$\alpha[G]$	Independence number
$\beta[G]$	Dominance number
$\gamma[G]$	Chromatic number

Γ Set of correspondences

$\Gamma(x_i)$ Set of vertices x_j such that $(x_i, x_j) \in A$

$\Gamma^{-1}(x_i)$ Set of vertices x_j such that $(x_j, x_i) \in A$

$\Theta = [\theta_{ij}]$ Matrix storing shortest paths

$\xi_{ij} = \xi(x_i, x_j)$ Flow in arc (x_i, x_j)

ξ_{ij}^e, ξ_{ij}^0 Entering and leaving flow in arc (x_i, x_j) respectively

Ξ Maximum flow pattern

Ξ Optimum flow pattern

Ξ Optimum maximum flow pattern

$\Phi = [\phi_{ij}]$ Circuit matrix

Φ Circuit

Contents

Chapter 7. Trees

Chapter 8. Shortest Paths

Chapter 9. Circuits, Cut-sets and Euler's Problem

Chapter 10. Hamiltonian Circuits, Paths and the Travelling Salesman Problem

Part I

Part II

Chapter 11. Network Flows

Chapter 12. Matchings, Transportation and Assignment Problems

Chapter 1

Introduction

1. Graphs—Definition

A *graph* G is a collection of points or *vertices* $x_1, x_2, \ldots, x_n$ (denoted by the set X), and a collection of lines $a_1, a_2, \ldots, a_m$ (denoted by the set A) joining all or some of these points. The graph G is then fully described and denoted by the doublet (X, A).

If the lines in A have a direction—which is usually shown by an arrow—they are called *arcs* and the resulting graph is called a *directed* graph (Fig. 1.1(a)). If the lines have no orientation they are called *links* and the graph is *nondirected* (Fig. 1.1(b)). In the case where $G = (X, A)$ is a directed graph but we want to disregard the direction of the arcs in A, the nondirected counterpart to G will be written as $\bar{G} = (X, \bar{A})$.

Throughout this book, when an arc is denoted by the pair of its *initial* and *final* vertices (i.e. by its two *terminal* vertices), its direction will be assumed to be from the first vertex to the second. Thus, in Fig. 1.1(a) (x_1, x_2) refers to arc a_1, and (x_2, x_1) to arc a_2.

An alternative and often preferable way to describe a directed graph G, is by specifying the set X of vertices and a *correspondence* Γ which shows how the vertices are related to each other. Γ is called a mapping of the set X in X and the graph is denoted by the doublet $G = (X, \Gamma)$.

In the example of Fig. 1.1(a) we have

$$\Gamma(x_1) = \{x_2, x_5\} \text{ i.e. } x_2 \text{ and } x_5 \text{ are the final vertices of arcs whose initial vertex is } x_1.$$

$$\Gamma(x_2) = \{x_1, x_3\}$$

$$\Gamma(x_3) = \{x_1\}$$

$$\Gamma(x_4) = \varnothing, \text{ the } null \text{ or empty set}$$

and

$$\Gamma(x_5) = \{x_4\}.$$

1

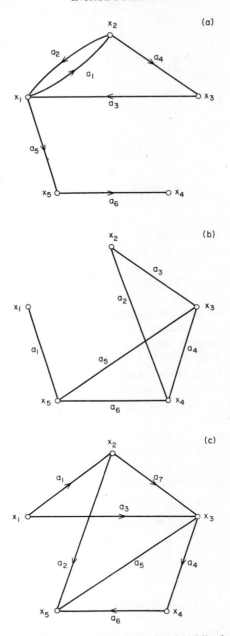

FIG. 1.1. (a) Directed, (b) Nondirected and (c) Mixed graphs

In the case of a nondirected graph, or of a graph containing both arcs and links (such as the graphs shown in Figs 1.1(b) and 1.1(a)), the correspondences Γ will be assumed to be those of an equivalent directed graph in which every link has been replaced by two arcs in opposite directions. Thus, for the graph of Fig. 1.1(b) for example, $\Gamma(x_5) = \{x_1, x_3, x_4\}$, $\Gamma(x_1) = \{x_5\}$ etc.

Since $\Gamma(x_i)$ has been defined to mean the set of those vertices $x_j \in X$ for which an arc (x_i, x_j) exists in the graph, it is natural to write as $\Gamma^{-1}(x_i)$ the set of those vertices x_k for which an arc (x_k, x_i) exists in G. The relation $\Gamma^{-1}(x_i)$ is then called the *inverse correspondence*. Thus for the graph shown in Fig. 1.1(a) we have:

$$\Gamma^{-1}(x_1) = \{x_2, x_3\}$$
$$\Gamma^{-1}(x_2) = \{x_1\}$$

etc.

It is quite obvious that for a nondirected graph $\Gamma^{-1}(x_i) = \Gamma(x_i)$ for all $x_i \in X$.

When the correspondence Γ does not operate on a single vertex but on a set of vertices such as $X_q = \{x_1, x_2, \ldots, x_q\}$, then $\Gamma(X_q)$ is taken to mean:

$$\Gamma(X_q) = \Gamma(x_1) \cup \Gamma(x_2) \cup \ldots \ldots \cup \Gamma(x_q)$$

i.e. $\Gamma(X_q)$ is the set of those vertices $x_j \in X$ for which at least one arc (x_i, x_j) exists in G, for some $x_i \in X_q$. Thus, for the graph of Fig. 1.1(a); $\Gamma(\{x_2, x_5\}) = \{x_1, x_3, x_4\}$ and $\Gamma(\{x_1, x_3\}) = \{x_2, x_5, x_1\}$.

The double correspondence $\Gamma(\Gamma(x_i))$ is written as $\Gamma^2(x_i)$. Similarly the triple correspondence $\Gamma(\Gamma(\Gamma(x_i)))$ is written as $\Gamma^3(x_i)$ and so on. Thus, the graph in Fig. 1.1(a):

$$\Gamma^2(x_1) = \Gamma(\Gamma(x_1)) = \Gamma(\{x_2, x_5\}) = \{x_1, x_3, x_4\}$$
$$\Gamma^3(x_1) = \Gamma(\Gamma^2(x_1)) = \Gamma(\{x_1, x_3, x_4\}) = \{x_1, x_2, x_5\}$$

etc.

Similarly for $\Gamma^{-2}(x_i)$, $\Gamma^{-3}(x_i)$ and so on.

2. Paths and Chains

A *path* in a directed graph is any sequence of arcs where the final vertex of one is the initial vertex of the next one.

Thus in Fig. 1.2 the sequence of arcs:

$$a_6, a_5, a_9, a_8, a_4 \tag{1.1}$$
$$a_1, a_6, a_5, a_9 \tag{1.2}$$
$$a_1, a_6, a_5, a_9, a_{10}, a_6, a_4 \tag{1.3}$$

are all paths.

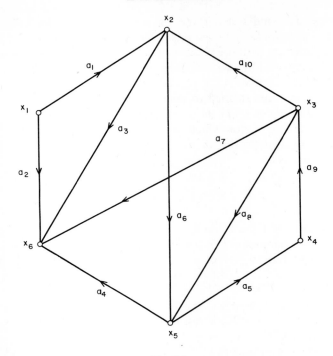

Fig. 1.2

Arcs $a = (x_i, x_j)$, $x_i \neq x_j$ which have a common terminal vertex are called *adjacent*. Also, two vertices x_i and x_j are called adjacent if either arc (x_i, x_j) or arc (x_j, x_i) or both exist in the graph. Thus in Fig. 1.2 arcs a_1, a_{10}, a_3 and a_6 are adjacent and so are the vertices x_5 and x_3; on the other hand arcs a_1 and a_5 or vertices x_1 and x_4 are not adjacent.

A *simple* path is a path which does not use the same arc more than once. Thus the paths (1.1) and (1.2) above are simple but path (1.3) is not, since it uses arc a_6 twice.

An *elementary* path is a path which does not use the same vertex more than once. Thus the path (1.2) is elementary but paths (1.1) and (1.3) are not. Obviously an elementary path is also simple but the reverse is not necessarily true. Note for example that path (1.1) is simple but not elementary, that path (1.2) is both simple and elementary and that path (1.3) is not simple and not elementary.

A *chain* is the nondirected counterpart of the path and applies to graphs with the direction of its arcs disregarded. Thus a chain is a sequence of links

$(\bar{a}_1, \bar{a}_2, \ldots, \bar{a}_q)$ in which every link $\bar{a}_i$, except perhaps the first and last links, is connected to the links $\bar{a}_{i-1}$ and $\bar{a}_{i+1}$ by its two terminal vertices.

$$\bar{a}_2', \bar{a}_4, \bar{a}_8, \bar{a}_{10} \tag{1.4}$$

$$\bar{a}_2, \bar{a}_7, \bar{a}_8, \bar{a}_4, \bar{a}_3 \tag{1.5}$$

and

$$\bar{a}_{10}, \bar{a}_7, \bar{a}_4, \bar{a}_8, \bar{a}_7, \bar{a}_2 \tag{1.6}$$

are all chains; where a bar above the symbol of an arc means that its direction is disregarded i.e. it is to be considered as a link.

In an exactly analogous way as we defined simple and elementary paths we can define simple and elementary chains. Thus, chain (1–4) is simple and elementary, chain (1–5) is simple but not elementary and chain (1–6) is both not simple and not elementary.

A path or a chain may also be represented by the sequence of vertices that are used. Thus, path (1.1) may also be represented by the sequence $x_2, x_5, x_4, x_3, x_5, x_6$ of vertices, and this representation is often more useful when one is concerned with finding elementary paths or chains.

2.1 Weights and the length of a path

A number c_{ij} may sometimes be associated with an arc (x_i, x_j). These numbers are called *weights*, *lengths* or *costs* and the graph is then called *arc-weighted*. Also a weight v_i may sometimes be associated with a vertex x_i and the resulting graph is then called *vertex-weighted*. If a graph is both arc and vertex weighted it is simply called *weighted*.

Considering a path μ represented by the sequence of arcs $(a_1, a_2, \ldots, a_q)$, the *length (or cost) of the path* $l(\mu)$ is taken to be the sum of the arc weights on the arcs appearing in μ, i.e.

$$l(\mu) = \sum_{(x_i, x_j) \text{ in } \mu} c_{ij}$$

Thus, the words "length", "cost" and "weight" when applied to arcs, are all considered to be equivalent, and in specific applications that word will be chosen which gives the best intuitive meaning and which is in agreement with the usage found in the literature.

The *cardinality* of the path μ is q i.e. the number of arcs appearing in the path.

3. Loops, Circuits and Cycles

A *loop* is an arc whose initial and final vertices are the same. In Fig. 1.3 for example arcs a_2 and a_5 form loops.

A *circuit* is a path $a_1, a_2, \ldots, a_q$ in which the initial vertex of a_1 coincides with the final vertex of a_q.

Thus in Fig. 1.3 the sequences:

$$a_3, a_6, a_{11} \tag{1.7}$$

$$a_{11}, a_3, a_4, a_7, a_1, a_{12}, a_9 \tag{1.8}$$

$$a_3, a_4, a_7, a_{10}, a_9, a_{11} \tag{1.9}$$

are all circuits.

Circuits (1.7) and (1.9) are elementary since they do not use the same vertex more than once (except for the initial and final vertices which are the same), but circuit (1.8) is not elementary since vertex x_1 is used twice.

An elementary circuit which passes through all the n vertices of a graph G is of special significance and is known as a *Hamiltonian* circuit. Of course not all graphs have a Hamiltonian circuit. Thus, circuit (1.9) is a Hamiltonian circuit of the graph of Fig. 1.3; but the graph of Fig. 1.2 has no Hamiltonian circuits as can be seen quite easily from the fact that there is no arc having x_1 as its final vertex.

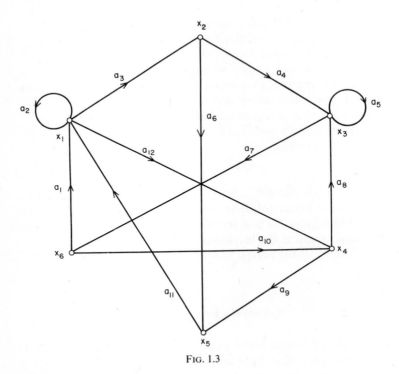

FIG. 1.3

A *cycle* is the nondirected counterpart of the circuit. Thus, a cycle is a chain $x_1, x_2, \ldots, x_q$ in which the beginning and end vertices are the same, i.e. in which $x_1 = x_q$.

In Fig. 1.3 the chains:

$$\bar{a}_1, \bar{a}_{12}, \bar{a}_{10} \tag{1.10}$$

and

$$\bar{a}_{10}, \bar{a}_1, \bar{a}_3, \bar{a}_4, \bar{a}_7, \bar{a}_1, \bar{a}_{12} \tag{1.11}$$

all form cycles.

4. Degrees of a Vertex

The number of arcs which have a vertex x_i as their initial vertex is called the *outdegree* of vertex x_i, and similarly the number of arcs which have x_i as their final vertex is called the *indegree* of vertex x_i.

Thus, referring to Fig. 1.3, the outdegree of x_6, denoted by $d_0(x_6)$, is $|\Gamma(x_6)| = 2$, and the indegree of x_6 denoted by $d_t(x_6)$, is $|\Gamma^{-1}(x_6)| = 1$.

It is quite obvious that the sum of the outdegrees or indegrees of all the vertices in a graph is equal to the total number of arcs of G i.e.

$$\sum_{i=1}^{n} d_0(x_i) = \sum_{i=1}^{n} d_t(x_i) = m \tag{1.12}$$

where n is the total number of vertices and m the total number of arcs of G.

For a nondirected graph $G = (X, \Gamma)$ the degree of a vertex x_i is similarly defined by $d(x_i) \equiv |\Gamma(x_i)|$, and when no confusion can arise it will be written as d_i.

5. Partial Graphs and Subgraphs

Given a graph $G = (X, A)$, a *partial* graph G_p of G is the graph (X, A_p) with $A_p \subset A$. Thus a partial graph is a graph with the same number of vertices but with only a subset of the arcs of the original graph.

If Fig. 1.4(a) represents the graph G, Fig. 1.4(b) is a partial graph G_p.

Given a graph $G = (X, \Gamma)$ a *subgraph* G_s is the graph (X_s, Γ_s) with $X_s \subset X$; and for every $x_i \in X_s$, $\Gamma_s(x_i) = \Gamma(x_i) \cap X_s$. Thus, a subgraph has only a subset X_s of the set of vertices of the original graph but contains all the arcs whose initial and final vertices are both within this subset. It is often very convenient to denote the subgraph G_s simply by $\langle X_s \rangle$ and when no confusion can arise we will use this latter symbolism.

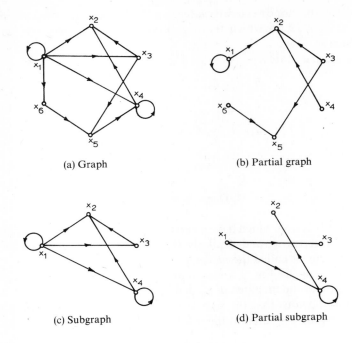

(a) Graph   (b) Partial graph

(c) Subgraph   (d) Partial subgraph

Fig. 1.4

Fig. 1.4(c) shows a subgraph of the graph in Fig. 1.4(a) containing only vertices x_1, x_2, x_3 and x_4, and those arcs which interconnect them.

The two definitions given above can be combined to define the *partial subgraph*, an example of which is given in Fig. 1.4(d) which shows a partial graph of the subgraph in Fig. 1.4(c).

If a graph represents an entire organization with the vertices representing people and the arcs representing, say, lines of communication, then the graph representing only the more important communication channels of the whole organization is a partial graph; the graph which represents the detailed lines of communication of only a part of the organization (say a division) is a subgraph; and the graph which represents only the important lines of communication within this division is a partial subgraph.

6. Types of Graphs

A graph $G = (X, A)$ is said to be *complete* if for every pair of vertices x_i and x_j in X, there exists a link $(\overline{x_i, x_j})$ in $\overline{G} = (X, \overline{A})$ i.e. there must be at least one arc joining every pair of vertices. The complete nondirected graph on n vertices is denoted by K_n.

A graph (X, A) is said to be *symmetric* if, whenever an arc (x_i, x_j) is one of the arcs in the set A of arcs, the opposite arc (x_j, x_i) is also in the set A.

An *antisymmetric* graph is a graph in which whenever an arc $(x_i, x_j) \in A$, the opposite arc $(x_j, x_i) \notin A$. Obviously an antisymmetric graph cannot contain any loops.

Fig. 1.5(a) shows a symmetric and Fig. 1.5(b) an antisymmetric graph.

For example if the vertices of a graph represents a group of people and an arc directed from vertex x_i to vertex x_j means that x_i is the friend or relative of x_j then the graph would be symmetric. On the other hand if an arc directed from x_i to x_j means that x_j is a subordinate of x_i then the resulting graph would be antisymmetric.

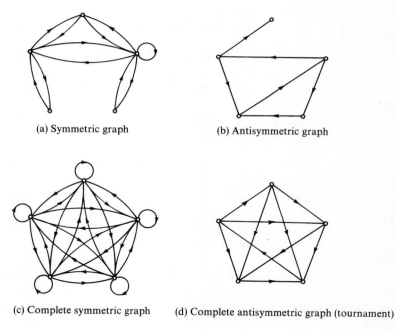

(a) Symmetric graph   (b) Antisymmetric graph

(c) Complete symmetric graph   (d) Complete antisymmetric graph (tournament)

FIG. 1.5

Combining the above definitions one can define the *complete symmetric graph*, an example of which is shown in Fig. 1.5(c), and the *complete antisymmetric* graph, with an example shown in Fig. 1.5(d). This last type of graph is also often referred to as a *tournament*.

A nondirected graph $G = (X, A)$ is said to be *bipartite*, if the set X of its vertices can be partitioned into two subsets X^a and X^b so that all arcs have one terminal vertex in X^a and the other in X^b. A directed graph G is said to be bipartite if its nondirected counterpart $\bar{G}$ is bipartite. It is quite easy to show that:

THEOREM 1. *A nondirected graph G is bipartite if and only if it contains no circuits of odd cardinality.*

Proof. Necessity. Since X is partitioned into X^a and X^b;

$$X^a \cup X^b = X \qquad \text{and} \qquad X^a \cap X^b = \varnothing \qquad (1.13)$$

Let an odd cardinality circuit $x_{i_1}, x_{i_2}, \ldots, x_{i_q}, x_{i_1}$ exist and without loss of generality take $x_{i_1} \in X^a$. Since, from the definition, two consecutive vertices on this circuit must belong one to X^a and the other to X^b it follows that $x_{i_2} \in X^b$, $x_{i_3} \in X^a$ etc. and, in general, $x_{i_k} \in X^a$ if k is odd and $x_{i_k} \in X^b$, if k is even. Since we assumed the cardinality of the circuit to be odd, $x_{i_q} \in X^a$ which implies $x_{i_1} \in X^b$. This is a contradiction since $X^a \cap X^b = \varnothing$ and no vertex can belong to both X^a and X^b.

Sufficiency. Assume that no circuits of odd cardinality exist. Choose any vertex x_i, say, and label it "$+$", the iteratively perform the operations:

Choose any labelled vertex x_i and label all vertices in $\Gamma(x_i)$ with the reverse sign of the label of x_i.

Continue applying this operation until either:

(i)   all vertices have been labelled and are consistent, i.e. for any two vertices joined by a link, one is labelled "$+$" and the other "$-$", or

(ii) some vertex, say x_{i_k}, which was labelled with one sign can now be labelled (from a different vertex) with the opposite sign, or

(iii) for all labelled vertices x_i, $\Gamma(x_i)$ is labelled but other unlabelled vertices exist.

In case (i) let all vertices labelled "$+$" be in the set X^a and those labelled "$-$" be in the set X^b. Since all links are between differently labelled vertices, the graph G is bipartite.

In case (ii) vertex x_{i_k} must have received its "$+$" label along some path (μ_1 say) of vertices, alternatingly labelled "$+$" and "$-$", starting from x_{i_1} and finishing at x_{i_k}. Similarly the "$-$" label on x_{i_k} was obtained along some path μ_2. Let x^* be the last but one (the last being x_{i_k}) vertex common to paths μ_1 and μ_2. If x^* is labelled "$+$" the part of path μ_2 from x^* to x_{i_k} must be of even, and the part of path μ_2 from x^* to x_{i_k} must be of odd cardinality respectively. The opposite is true if x^* is labelled "$-$". Hence the circuit consisting of path μ_1 from x^* to x_{i_k} and the reverse part of path μ_2 from x_{i_k} back to x^* is always of odd cardinality. This contradicts the assumption that no odd cardinality circuits exist in G and hence case (ii) is impossible.

Case (iii) implies that there is no link between any labelled and unlabelled

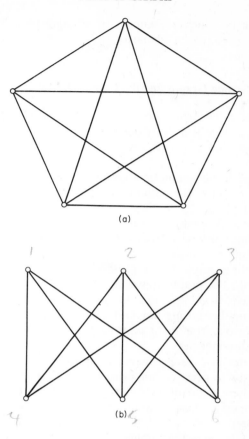

FIG. 1.6. The Kuratowski nonplanar graphs. (a) K_5 and (b) $K_{3,3}$

vertex which means that G is disconnected into two or more parts and each part can then be considered in isolation. Thus, only case (i) is eventually possible, and hence the theorem.

When a graph is bipartite and this property needs to be emphasized we will denote the graph as $(X^a \cup X^b, A)$ with equations (1.13) being implied.

A bipartite graph $G = (X^a \cup X^b, A)$ is said to be *complete* if for every two vertices $x_i \in X^a$ and $x_j \in X^b$ there exists a link $\overline{(x_i, x_j)}$ in $\overline{G} = (X, \overline{A})$. If $|X^a|$, the number of vertices in set X^a, is r and $|X^b| = s$, then the complete non-directed bipartite graph $G = (X^a \cup X^b, A)$ is denoted by K_{rs}.

A graph $G = (X, A)$ is said to be *planar*, if it can be drawn on a plane (or sphere) in such a way so that no two arcs intersect each other. Figure 1.6(a) shows the complete graph K_5 and Fig. 1.6(b) shows the complete bipartite graph $K_{3,3}$ both of which are known to be *nonplanar* [1, 3]. These two graphs play a central role in planarity and are known as the Kuratowski graphs.

7. Strong Graphs and the Components of a Graph

A graph is said to be strongly connected or *strong* if for any two distinct vertices x_i and x_j, there is at least one path going from x_i to x_j. The above definition implies that any two vertices of a strong graph are mutually reachable.

A graph is said to be unilaterally connected or *unilateral* if for any two distinct vertices x_i and x_j, there is at least one path going from either x_i to x_j, or from x_j to x_i, or both.

A graph is said to be weakly connected or *weak* if there is at least one chain joining every pair of distinct vertices. If for a pair of vertices such a chain does not exist, the graph is said to be *disconnected* [2].

Considering, for example, the graph shown in Fig. 1.7(a) it is easy to check that this is strong. The graph shown in Fig. 1.7(b) on the other hand is not strong (there is no path going from x_1 to x_3), but is unilateral. The graph shown in Fig. 1.7(c) is neither strong nor unilateral—since there is no path going from x_2 to x_5 nor one from x_5 to x_2. It is, however, weak. Finally, the graph of Fig. 1.7(d) is disconnected.

Given any property P to characterize a graph, a *maximal subgraph* $\langle \hat{X}_s \rangle$ if a graph G with respect to that property, is a subgraph which has this property, and there is no other subgraph $\langle X_s \rangle$ with $X_s \supset \hat{X}_s$ and which also has the property. Thus, if the property P is strong-connectedness, then a maximal strong subgraph of G is a strong subgraph of G which is not contained in any other strong subgraph. Such a subgraph is called a *strong component* of G. Similarly a *unilateral component* is a maximal unilateral subgraph and a *weak component* is a maximal weak subgraph.

For example, in the graph G shown in Fig. 1.7(b), the subgraph $\langle \{x_1, x_4, x_5, x_6\} \rangle$ is a strong component of G. On the other hand $\langle \{x_1, x_6\} \rangle$ and $\langle \{x_1, x_5, x_6\} \rangle$ are not strong components—even though they are strong subgraphs—since these are contained in $\langle \{x_1, x_4, x_5, x_6\} \rangle$ and are, therefore, not maximal. In the graph shown in Fig. 1.7(c) the subgraph $\langle \{x_1, x_4, x_5\} \rangle$ is a unilateral component. In the graph shown in Fig. 1.7(d) the subgraphs $\langle \{x_1, x_5, x_6\} \rangle$ and $\langle \{x_2, x_3, x_4\} \rangle$ are both weak components and are the only two such components.

It is quite obvious from the definitions that unilateral components are not necessarily pairwise vertex-disjoint. A strong component must be contained in at least one unilateral component and a unilateral component contained in a weak component for any given graph G.

8. Matrix Representations

A convenient way of representing a graph algebraically is by the use of matrices, as follows.

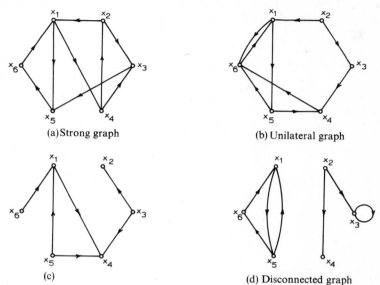

(a) Strong graph

(b) Unilateral graph

(c)

(d) Disconnected graph

FIG. 1.7.

8.1 The adjacency matrix

Given a graph G, its *adjacency* matrix is denoted by $\mathbf{A} = [a_{ij}]$ and is given by:

$$a_{ij} = 1 \text{ if arc } (x_i, x_j) \text{ exists in } G$$

$$a_{ij} = 0 \text{ if arc } (x_i, x_j) \text{ does not exist in } G.$$

Thus, the adjacency matrix of the graph shown in Fig. 1.8 is:

$$
\mathbf{A} =
\begin{array}{c|cccccc}
 & x_1 & x_2 & x_3 & x_4 & x_5 & x_6 \\
\hline
x_1 & 0 & 1 & 1 & 0 & 0 & 0 \\
x_2 & 0 & 1 & 0 & 0 & 1 & 0 \\
x_3 & 0 & 0 & 0 & 0 & 0 & 0 \\
x_4 & 0 & 0 & 1 & 0 & 0 & 0 \\
x_5 & 1 & 0 & 0 & 1 & 0 & 0 \\
x_6 & 1 & 0 & 0 & 0 & 1 & 1 \\
\end{array}
$$

The adjacency matrix defines completely the structure of the graph. For example, the sum of all the elements in row x_i of the matrix gives the out-degree of vertex x_i and the sum of the elements in column x_i gives the in-degree of vertex x_i. The set of columns which have an entry of 1 in row x_i is the set $\Gamma(x_i)$, and the set of rows which have an entry of 1 in column x_i is the set $\Gamma^{-1}(x_i)$.

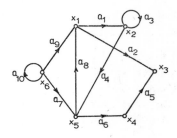

FIG. 1.8.

Consider the adjacency matrix raised to the second power. An element, $a_{ik}^{(2)}$ say, of matrix $\mathbf{A}^2$ is given by:

$$a_{ik}^{(2)} = \sum_{j=1}^{n} a_{ij} a_{jk} \tag{1.14}$$

Each term in the summation of equation (1.14) has the value 1 only if a_{ij} and a_{jk} are both 1, otherwise it has the value 0. Since $a_{ij} = a_{jk} = 1$ implies a path of cardinality 2 from vertex x_i to vertex x_k via vertex x_p, the term $a_{ik}^{(2)}$ is then the total number of cardinality 2 paths from x_i to x_k.

Similarly, if $a_{ik}^{(p)}$ is an element of $\mathbf{A}^p$, then $a_{ik}^{(p)}$ is the number of paths (not necessarily simple or elementary) of cardinality p from x_i to x_k.

8.2 The incidence matrix

Given a graph G of n vertices and m arcs, the *incidence* matrix of G is denoted by $\mathbf{B} = [b_{ij}]$ and is an $n \times m$ matrix defined as follows.

$$b_{ij} = 1 \text{ if } x_i \text{ is the initial vertex of arc } a_j$$

$$b_{ij} = -1 \text{ if } x_i \text{ is the final vertex of arc } a_j$$

and $b_{ij} = 0$ if x_i is not a terminal vertex of arc a_j or if a_j is a loop.

For the graph shown in Fig. 1.8, the incidence matrix is:

$$B = \begin{array}{c} \\ x_1 \\ x_2 \\ x_3 \\ x_4 \\ x_5 \\ x_6 \end{array} \begin{array}{|cccccccccc|} a_1 & a_2 & a_3 & a_4 & a_5 & a_6 & a_7 & a_8 & a_9 & a_{10} \\ \hline 1 & 1 & 0 & 0 & 0 & 0 & 0 & -1 & -1 & 0 \\ -1 & 0 & 0 & 1 & 0 & 0 & 0 & 0 & 0 & 0 \\ 0 & -1 & 0 & 0 & -1 & 0 & 0 & 0 & 0 & 0 \\ 0 & 0 & 0 & 0 & 1 & -1 & 0 & 0 & 0 & 0 \\ 0 & 0 & 0 & -1 & 0 & 1 & -1 & 1 & 0 & 0 \\ 0 & 0 & 0 & 0 & 0 & 0 & 1 & 0 & 1 & 0 \end{array}$$

Since each arc is adjacent to exactly two vertices, each column of the incidence matrix contains one 1 and one −1 entry, except when the arc forms a loop in which case it contains only zero entries.

If G is a nondirected graph, then the incidence matrix is defined as above except that all entries of −1 are now changed to +1.

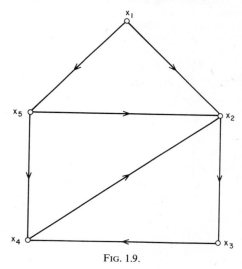

FIG. 1.9.

9. Problems P1

1. For the graph of Fig. 1.9 find:
   (a) $\Gamma(x_2)$
   (b) $\Gamma^{-1}(x_2)$
   (c) $\Gamma^2(x_2)$
   (d) $\Gamma^{-2}(x_2)$
   (e) $d_0(x_2)$
   (f) $d_t(x_2)$
   (g) the adjacency matrix A
   (h) the incidence matrix B

2. For the graph $G = (X, A)$ of Fig 1.9 draw:
    (a) The subgraph $\langle\{x_1, x_2, x_4, x_5\}\rangle$
    (b) The partial graph (X, A'), where $(x_i, x_j) \in A'$ if and only if $i + j$ is odd.
    (c) The partial graph as defined in (b) of the subgraph in (a).
3. For a nondirected graph prove that the number of vertices of odd degree is even. (Zero is an even number).
4. Show that any complete symmetric graph contains a Hamiltonian circuit.
5. Show that the rank of the incidence matrix **B** of a connected graph is $n - 1$, and hence prove that the rank of **B** for a graph with P (weak) components is $n - P$.
6. Prove that a nondirected connected graph remains connected after the removal of a link if and only if that link is part of some circuit.
7. Prove that a nondirected connected graph with n vertices
    (a) Contains at least $n - 1$ links
    (b) If it contains more than $n - 1$ links then it has at least one circuit.

10. References

1. Berge, C., (1962). "The theory of graphs", Methuen, London.
2. Harary, F., Norman, R. Z. and Cartwright, D. (1965). "Structural models: An introduction to the theory of directed graphs," Wiley, New York.
3. Kuratowski, G. (1930). Sur le problème des courbes gauches en topologie, *Fund. Match.*, 15–16, p. 271.
4. Roy, B. (1969, 1970). Algebre moderne et theorie des graphes, Vol. 1 (1969), Vol. 2 (1970), Dunod, Paris.

Chapter 2

Reachability and Connectivity

1. Introduction

In the previous chapter it was mentioned that the communication system of an organization can be considered in terms of a graph where people are represented by vertices, and communication channels by arcs. A natural question to ask of such a system, is as to whether information held by a given individual x_i can be communicated to another individual x_j; that is, whether there is a path leading from vertex x_i to vertex x_j. If such a path exists we say that x_j is *reachable* from x_i. If the reachability is restricted to paths of limited cardinality, we would like to know if x_j is still reachable from x_i. The purpose of the present Chapter is to discuss some fundamental concepts relating to the reachability and connectivity properties of graphs and introduce some very basic algorithms.

In terms of the graph representing an organization, the present chapter considers the questions:

(i)   What is the least number of people from which every other person in the organization can be reached?
(ii) What is the largest number of people who are mutually reachable?
(iii) How are (i) and (ii) above related?

2. Reachability and Reaching Matrices

The reachability matrix $\mathbf{R} = [r_{ij}]$ is defined as follows:

$$r_{ij} = 1 \text{ if vertex } x_j \text{ is reachable from vertex } x_i.$$
$$= 0 \text{ otherwise.}$$

The set of vertices $R(x_i)$ of the graph G which can be reached from a given vertex x_i consists of those elements x_j for which there is an entry of 1 in cell (x_i, x_j) of the reachability matrix. Obviously all the diagonal elements of $\mathbf{R}$ are 1 since every vertex is reachable from itself, by a path of cardinality 0.

17

Since $\Gamma(x_i)$ is that set of vertices x_j which are reachable from x_i along a path of cardinality 1 (i.e. that set of vertices for which the arcs (x_i, x_j) exist in the graph) and since $\Gamma(x_j)$ is that set of vertices reachable from x_j along a path of cardinality 1; the set of vertices $\Gamma(\Gamma(x_i)) = \Gamma^2(x_i)$ consists of those vertices reachable from x_i along a path of cardinality 2. Similarly $\Gamma^p(x_i)$ is that set of vertices which are reachable from vertex x_i along a path of cardinality p.

Since any vertex of the graph G which is reachable from x_i must be reachable along a path (or paths) of cardinality 0 or 1 or 2, ... or p for some finite but suitably large value of p, the reachable set of vertices from vertex x_i is:

$$R(x_i) = \{x_i\} \cup \Gamma(x_i) \cup \Gamma^2(x_i) \cup \ldots \cup \Gamma^p(x_i) \qquad (2.1)$$

Thus, the reachable set $R(x_i)$ can be obtained from eqn (2.1) by performing the union operations from left to right until such time when the current total set is not increased in size by the next union. When this occurs any subsequent unions will obviously not add any new members to the set and this is then the reachable set $R(x_i)$. The number of unions that may have to be performed depends on the graph, although it is quite clear that p is bounded from above by the number of vertices in the graph minus one.

The reachability matrix can therefore be constructed as follows. Find the reachable sets $R(x_i)$ for all vertices $x_i \in X$ as mentioned above. Set $r_{ij} = 1$ if $x_j \in R(x_i)$, otherwise set $r_{ij} = 0$. The resulting matrix **R** is the reachability matrix.

The reaching matrix $\mathbf{Q} = [q_{ij}]$ is defined as follows:

$$q_{ij} = 1 \text{ if vertex } x_j \text{ can reach vertex } x_i$$

$$= 0 \text{ otherwise.}$$

The reaching set $Q(x_i)$ of the graph G is that set of vertices which can reach

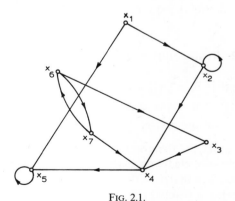

FIG. 2.1.

vertex x_i. In a manner analogous to the calculation of the reachable set $R(x_i)$ from eqn (2.1), the set $Q(x_i)$ can be calculated as:

$$Q(x_i) = \{x_i\} \cup \Gamma^{-1}(x_i) \cup \Gamma^{-2}(x_i) \cup \ldots \cup \Gamma^{-p}(x_i) \qquad (2.2)$$

where $\Gamma^{-2}(x_i) = \Gamma^{-1}(\Gamma^{-1}(x_i))$ etc.

The operations are performed from left to right until the next set union operation does not change the set $Q(x_i)$.

It is quite apparent from the definitions that column x_i of the matrix $\mathbf{Q}$ (found by setting $q_{ij} = 1$ if $x_j \in Q(x_i)$, and $q_{ij} = 0$ otherwise), is the same as row x_i of the matrix $\mathbf{R}$; i.e. $\mathbf{Q} = \mathbf{R}^t$, the transpose of the reachability matrix.

2.1 Example

Find the reachability and reaching matrices of the graph G shown in Fig. 2.1. The adjacency matrix of G is:

$$\mathbf{A} = x_4 \quad \begin{array}{c} \\ x_1 \\ x_2 \\ x_3 \\ x_4 \\ x_5 \\ x_6 \\ x_7 \end{array} \begin{bmatrix} \begin{array}{ccccccc} x_1 & x_2 & x_3 & x_4 & x_5 & x_6 & x_7 \\ 0 & 1 & 0 & 0 & 1 & 0 & 0 \\ 0 & 1 & 0 & 1 & 0 & 0 & 0 \\ 0 & 0 & 0 & 1 & 0 & 0 & 0 \\ 0 & 0 & 0 & 0 & 1 & 0 & 0 \\ 0 & 0 & 0 & 0 & 1 & 0 & 0 \\ 0 & 0 & 1 & 0 & 0 & 0 & 1 \\ 0 & 0 & 0 & 1 & 0 & 1 & 0 \end{array} \end{bmatrix}$$

The reachable sets are calculated from eqn (2.1) as:

$$R(x_1) = \{x_1\} \cup \{x_2, x_5\} \cup \{x_2, x_4, x_5\} \cup \{x_2, x_4, x_5\}$$
$$= \{x_1, x_2, x_4, x_5\}$$

$$R(x_2) = \{x_2\} \cup \{x_2, x_4\} \cup \{x_2, x_4, x_5\} \cup \{x_2, x_4, x_5\}$$
$$= \{x_2, x_4, x_5\}$$

$$R(x_3) = \{x_3\} \cup \{x_4\} \cup \{x_5\} \cup \{x_5\}$$
$$= \{x_3, x_4, x_5\}$$

$$R(x_4) = \{x_4\} \cup \{x_5\} \cup \{x_5\}$$
$$= \{x_4, x_5\}$$

$R(x_5) = \{x_5\} \cup \{x_5\}$

$\quad = \{x_5\}$

$R(x_6) = \{x_6\} \cup \{x_3, x_7\} \cup \{x_4, x_6\} \cup \{x_3, x_5, x_7\} \cup \{x_4, x_5, x_6\}$

$\quad = \{x_3, x_4, x_5, x_6, x_7\}$

$R(x_7) = \{x_7\} \cup \{x_4, x_6\} \cup \{x_3, x_5, x_7\} \cup \{x_4, x_5, x_6\}$

$\quad = \{x_3, x_4, x_5, x_6, x_7\}$

The reachability matrix is therefore given by:

$$
\mathbf{R} = \begin{array}{c} \\ x_1 \\ x_2 \\ x_3 \\ x_4 \\ x_5 \\ x_6 \\ x_7 \end{array}
\begin{array}{c} \begin{array}{ccccccc} x_1 & x_2 & x_3 & x_4 & x_5 & x_6 & x_7 \end{array} \\
\left[\begin{array}{ccccccc}
1 & 1 & 0 & 1 & 1 & 0 & 0 \\
0 & 1 & 0 & 1 & 1 & 0 & 0 \\
0 & 0 & 1 & 1 & 1 & 0 & 0 \\
0 & 0 & 0 & 1 & 1 & 0 & 0 \\
0 & 0 & 0 & 0 & 1 & 0 & 0 \\
0 & 0 & 1 & 1 & 1 & 1 & 1 \\
0 & 0 & 1 & 1 & 1 & 1 & 1
\end{array} \right] \end{array}
$$

and the reaching matrix is given by:

$$
\mathbf{Q} = \mathbf{R}^t = \begin{array}{c} \\ x_1 \\ x_2 \\ x_3 \\ x_4 \\ x_5 \\ x_6 \\ x_7 \end{array}
\begin{array}{c} \begin{array}{ccccccc} x_1 & x_2 & x_3 & x_4 & x_5 & x_6 & x_7 \end{array} \\
\left[\begin{array}{ccccccc}
1 & 0 & 0 & 0 & 0 & 0 & 0 \\
1 & 1 & 0 & 0 & 0 & 0 & 0 \\
0 & 0 & 1 & 0 & 0 & 1 & 1 \\
1 & 1 & 1 & 1 & 0 & 1 & 1 \\
1 & 1 & 1 & 1 & 1 & 1 & 1 \\
0 & 0 & 0 & 0 & 0 & 1 & 1 \\
0 & 0 & 0 & 0 & 0 & 1 & 1
\end{array} \right] \end{array}
$$

as can easily be checked.

It should be mentioned here that since all entries in the $\mathbf{R}$ and $\mathbf{Q}$ matrices are either 0 or 1, each row can be stored in binary form using one (or more) computer words. Thus, finding the $\mathbf{R}$ and $\mathbf{Q}$ matrices becomes a computationally very simple task, since the union of sets indicated by eqns (2.1) and (2.2) and the comparison after each union—in order to determine if it is necessary to continue—can then each be done by a single logical operation on a computer.†

Since $R(x_i)$ is the set of vertices which can be reached from x_i and $Q(x_j)$ the set of vertices which can reach x_j, the set $R(x_i) \cap Q(x_j)$ is then the set of vertices which are on at least one path going from x_i to x_j. These vertices are called *essential* with respect to the two end vertices x_i and x_j [7]. All other vertices $x_k \notin R(x_i) \cap Q(x_j)$ are called *inessential* or *redundant* since their removal does not affect any path from x_i to x_j.

The reachability and reaching matrices defined above are absolute, in the sense that the number of steps in the paths reaching from x_i to x_j was not restricted. On the other hand a *limited* reachability and reaching matrices can be defined where the cardinality of the path must not exceed a certain number. These limited matrices can also be calculated from eqns (2.1) and (2.2) in an exactly analogous manner to that described previously where p would now be the upper bound on the allowed path cardinality.

A graph is said to be *transitive* if the existence of arcs (x_i, x_j) and (x_j, x_k) implies the existence of arc (x_i, x_k). The *transitive closure* of a graph $G = (X, A)$ is the graph $G_{tc} = (X, A \cup A')$ where A' is the minimal number of arcs necessary to make G_{tc} transitive. It is then quite apparent that, since a path from x_i to x_j in G must correspond to an arc (x_i, x_j) in G_{tc}, the reachability matrix **R** of G is the same as the adjacency matrix **A** of G_{tc} with all its diagonal elements set to 1.

3. Calculation of Strong Components

A strong component (SC) of a graph G has been defined in the previous chapter as being a maximal strongly connected subgraph of G. Since, for a strongly connected graph, vertex x_j is reachable from vertex x_i and *vice versa* for any x_i and x_j, the SC containing a given vertex x_i is unique and x_i will appear in the set of vertices of one and only one SC. This statement is quite obvious since if x_i appears in two or more strong components, then a path from any vertex in one SC to any other vertex in another SC would always exist (via x_i), and hence the union of the strong components would be strongly connected, a fact which is contrary to the definition of the SC.

If vertex x_i is taken to be both the initial and terminal vertex then the set of vertices essential with respect to these two identical ends (i.e. the set of vertices on some circuit containing x_i) is given by $R(x_i) \cap Q(x_i)$. Since all these essential vertices can reach and be reached from x_i, they can also reach and be reached from each other. Moreover, as no other vertex is

† Alternative ways of constructing the sets $R(x_i)$ and $Q(x_i)$ using a vertex labelling procedure starting from x_i, are indicated in Chapter 7 dealing with *trees*.

essential with respect to the ends x_i and x_j, the set $R(x_i) \cap Q(x_i)$—which can be calculated by eqns (2.1) and (2.2)—defines the vertices of the unique SC of G containing vertex x_i.

If these vertices are removed from the graph $G = (X, \Gamma)$, the remaining subgraph $G' = \langle X - R(x_i) \cap Q(x_i) \rangle$ can be treated in the same way by finding a SC containing a vertex $x_j \in X - R(x_i) \cap Q(x_i)$. This process can be repeated until all the vertices of G have been allocated to one SC. When this is done the graph G is said to have been *partitioned* into its strong components [3].

The graph $G^* = (X^*, \Gamma^*)$ defined so that each of its vertices represents the vertex set of a strong component of G, and an arc (x_i^*, x_j^*) exists if and only if there exists an arc (x_i, x_j) in G for some $x_i \in x_i^*$ and some $x_j \in x_j^*$; is called the *condensed* graph of G.

It is quite apparent that the condensed graph G^* contains no circuits, since a circuit would mean that any vertex on that circuit could reach (and be reached from) any other vertex, and hence the union of these vertices of G^* would be a SC of G^* and therefore also a SC of G, a fact which is contrary to the definition that the vertices of G^* correspond to the SC's of G.

3.1 Example

For the graph G given in Fig. 2.2, find the strong components, and the condensed graph G^*.

Let us find the SC of G containing vertex x_1.

From eqns (2.1) and (2.2) we find:

$$R(x_1) = \{x_1, x_2, x_4, x_5, x_6, x_7, x_8, x_9, x_{10}\}$$

and

$$Q(x_1) = \{x_1, x_2, x_3, x_5, x_6\}$$

Therefore the SC containing vertex x_1 is the subgraph

$$\langle R(x_1) \cap Q(x_1) \rangle = \langle \{x_1, x_2, x_5, x_6\} \rangle$$

Similarly, the SC containing vertex x_8, say, is the subgraph $\langle \{x_8, x_{10}\} \rangle$, the SC containing x_7 is $\langle \{x_4, x_7, x_9\} \rangle$, the SC containing vertex x_{11} is $\langle \{x_{11}, x_{12}, x_{13}\} \rangle$ and the SC containing vertex x_3 is $\langle \{x_3\} \rangle$. It should be noted that this last SC contains just a single vertex of G.

The condensed graph G^* is then given by the graph of Fig. 2.3.

The operations described above in order to find the SC's of a graph can also be done most conveniently by the direct use of the **R** and **Q** matrices defined in the previous section. Thus, if we write $\mathbf{R} \otimes \mathbf{Q}$ to mean the element-by-element multiplication of the two matrices, then it is immediately apparent that row x_i of the matrix $\mathbf{R} \otimes \mathbf{Q}$ contains values of 1 in those columns x_j for

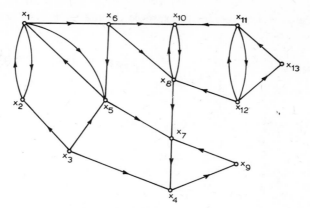

FIG. 2.2. The graph G

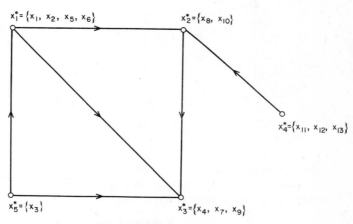

FIG. 2.3. The condensed graph G^*

which x_i and x_j are mutually reachable, and values of 0 in all other places. Thus, two vertices are in the same SC if and only if their corresponding rows (or columns) are identical. The vertices whose corresponding rows contain an entry of 1 under column x_j, then form the vertex set of the SC containing x_j. It is quite apparent that the matrix $\mathbf{R} \otimes \mathbf{Q}$ can be transformed by transposition of rows and columns into block diagonal form; each of the diagonal submatrices corresponding to a SC of G and having entries of 1's, all other entries being 0. For the above example the matrix $\mathbf{R} \otimes \mathbf{Q}$ arranged in this form becomes:

$$\mathbf{R} \otimes \mathbf{Q} =$$

	x_1	x_2	x_5	x_6	x_8	x_{10}	x_4	x_7	x_9	x_{11}	x_{12}	x_{13}	x_3
x_1	1	1	1	1									
x_2	1	1	1	1		0		0			0		0
x_5	1	1	1	1									
x_6	1	1	1	1									
x_8		0			1	1		0			0		0
x_{10}					1	1							
x_4							1	1	1				
x_7		0				0	1	1	1		0		0
x_9							1	1	1				
x_{11}										1	1	1	
x_{12}		0				0		0		1	1	1	0
x_{13}										1	1	1	
x_3		0				0		0			0		1

4. Bases

A *basis B* of a graph is a set of vertices from which every vertex of the graph can be reached and which is minimal in the sense that no subset of B has this property. Thus, if we write $R(B)$—the reachable set of B—to mean:

$$R(B) = \bigcup_{x_i \in B} R(x_i) \tag{2.3}$$

then B is a basis if and only if:

$$R(B) = X \qquad \text{and} \qquad \forall S \subset B, R(S) \neq X \tag{2.4}$$

The second condition ($R(S) \neq X, \forall S \subset B$) of eqn (2.4) is equivalent to the condition $x_j \notin R(x_i)$ for any two distinct $x_i, x_j \in B$, i.e. a vertex in B cannot be reached from another vertex also in B. This can be shown as follows:

Since for any two sets H and $H' \subseteq H$ we have $R(H') \subseteq R(H)$, the condition $R(S) \neq X, \forall S \subset B$ is equivalent to $R(B - \{x_j\}) \neq X$ for all $x_j \in B$, in other words $R(x_j) \nsubseteq R(B - \{x_j\})$. This last condition can be satisfied if and only if $x_j \notin R(B - \{x_j\})$, i.e. if and only if $x_j \notin R(x_i)$ for any $x_i, x_j \in B$.

A basis is, therefore, a set B of vertices which satisfies the following two conditions:

(i) All vertices of G are reachable from some vertex in B and

(ii) No vertex in B can reach another vertex also in B.

From the above two conditions we can immediately state that:

(a) No two vertices in B can be in the same SC of the graph G.

and (b) For any graph without circuits there is a unique basis consisting of of the set of points with indegree 0.

The proofs of these statements are very simple and follow directly from the definitions. (See problems P2.3 and P2.4).

Thus, according to (a) and (b) above, since the condensed graph G^* of a graph G has no circuits, its basis B^* say, is the set of vertices of G^* with indegree 0. The bases of G itself can then be found by forming sets containing one vertex from each one of the vertex-sets in B^*, i.e. if $B^* = \{S_1, S_2, \ldots, S_m\}$ —m being the number of vertex-sets S_j in the basis B^* of G^*—then B is any set $\{x_{i_1}, x_{i_2}, \ldots, x_{i_m}\}$ where $x_{i_j} \in S_j$.

4.1 Example

For the graph G shown in Fig. 2.2 the condensed graph G^* is given in Fig. 2.3. The basis of this graph is $\{x_4^*, x_5^*\}$ since x_4^* and x_5^* are the only two vertices of G^* with indegree 0. The bases of G are then given by $\{x_3, x_{11}\}$, $\{x_3, x_{12}\}$ and $\{x_3, x_{13}\}$.

A concept dual to that of the basis can be defined in terms of the reaching sets $Q(x_i)$ as follows:

A *contra-basis* $\bar{B}$ is a set of vertices of the graph $G = (X, \Gamma)$, so that

and
$$\left. \begin{array}{c} Q(\bar{B}) = \bigcup_{x_i \in \bar{B}} Q(x_i) = X \\ \forall S \subseteq \bar{B}, \quad Q(S) \neq X \end{array} \right\} \tag{2.5}$$

i.e. $\bar{B}$ is a minimal set of vertices which can be reached from every other vertex. The properties of $\bar{B}$ are analogous to those of B where directed concepts are replaced by the opposite counterparts. For example, the definition of eqn (2.5) is equivalent to two conditions similar to (i) and (ii) above but with B replaced by $\bar{B}$ and the words "are reachable from" replaced by "can reach" and *vice versa*.

Thus, the contra-basis of a condensed graph G^* is the set of vertices of G^* with outdegree 0, and the contra-bases of G itself can then be found by constructing sets taking one vertex from each vertex-set in the contra-basis of G^*, similar to what was done previously for the bases.

In the example of the graph G shown in Fig. 2.2, the condensed graph G^*, (shown in Fig. 2.3), contains only one vertex x_3^* with out-degree 0. Thus the contra-basis of G^* is $\{x_3^*\}$ and the contra-bases of G are $\{x_4\}$, $\{x_7\}$ and $\{x_9\}$.

4.2 Application to organizational structure

If the graph G represents the authority or influence structure of an organization, then members of a strong component of G would have equal authority and influence over each other such as could, for example, exist in a committee. A basis of G could then be interpreted as a "coalition" involving the least number of people which would have authority over every person in the organization [2, 3].

On the set of vertices representing people of the same organization, let a new graph G' be constructed to represent channels of communication, so that arc (x_i, x_j) implies that x_i can communicate with x_j. G' will of course be related to G but in which way is not necessarily obvious. The least number of people who know or can obtain all the facts about the organization then form one of the contra-bases of G'. One may then be justified in saying that an effective coalition to control the organization would be the set H of people given by:

$$H = B(G) \cup \bar{B}(G') \tag{2.6}$$

where $B(G)$ and $\bar{B}(G')$ are one of the bases and contra-bases of G and G' respectively chosen so that $|H|$, the number of people in H, is a minimum.

The above graph-theoretic description of an organization is, of course, greatly oversimplified. One of the shortcomings which spring immediately to mind is that it may not be so desirable for a person outside the basis to have authority over a person who is inside.

One can, therefore, define a *power-basis* as the set of vertices $B_p \subseteq X$, so that,

$$R(B_p) = X, \qquad Q(B_p) = B_p \tag{2.7a}$$

and

$$R(S) \neq X \qquad \forall S \subset B_p \tag{2.7b}$$

The second part of condition (2.7a) expresses the fact that only people within B_p can have authority over other people also in B_p, and could be replaced by the equivalent condition $R(X - B_p) \cap B_p = \varnothing$. This condition implies that if a vertex in a SC of G is in B_p, then every other vertex in the same SC must also be in B_p. Since the basis of G^* is the set of those of its vertices with indegree 0 i.e. those not reachable from any other vertex, the power-basis of G is then the union of the vertex-sets in the basis of G^* i.e.

$$B_p = \bigcup_{S_i \in B^*} S_i \tag{2.8}$$

For the graph of example 4.1 (Figs. 2.2 and 2.3), the power-basis of G is $\{x_3, x_{11}, x_{12}, x_{13}\}$. It may be noted that if this graph represents an organization, x_3 may be regarded as its top boss having authority over the sets of people x_1^*, x_2^* and x_3^*, whereas $\{x_{11}, x_{12}, x_{13}\}$ may be regarded as a committee having authority over the two sets of people x_2^* and x_3^*.

5. Problems Associated with Limiting the Reachability

The basis was defined in the last section from the unrestricted reachability sets. If the reachability is restricted to those vertices which are reached along paths of limited cardinality (as mentioned earlier in Section 2), and a restricted basis is defined in terms of these reachabilities, two complications arise.

(a) The concepts of strong components and condensed graphs, which simplified the problem of finding bases in the case of unrestricted reachability, cannot now be used. Extensions of these concepts to the case of restricted reachability do not lead to any significant reduction in problem complexity.

In the case where reachability is limited to single arcs, the restricted bases are called *minimal dominating sets*, and they are considered in greater detail in Chapter 3. In the case where reachability is limited to, say, q arcs, a graph G' may be defined whose vertex set is the same as that of G and where arc (x_i, x_j) exists if and only if there is a path of cardinality less than or equal to q leading from x_i to x_j (see eqn 2.1). The restricted reachability matrix of G will then correspond to the adjacency matrix of G' which is a graph that can be called the restricted transitive closure of G: according to the definition of the transitive closure given in Section 2. The problem of finding the restricted bases of G is then equivalent to the problem of finding the minimal dominating sets of G'.

(b) In the unrestricted case, all the different bases contain the same number of elements; this number being given by the number of vertices of the condensed graph having indegree 0. In the restricted case, however, the restricted bases may contain different numbers of elements and what may now be required is that particular restricted basis with the smallest number of elements. Alternatively, if the restriction on the reachability is not given, one may require that restricted basis which contains exactly p (say) vertices and which can reach all other vertices of G within the smallest possible restricted reachability. These problems are very closely related to the problem of finding a *p-centre* and which is discussed in much greater detail in Chapter 4.

6. Problems P2

1. For the graph of Fig. 2.4 find the reachability and reaching matrices.

2. For the graph of Fig. 2.4 calculate the strong components, draw the condensed graph and find all the bases.

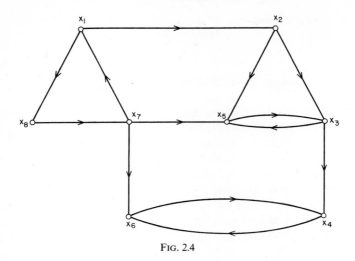

FIG. 2.4

3. Prove that in any graph G all vertices of indegree 0 are in every basis.

4. Prove that in any graph without circuits there is a unique basis consisting of all points of indegree 0.

5. Show that any two bases of a graph G have the same number of vertices.

6. Prove that a vertex x_i is in both the basis B and contrabasis $\bar{B}$ of a graph G if and only if the strong component containing x_i corresponds to an isolated vertex of the condensed graph G^*.

7. Show that if $\hat{x}_i$ is a vertex of the graph (X, A) which produces a maximum to the expression:

$$\max_{x_i \in X} |R(x_i)|,$$

then $\hat{x}_i$ is in a basis.

8. With all arithmetic operations taken to be boolean, (i.e. $0 + 0 = 0$, $0 + 1 = 1 + 1 = 1$, $0 \cdot 0 = 0 \cdot 1 = 0$ and $1 \cdot 1 = 1$), show that—with reachability restricted to paths of q arcs or less—the restricted reachability matrix $\mathbf{R}_q$ is given by:

$$\mathbf{R}_q = \mathbf{I} + \mathbf{A} + \mathbf{A}^2 + \ldots + \mathbf{A}^q = (\mathbf{I} + \mathbf{A})^q$$

where $\mathbf{I}$ is the $n \times n$ unity matrix.

9. Show that the entry $r_{ii}^{(2)}$ of the matrix $\mathbf{R}^2$ is equal to the number of vertices in the strong component containing vertex x_i.

7. References

1. Chen, Y. C. and O. Wing (1964), Connectivity of directed graphs, Proc. of Allerton Conference on Circuit and System Theory, Univ. of Illinois.
2. Flament, C. (1963), "Applications of graph theory to group structure", Prentice-Hall, Engelwood Cliffs, New Jersey.

3. Harary, F., Norman, R. Z. and Cartwright, D. (1965), "Structural models: An introduction to the theory of directed graphs", Wiley, New York.
4. Leifman, L. Ya. (1966), An efficient algorithm for partitioning an oriented graph into bicomponents, *Kibernetika*, **2**, p. 19 (in Russian). English translation: Cybernetics, Plenum Press, New York.
5. Marimont, R. B. (1959), A new method of checking the consistency of precedence matrices, *Jl. of ACM*, **6**, p. 164.
6. Moyles, D. M. and Thompson, G. L. (1969), An algorithm for finding a minimum equivalent graph of a digraph. *Jl. of ACM*, **16**, p. 455.
7. Ramamoorthy, C. V. (1966), Analysis of graphs by connectivity considerations, *Jl. of ACM*, **13**, p. 211.

Chapter 3

Independent and Dominating Sets
—The Set Covering Problem

1. Introduction

Given a graph $G = (X, \Gamma)$, one is often interested in finding subsets of the set X of vertices of G which possess some predefined property. For example, what is the maximum cardinality of a subset $S \subseteq X$ so that the subgraph $\langle S \rangle$ is complete? or what is the maximum cardinality of a subset S so that $\langle S \rangle$ is totally disconnected? The answer to the first question is known as the *clique number* of G and the answer to the second question as the *independence number*. Another problem is to find the minimum cardinality of a subset S so that every vertex of $X–S$ can be reached from a vertex of S by a single arc. The answer to this problem is known as the *dominance number* of G.

These numbers and the subsets of vertices from which they are derived, describe important structural properties of the graph, and have a variety of direct applications in project scheduling [3], cluster analysis and numerical taxonomy [36, 2], parallel processing on a computer [17] and in facilities location and placement of electrical components [66, 56].

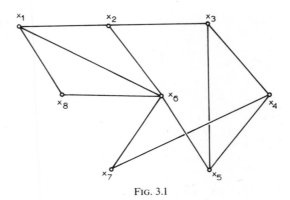

FIG. 3.1

30

In the present Chapter algorithms are given for the determination of the above numbers and several of the application areas are discussed. The *Set Covering Problem*—which is a generalization of the problem of determining the dominance number of graph—is introduced and a method for its solution is described. This last problem is very important not only because of the large number of its direct applications, but also in that it appears often as a subproblem in other areas of graph theory covered by this book. In particular it plays an important role in the determination of the "chromatic number" (Chapter 4), graph "centres" (Chapter 5) and "matchings" (Chapter 12).

2. Independent Sets

Consider a nondirected graph $G = (X, \Gamma)$. An *independent vertex set* (also known as an *internally stable set* [11]) is a set of vertices of G so that no two vertices of the set are adjacent; i.e. no two vertices are joined by a link. Hence, any set $S \subset X$ which satisfies the relation:

$$S \cap \Gamma(S) = \phi \qquad (3.1)$$

is an independent vertex set. For the graph of Fig. 3.1 for example, the sets of vertices: $\{x_7, x_8, x_2\}, \{x_3, x_1\}, \{x_7, x_8, x_2, x_5\}$ etc. are all independent vertex sets. When no confusion arises the word "vertex" will be dropped and these will simply be referred to as independent sets.

An independent set is called maximal when there is no other independent set that contains it. Thus, a set S is a *maximal independent set* if it satisfies (3.1) together with the relation

$$H \cap \Gamma(H) \neq \phi, \qquad \forall H \supset S \qquad (3.2)$$

hence for the graph of Fig. 3.1 the set $\{x_7, x_8, x_2\}$ is not maximal but the set $\{x_7, x_8, x_2, x_5\}$ is. The sets $\{x_1, x_3, x_7\}$ or $\{x_4, x_6\}$ are also maximal independent sets of the graph in Fig. 3.1 and, therefore, in general there is more than one maximal independent set to a given graph. One should also note that the number of elements (vertices) in the various maximal independent sets is not the same for all the sets as can be seen from the above example.

If Q is the family of internally stable sets then the number

$$\alpha[G] = \max_{S \in Q} |S| \qquad (3.3)$$

is called the *independence number* of a graph G, and the set S^* from which it is derived is called a *maximum independent* set.

In the graph of Fig. 3.1 the family of maximal independent sets is composed of the sets:

$$\{x_8, x_7, x_2, x_5\}, \{x_1, x_3, x_7\}, \{x_2, x_4, x_8\}, \{x_6, x_4\}, \{x_6, x_3\}$$
$$\{x_7, x_5, x_1\}, \{x_1, x_4\}, \{x_3, x_7, x_8\}.$$

The largest of these sets has 4 elements and hence $\alpha[G] = 4$. The set $\{x_8, x_7, x_2, x_5\}$ is a maximum independent set.

2.1 Example: project selection

Consider n projects which must be executed, and suppose that each project x_i requires some subset $R_i \subseteq \{1, \ldots, p\}$ of p available resources for its execution. Further assume that each project (given its resource requirements), can be executed in a single time period. Now form a graph G each vertex of which corresponds to a project and with links (x_i, x_j) added whenever x_i and x_j have some resource requirement in common, i.e. whenever $R_i \cap R_j \neq \phi$. A maximal independent set of G then represents a maximal set of projects that can be executed simultaneously during the single time period.

In a dynamic system new projects become available for execution after each time period, so that the family of maximal independent sets of G has to be found repeatedly in order to choose which maximal set of projects to execute during the current period. In a practical system, it is not sufficient simply to execute the set of projects corresponding to the maximum independent set at every period, since some projects may then be delayed indefinitely. A better way would be to give a penalty p_i to every project (vertex) x_i which increases as the delay in the execution of x_i increases, and then at every point choose from the family of maximal independent sets that set which maximizes some function of the penalties on the vertices that it contains.

2.2 Maximal complete subgraphs (cliques)

A concept which is the opposite of that of the maximal independent set is that of a maximal complete subgraph. Thus, a *maximal complete subgraph* (*clique*) of G is a subgraph based on the set S of vertices which is complete and which is maximal in the sense that any other subgraph of G based on a set $H \supset S$ of vertices is not complete. Hence, in contrast to a maximal independent set for which no two vertices are adjacent, the set of vertices of a clique are all adjacent to each other. It is quite obvious, therefore, that the maximal independent set of a graph G corresponds to a clique of the graph $\tilde{G}$ and *vice versa*, where $\tilde{G}$ is the graph complementary to G.

It is also quite apparent, that for a nondirected graph, a clique is a concept which is similar, but more stringent, than that of a strong component of a graph (see Chapter 2). A clique may in fact also be considered as a strong component where reachabilities are restricted to single arcs as mentioned in Section 5 of Chapter 2.

In a way analogous to the definition of the independence number by equation (3.3) we can define the *clique number* (also known as the *density*) as being the maximum number of vertices in a clique. Then, roughly speaking, for a "dense" graph the clique number is likely to be high and the independence number low, whereas for a "sparse" graph the opposite is likely to be the case.

2.3 The computation of all maximal independent sets

Because of the relationship between cliques and maximal independent sets mentioned above, the methods of computation presented in this section and which are described in terms of the maximal independent sets can also be used directly for the computation of cliques. One may, at first sight, suppose that the computation of all the maximal independent sets of a graph is a very simple task, which could be done by systematically enumerating independent sets, testing if they are the largest possible (by attempting to add any other vertex to the set whilst preserving independence) and storing the maximal ones. The supposition is certainly true for small graphs of, say, up to 20 vertices, but as the number of vertices increases this method of generation becomes computationally unwieldy. This is so, not so much because the number of maximal independent sets becomes excessive, but owing to the fact that during the process a very large number of independent sets are formed and subsequently rejected because they are found to be contained in other previously generated sets and are, therefore, not maximal in themselves.

In the current section we will describe a systematic enumerative method due to Bron and Kerbosch [14], which substantially overcomes this difficulty so that independent sets—once generated—need not be checked for maximality against the previously generated sets.

2.3.1 THE BASIS OF THE ALGORITHM. The algorithm is essentially a simple enumerative tree search during which—at some stage k—an independent set of vertices S_k is augmented by the addition of another suitably chosen vertex to produce an independent set S_{k+1}, at stage $k + 1$, until no further augmentation is possible and the set becomes a maximal independent set. At stage k let Q_k be the largest set of vertices for which $S_k \cap Q_k = \phi$, i.e. any vertex from Q_k added to S_k produces a set S_{k+1} which is independent.

At some point during the algorithm, Q_k consists of two types of vertices: vertices in Q_k^- (say) which have already been used earlier in the search to augment S_k and vertices in Q_k^+ (say) which have not yet been used. A forward branching during the tree search then involves choosing a vertex $x_{i_k} \in Q_k^+$, appending it to S_k to produce

$$S_{k+1} = S_k \cup \{x_{i_k}\} \tag{3.4}$$

and creating new sets Q_{k+1} as:

$$Q_{k+1}^- = Q_k^- - \Gamma(x_{i_k}) \tag{3.5}$$

and

$$Q_{k+1}^+ = Q_k^+ - \Gamma(x_{i_k}) - \{x_{i_k}\} \tag{3.6}$$

A backtracking step of the algorithm involves the removal of x_{i_k} from S_{k+1} to revert back to S_k, and the removal of x_{i_k} from the old set Q_k^+ and its addition to the old set Q_k^- to form two new sets Q_k^+ and Q_k^-.

The following observations are quite obvious. A set S_k is a maximal independent set only if it cannot be augmented further, i.e. only if $Q_k^+ = \phi$. If $Q_k^+ \neq \phi$, it immediately follows that the current S_k was at some previous stage augmented by some vertex in Q_k^- and is therefore not maximal. If $Q_k^- = \phi$, the current S_k has not been previously augmented, and since sets are generated without duplication, S_k is a maximal independent set. Thus, the necessary and sufficient conditions for S_k to be a maximal independent set are:

$$Q_k^+ = Q_k^- = \phi \tag{3.7}$$

It is now quite apparent that if a stage is reached when some vertex $x \in Q_k^-$ exists for which $\Gamma(x) \cap Q_k^+ = \phi$, then regardless of which vertex from Q_k^+ is used to augment S_k in any number of forward branchings, vertex x can never be removed from Q_p^- at any future step $p > k$. Thus, the condition:

$$\exists x \in Q_k^- \quad \text{so that} \quad \Gamma(x) \cap Q_k^+ = \phi \tag{3.8}$$

is sufficient for a backtracking step to be taken since no maximal independent set can result from any forward branching from S_k.

As in all tree search methods, it is obviously beneficial to aim for early backtracking steps since these tend to limit the unnecessary part of the tree search. It is therefore worth while to aim at forcing condition (3.8) as early as possible by a suitable choice of the vertices used in augmenting the set S_k. During a forward step one can choose any vertex $x_{i_k} \in Q_k^+$ with which to augment set S_k, and on backtracking x_{i_k} will be removed from Q_k^+ and inserted into Q_k^-. Now if x_{i_k} is chosen to be a vertex $\in \Gamma(x)$ for some $x \in Q_k^-$, then at the corresponding backtracking step the number:

$$\Delta(x) = |\Gamma(x) \cap Q_k^+| \tag{3.9}$$

will be decreased by one (from its value prior to the completion of the forward and backward steps) so that condition (3.8) is now more nearly true.

Thus, one possible way of choosing the vertex x_{i_k} with which to augment S_k, is to first determine the vertex $x^* \in Q_k^-$ with the smallest possible value of $\Delta(x^*)$ and then choose x_{i_k} from the set $\Gamma(x^*) \cap Q_k^+$. Such a choice of x_{i_k} at every step will cause $\Delta(x^*)$ to decrease by one every time a backtracking step at level k is made, until eventually vertex x^* satisfies condition (3.8), at which time backtracking can occur.

It should be noted that since at a backtracking step x_{i_k} enters Q_k^-, it may be that this new entry has a smaller value of Δ than the previously fixed vertex x^* so that this new vertex should now become the target for the attempt to force condition (3.8). This is particularly important in the first forward branchings when $Q_k^- = \phi$.

2.3.2 DESCRIPTION OF THE ALGORITHM

Initialisation

Step 1. Set $S_0 = Q_0^- = \phi$, $Q_0^+ = X$, $k = 0$.

Forward step

Step 2. Choose a vertex $x_{i_k} \in Q_k^+$ as mentioned earlier and form S_{k+1}, Q_{k+1}^- and Q_{k+1}^+ keeping Q_k^- and Q_k^+ intact. Set $k = k + 1$.

Test

Step 3. If condition (3.8) is satisfied to go to step 5, else go to step 4.

Step 4. If $Q_k^+ = Q_k^- = \phi$ print out the maximal independent set S_k and go to step 5. If $Q_k^+ = \phi$ but $Q_k^- \neq \phi$ go to step 5. Otherwise go to step 2.

Backtrack

Step 5. Set $k = k - 1$. Remove x_{i_k} from S_{k+1} to produce S_k. Retrieve Q_k^- and Q_k^+, remove x_{i_k} from Q_k^+ and add it to Q_k^-. If $k = 0$ and $Q_0^- = \phi$, Stop. (All maximal independent sets have been printed out). Otherwise goto step 3.

Comments on the implementation of the above algorithm are given in [14] together with an Algol program listing. This program has been tested on a number of graphs (including the Moon–Moser graphs—see problem 10), and the computation times per maximal independent set generated were found to be almost constant and independent of the size of the graph itself, which makes the algorithm close to the best possible.

3. Dominating Sets

For a graph $G = (X, \Gamma)$ a *dominating vertex set* (also known as an *externally stable set* [11]), is a set of vertices $S \subseteq X$ chosen so that for every vertex x_j not in S, there is an arc from a vertex in S to x_j.

Thus S is a dominating vertex set (or simply a dominating set—when no confusion arises) if:

$$S \cup \Gamma(S) = X \qquad (3.10)$$

For the graph of Fig. 3.2 the vertex sets $\{x_1, x_4, x_6\}$, $\{x_1, x_4\}$, $\{x_3, x_5, x_6\}$ etc. are all dominating sets.

A dominating set is called minimal if there is no other dominating set which is contained in it. Thus, a set S is a minimal dominating set if it satisfies eqn (3.10) and there is no other set $H \subset S$ which also satisfies eqn (3.10). Hence, for the graph of Fig. 3.2 the set $\{x_1, x_4\}$ is minimal but the set $\{x_1, x_4, x_6\}$ is not.

The set $\{x_3, x_5, x_6\}$ is also a minimal dominating set and other such sets exist for this example. Thus, as in the case of maximal independent sets, there are in general more than one minimal dominating set and they do not necessarily all contain the same number of vertices.

If P represents the family of minimal dominating sets then the number

$$\beta[G] = \min_{S \in P} |S| \qquad (3.11)$$

is called the *dominance number* of a graph G, and the set S^* from which it is derived is called the *minimum* dominating set.

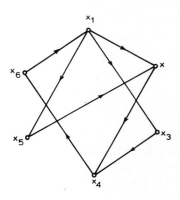

FIG. 3.2

For the graph of Fig. 3.2 the minimal dominating set containing the smallest number of elements is the set $\{x_1, x_4\}$ and hence $\beta[G] = 2$.

3.1 Example: Location of "centres" to cover a region

A very wide range of problems fall into this general category. These include:

(a) The location of T.V. or radio transmitters to broadcast to a specified area.

(b) The location of army posts to control a region.

(c) The location of home bases for salesmen to serve a region.

Suppose that a region—shown by the large square in Fig. 3.3(a)—is divided into 16 smaller areas as shown. An army post located in an area is assumed to be able to control not only the square that it is located in, but also those areas which have common borders with that square. What is then required is to choose the smallest possible number and the locations of army posts so that the whole region is under control.

If we represent each small area by a vertex and add a link between two vertices whenever the corresponding areas have a common border, a graph

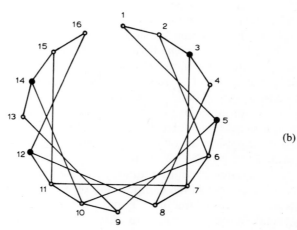

FIG. 3.3

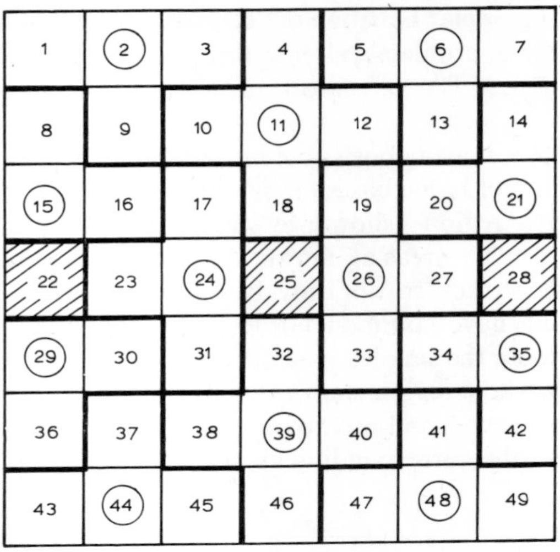

FIG. 3.4

is formed as shown in Fig. 3.3(b). The problem then reduces to one of finding (for this graph) a minimum dominating set. The number $\beta[G]$ is then the smallest number of posts to cover the entire region. For the example of Fig. 3.3, $\beta[G] = 4$ and the posts should be in locations $\{3, 5, 12, 14\}$ or in locations $\{2, 9, 15, 8\}$.

Similarly, for the region shown in Fig. 3.4 the dominance number of the corresponding graph is 12 and the location of the posts should be $\{2, 6, 11, 15, 21, 24, 26, 29, 35, 39, 44, 48\}$. Only the three shaded squares are protected by two posts simultaneously.

The relationship between the concept of a minimal dominating set and that of a basis is quite apparent; as is the relationship between a minimal dominating set and a p-centre (see Chapter 5). These relationships have been discussed in Section 5 of Chapter 2, and in Chapter 5 an algorithm for the computation of the minimum dominating set is used as a basic step in the more general algorithm for the calculation of p-centres.

In the case of maximal independent sets we have given an algorithm which generates the complete list of such sets because this list is required in many practical applications [36, 3, 2]. In the case of dominating sets, however, what is invariably required is simply a minimum dominating set and we, therefore, restrict ourselves here to the computation of such a set. In the next section we cast the problem of finding a minimum dominating set into a slightly more general framework, and this helps to highlight relationships with concepts discussed in other parts of this book.

4. The Set Covering Problem

If one considers the transpose of the adjacency matrix A^t, with all diagonal elements set to 1, then the problem of finding the minimum dominating set is equivalent to the problem of choosing the least number of columns so that every row contains an entry of 1 under at least one of the chosen columns. This last problem of finding the minimum number of columns to "cover" all the rows has been studied extensively under the name of the *Set Covering Problem* (SCP).

In the general SCP the matrix of 0–1 coefficients need not be square as is the case with the adjacency matrix. Moreover, associated with each column j, (vertex x_j in our case), is a cost c_j, and what is required is to choose a cover (dominating vertex set in the graph case) with the least total cost. One may— at first sight—suppose that since the problem of finding the minimum dominating vertex set is a very special covering problem with $c_j = 1$ for all $j = 1, 2, \ldots, n$, the finding of this set may in fact be much simpler than the solution of the general SCP. However, this is not generally the case and in this section we will, therefore, be concerned with the solution of the general SCP.

4.1 Problem formulation

The SCP owes its name to the following set-theoretic interpretation. Given a set $R = \{r_1, \ldots, r_M\}$, and a family $\mathscr{S} = \{S_1, \ldots, S_N\}$ of sets $S_j \subset R$ any subfamily $\mathscr{S}' = \{S_{j_1}, S_{j_2}, \ldots, S_{j_k}\}$ of $\mathscr{S}$ such that

$$\bigcup_{i=1}^{k} S_{j_i} = R \tag{3.12}$$

is called a *set-covering* of R, and the S_{j_i} are called the *covering sets*. If, in addition to equation (3.12), $\mathscr{S}'$ also satisfies

$$S_{j_h} \cap S_{j_l} = \phi, \qquad \forall h, l \in \{1, \ldots, k\}, \qquad h \neq l \tag{3.13}$$

i.e. the S_{j_i} $(i = 1, \ldots, k)$ are pairwise disjoint, then $\mathscr{S}'$ is called a *set-partitioning* of R.

If, with each $S_j \in \mathscr{S}$, there is associated a (positive) cost c_j, the SCP is to find that set-covering of R which has minimum cost, the cost of $\mathscr{S}' = \{S_{j_1}, \ldots, S_{j_k}\}$ being $\sum_{i=1}^{k} c_{j_i}$. The *Set Partitioning Problem* (SPP) is defined correspondingly.

In the matrix form referred to earlier—where the rows of an M-row N-column 0–1 matrix $[t_{ij}]$ are to be covered by columns—the SCP can be formulated as a 0–1 linear program:

Minimize:
$$z = \sum_{j=1}^{N} c_j \xi_j$$

subject to: $$\sum_{j=1}^{N} t_{ij}\xi_j \geq 1 \qquad i = 1, 2, \ldots, M. \tag{3.14}$$

where $c_j \geq 0$, and ξ_j is 1, (0) depending on $S_j \in \mathscr{S}'$, $(S_j \notin \mathscr{S}')$ and where $t_{ij} = 1$, (0) depending on $r_i \in S_j$, $(r_i \notin S_j)$.

For the SPP, the inequalities (3.14) become

$$\sum_{j=1}^{N} t_{ij}\xi_j = 1, \qquad i = 1, 2, \ldots, M \tag{3.15}$$

4.2 Problem reduction

Due to the nature of the SCP it is often possible to make certain well-known *a priori* deductions and reductions. [6, 26, 27, 28, 30, 50, 51]

Thus:

(1) If for some $r_i \in R$, $r_i \notin S_j \forall j = 1, \ldots, N$, then r_i cannot be covered and the problem has no solution.

(2) If $\exists r_i \in R$ so that $r_i \in S_k$, and $r_i \notin S_j$, $\forall j \neq k$, then S_k must be in all solutions and the problem may be reduced by setting $R = R - \{r_i\}$ and $\mathscr{S} = \mathscr{S} - \{S_k\}$.

(3) Let $V_i = \{j \mid r_i \in S_j\}$. Then, if $\exists p, q \in \{1, \ldots, M\}$ with $V_p \subseteq V_q$, r_q may be deleted from R, since any set that covers r_p must also cover r_q, i.e. r_q is dominated by r_p.

(4) If, for some family of sets $\bar{\mathscr{S}} \subset \mathscr{S}$ we have $\bigcup\limits_{S_j \in \bar{\mathscr{S}}} S_j \supseteq S_k$ and $\sum\limits_{S_j \in \bar{\mathscr{S}}} c_j \leq c_k$ for any $S_k \in \mathscr{S} - \bar{\mathscr{S}}$, then S_k may be deleted from $\mathscr{S}$ since it is dominated by $\bigcup\limits_{S_j \in \bar{\mathscr{S}}} S_j$.

Let us now assume that these reductions, if applicable, have been applied and the original SCP reformulated in its irreducible form.

4.3 A tree search algorithm for the SPP

As noted earlier, the SPP is closely related to the SCP, being essentially an SCP with an additional (no-overcovering) restriction. This restriction is advantageous when we attempt to solve the problem by a tree search method, since it may allow the early abandonment of potential branches of the tree. Therefore, we shall first discuss a tree search algorithm for solving the SPP, and then show how it can be extended to solve the SCP.

Simple tree search methods for solving the SPP were proposed by Garfinkel and Nemhauser [27] and Pierce [48], and are essentially as follows:

At the start of the process, we make up M "blocks" of columns, one block for each element r_k of R. The kth block consists of those sets of $\mathscr{S}$ (represented by columns) which contain element r_k but which do not contain any lower-numbered elements $r_1, \ldots, r_{k-1}$. Each set (column), therefore, appears in exactly one block, and the totality of blocks can in general be arranged in tableau form as shown in Table 3.1 although some block(s) may be nonexistent in a particular problem.

During the course of the algorithm, blocks are searched sequentially, with block k not being searched unless every element r_i, $1 \leqslant i \leqslant k - 1$, has already been covered by the partial solution. Thus, if any set in block k were to contain elements indexed less than k, the set would have to be discarded (at this stage) due to the no-overcovering requirement.

The sets within each block are arranged (heuristically) in ascending order of their cost and the sets are now renumbered so that S_j will from now on be used to mean the set corresponding to the jth column of the tableau.

TABLE 3.1. Initial tableau

	Block 1	Block 2	Block 3	Block 4
r_1	1111	0		
r_2		1111	0	
r_3	0 or 1		1111	0
r_4		0 or 1		1111
.			0 or 1	
.				0 or 1
.				etc.
r_M				

At any stage of the tree search a current "best" solution $\check{B}$ of cost $\check{z}$ is known, where $\check{B}$ represents the family of corresponding covering sets. If B and z represent the corresponding values in the current state of the search, and E is a set representing those elements (i.e. rows) r_i covered by the sets in B, then a simple tree search algorithm proceeds as follows:

Initialization
Step 1. Set up the initial tableau and begin with the partial solution $B = \phi$, $E = \phi$, $z = 0$ and $\check{z} = \infty$.

Augmention:
Step 2. Find $p = \min [i \,|\, r_i \notin E]$. Set a marker at the top (lowest cost set) of block p.

Step 3. Beginning at the marked position in block p, examine its sets S_j^p, say, in increasing order of j.

(i) If set S_j^p is found such that $S_j^p \cap E = \phi$, and $z + c_j^p < \check{z}$ (where c_j^p is the cost of S_j^p), go to step 5.

(ii) Otherwise, if block p is exhausted or a set S_j^p reached for which $z + c_j^p \geqslant \check{z}$ go to step 4.

Backtrack
Step 4. B cannot lead to a better solution.
If $B = \phi$ (i.e. block 1 has been exhausted), terminate with the optimal solution $\check{B}$. Otherwise remove the last set, S_k^l say, added into B, put $p = l$, place a marker on set S_{k+1}^l, remove the previous marker in block l and go to step 3.

Test for new solution
Step 5. Update $B = B \cup \{S_j^p\}$, $E = E \cup S_j^p$, $z = z + c_j^p$.
If $E = R$ a better solution has been found. Set $\check{B} = B$, $\check{z} = z$ and go to step 4. Otherwise goto step 2.

As the search terminates with the exhaustion of block 1 (see step 4 above), it would seem worthwhile to arrange the blocks in ascending order according to the number of columns (sets) in each. This can be achieved by renumbering the elements (rows) $r_1, \ldots, r_M$ in increasing order of number of sets in S containing that element, before setting up the initial tableau.

Before showing how this SPP algorithm can be extended for the case of the SCP, we briefly mention some of the other methods proposed for solving the SPP. In [49], Pierce and Lasky give some modifications to the above basic algorithm, including subsidiary use of a linear program; while, in [45], Michaud describes another implicit enumeration algorithm which is based on a linear programming problem corresponding to the SPP, with the block structure given above being used in a secondary role.

Other algorithms involving simplex-type iterations have been proposed, both primal, [4, 5], and dual, [58, 37] whereas Jensen [37] has had success with a dynamic programming method for a certain type of SPP's.

4.4 A tree search algorithm for the SCP

The only step of the algorithm described in Section 4.3 above which is particular to the SPP, is step 3(i). If the "no-overcovering" requirement $S_j^p \cap E = \phi$ in that step were to be removed, then that algorithm could be used for the SCP. However, the setting up of the initial tableau would now have to be somewhat different from that shown in Table 3.2. Thus, let $S_j = \{r_{j_1}, r_{j_2}, r_{j_3}, \ldots\}$, with $j_1 < j_2 < j_3 \ldots$. It is not now sufficient to include S_j only in block j_1, because with the "no-overcovering" requirement removed, it is not now possible to exclude S_j from consideration at, say, block j_2 if r_{j_1} has already been covered by the partial solution. Hence S_j must be entered once in each of the blocks $j_1, j_2, j_3, \ldots$ On the other hand, since an element r_{j_α} of S_j would, by the sequential nature of the search, be covered before branching on the sets of block β ($\beta > j_\alpha$) is considered, it is now possible to remove all the elements x_{j_α} from a set S_j when entering S_j into any block $\beta > j_\alpha$ without affecting the solution of the problem in any way. This apparently trivial removal is in fact computationally very beneficial, since the initial tableau of the SCP now set up can be reduced by the preliminary reduction tests of Section 4.2 to much smaller dimensions that would have been possible without the removal of the elements x_{j_α} ($j_\alpha < \beta$) from the sets of block β [48].

It should be pointed out here that the algorithm of Section 4.3 is a primitive tree search devoid of any sophistication, and although with careful coding it can be made quite effective for the SPP [27] it is not, as it stands, an efficient algorithm for solving SCP's of even moderate size. We will now discuss some dominance tests and the calculation of lower bounds which can be used to limit the tree search and improve the efficiency of the basic algorithm.

4.4.1 SOME DOMINANCE TESTS.

We will first illustrate these dominance tests by an example before stating the general results. Consider a case where block 1 contains (among others), sets $S_1 = \{r_1, r_4, r_6\}$ and $S_2 = \{r_1, r_3, r_4\}$ of costs 3 and 4 respectively, and where block 2 contains (among others), sets $S_3 = \{r_2, r_3, r_5\}$ and $S_4 = \{r_2, r_4, r_5\}$ each of cost 2.

During the course of the algorithm of Section 4.3 a stage will be reached where $B_a = \{S_1, S_3\}$, $E_a = \{r_1, r_2, \ldots, r_6\}$, $z_a = 5$; and we will branch forward from this stage until either we find a better solution than the current $\check{B}$, or we realize that S_1 and S_3 cannot both appear in an optimal solution.

Many steps later, we will reach a stage where

$$B_b = \{S_2, S_3\}, \qquad E_b = \{r_1, \ldots, r_5\}, \qquad z_b = 6.$$

At this point it becomes apparent that any further branching from this stage is unnecessary, since $E_b \subseteq E_a$ and $z_b \geqslant z_a$. A similar situation occurs when the stage is reached where

$$B_c = \{S_2, S_4\}, \qquad E_c = \{r_1, \ldots, r_5\}, \qquad z_c = 6.$$

Thus, it may be worthwhile to keep, for each value of $z = 1, 2, \ldots \check{z}$ some (perhaps incomplete) list of the maximal E's which have already been achieved for this z, (where by maximal is meant a set not included in another set also in the list). These lists of E's can then be used to limit the search by eliminating branches that later on prove fruitless. It is generally impractical, however, to save *all* the maximal sets E, at any cost level. In fact if this is done, the method would be similar to a Dynamic Programming approach to the problem as a whole (or, alternatively, could be considered as a full breadth-first tree search).

Let us now assume that we have saved some list $(L(z_i)$ of previous sets E that have been attained during the course of the algorithm at a summed-cost level z_i. Suppose a stage is reached where $E = E'$, $B = B'$, $z = z'$, and we are about to examine block k (i.e. $k = \min \{i | r_i \notin E'\}$ at step 2 of the algorithm of Section 4.3), and to consider set S_j^k of cost c_j^k for the next branching. If $z' + c_j^k < \check{z}$, the original algorithm would have branched forward from this stage, and updated

$$E = E' \cup S_j^k, \qquad B = B \cup \{S_j^k\}, \qquad z = z + c_j^k,$$

regardless of any other considerations.

However, one could now also ensure (at step 3), that:

$$E' \cup S_j^k \nsubseteq E_l; \quad \forall E_l \in L(z_i), \quad \text{and for all } z_i, \quad z' < z_i \leqslant z' + c_j^k \quad (3.16)$$

before branching forward. If the set S_j^k fails the above test, then it is rejected and instead the next set, S_{j+1}^k, of block k is considered, etc. If S_j^k passes test (3.16), one could branch forward on S_j^k, and continue as previously, after updating $L(z)$ by introducing the set $E' \cup S_j^k$ into $L(z' + c_j^k)$.

Since, as mentioned earlier, it is practically impossible to store the complete lists $L(z_i)$, some heuristic criterion has to be used to decide on the size of these lists, and how they should be updated during the search.

(*A*) *List size.* It is obviously better to eliminate sets S_j^k which are potential tree branchings at an early level in the tree, as this would discard larger portions of the potential search-tree. It would thus seem intuitively worthwhile to keep larger-sized lists $L(z_i)$ for the smaller z_i. This argument is further reinforced by the fact that test (3.16) is more likely to be effective when the sets T contain only a few elements, which is generally the case for small z_i-levels.

(*B*) *Updating of the Lists.* At a particular z_i-level, the current E-set (E' say) is more likely to be a subset of a set E'' (in the list $L(z_i)$), which was derived from the same general part of the search-tree. This is so because E' and E'' have in common a large section of the path from the start-node to their respective tree nodes. Thus, it would seem to be advantageous to arrange the lists $L(z_i)$ as stacks and to use a FIFO (first in, first out) policy, with the bottom-most set being discarded in the event of stack overflow.

4.4.2 THE CALCULATION OF A LOWER BOUND. At some stage of the search, given by B', E', z', and where block k is the next block to be considered, a lower bound h on the minimum value of z, can be calculated and used to limit the tree search as follows:

Consider an uncovered element $r_i \in R - E'$ which does not appear in any of the sets of those blocks $k, k + 1, \ldots, i - 1$, corresponding to elements not yet covered by the partial solution. The element r_i cannot then be covered unless some set S_j^i of block i is chosen to add to B' at a future stage. Thus, for each such element r_i, construct a row for matrix $D = [d_{qs}]$ and a row for a second matrix $D' = [d'_{qs}]$, where d_{qs} is the number of elements in set S_s^q and d'_{qs} is its cost.

In addition, append an extra row, θ say, to D and D' with $d_{\theta s} = s$ for all $s = 0, 1, \ldots, M - |E'|$ and $d'_{\theta s} = s \cdot \min [c_j^i / |S_j^i|]$, the minimum being taken over all sets S_j^i with $r_i \notin E'$. Note that although the number of elements in a row q_1, of D (or D') may be different from that of another row q_2, we assume here that 0's and ∞'s are entered at the end of some rows of D and D' respectively, so that all rows have f (say) elements and the matrices become rectangular.

The following observations can now be made: since the optimal solution to the current subproblems must cover $M - |E'|$ elements, a choice of one entry from each row i of D, which satisfies

$$(d_{1s_1} + d_{2s_2} + \ldots + d_{\theta s_\theta}) \geqslant M - |E'|$$

and which minimizes the corresponding cost,

$$v = (d'_{1s_1} + \ldots + d'_{\theta s_\theta})$$

gives v as a lower bound to the optimal cost of the (covering) subproblem. Note that the assumption is made that the sets corresponding to the allocations in the rows of D are disjoint, which is obviously the best possible situation. The last row θ simply ensures that, if

$$\sum_{i=1}^{\theta - 1} d_{is_i} < M - |E'|,$$

the remaining elements are covered in the best possible way, i.e. at minimal cost-per-additional-element-covered.

The minimum value of

$$\sum_{i=1}^{\theta} d'_{is_i} \quad \text{subject to} \quad \sum_{i=1}^{\theta} d_{is_i} \geqslant M - |E'|$$

can easily be derived by a dynamic programming algorithm as follows:

Let $g_\rho(v)$ be the maximum number of elements that can be covered using only the first ρ rows of D (i.e. only ρ of the blocks in the subproblem), and whose total cost does not exceed v. $g_\rho(v)$ can then be calculated iteratively as:

$$g_\rho(v) = \max_{s=1,\ldots,f} [d_{\rho s} + g_{\rho-1}(v - d'_{\rho s})], \tag{3.17}$$

where $g_0(v)$ is initialized to 0 for all v.

Hence, the lowest value, v^*, of v, for which $g_\theta(v) \geqslant M - |E'|$ is then the required lower bound, h, and can be easily obtained from the dynamic programming tableau derived from the above iterative equation. One should note that only that range of v where $0 \leqslant v < \check{z} - z'$ need be considered, since if $h \geqslant z - z'$ (i.e. $g_\theta(v) < M - |E'|$ for $v \geqslant \check{z} - z'$), a backtracking step may be taken immediately.

4.5 Computational performance of the SCP algorithm

The computational performance of the algorithm described in Section 4.4 for the SCP is shown in Table 3.2 The numbers of rows and columns shown are those resulting after the reduction tests have been applied. All problems are randomly generated and have costs $c_j = 1$ for all columns j.

TABLE 3.2 Computing times of the SCP algorithm for random problems

Problem	Number of rows	Number of columns	Density	Computing time†	Nodes in tree search‡
1	15	20	0·22	0·05	0·02
2	15	25	0·25	0·04	0·01
3	15	25	0·30	0·06	0·02
4	20	60	0·18	0·25	0·07
5	20	50	0·20	0·11	0·02
6	20	75	0·23	0·35	0·05
7	25	110	0·17	0·60	0·06
8	25	120	0·19	1·30	0·16
9	25	170	0·22	1·50	0·11
10	30	80	0·12	1·00	0·20
11	30	180	0·15	10·50	1·60
12	30	250	0·17	6·00	0·52
13	30	340	0·21	9·20	0·30
14	30	475	0·20	11·00	0·09
15	30	500	0·23	22·00	0·40
16	30	700	0·23	105·50	3·02
17	30	725	0·25	32·50	0·12
18	30	875	0·26	57·20	0·18
19	30	1000	0·27	72·80	0·19
20	35	160	0·09	18·50	2·18
21	35	290	0·13	28·00	2·03
22	35	460	0·16	75·50	2·95
23	35	585	0·18	129·50	3·96
24	35	780	0·19	141·20	5·03
25	35	1075	0·20	110·00	0·55

† CDC 6600 seconds.
‡ Number of nodes generated by the tree (thousands).

Other methods of solving the SCP are given by Lemke, Salkin and Spielberg [41] who use a different type of tree search with an embedded linear program. The solution of this LP is used as a lower bound during the search and is also used in determining the next forward tree branching from the current node. Cutting-plane approaches, similar in principle to those used for general 0–1 programming [32] are given by House, Nelson and Rado [35] and by Bellmore and Ratliff [9]. A computational survey and comparison of these methods is given by Christofides and Korman [19].

5. Applications of the Covering Problem

5.1 Choice of interpreters

Let us assume that an organization wants to hire interpreters to translate from French, German, Greek, Italian, Spanish, Russian and Chinese into English, and suppose there are five prospective applicants A, B, C, D and E. Each applicant requires a certain salary and speaks only a subset of the above set of languages. What is then required is to find what candidates to hire in order to be able to translate from all the above languages into English with the least cost of hiring. This is quite obviously a covering problem.

If, for example, the salary requirements of all the applicants are the same and the languages they speak are as shown in the T matrix below, then the solution to the above problem is to hire persons B, C and D.

Language ↓	A	B*	C*	D*	E
French	1	0	1	1	0
German	1	1	0	0	0
Greek	0	1	0	0	0
Italian	1	0	0	1	0
Spanish	0	0	1	0	0
Russian	0	1	1	0	1
Chinese	0	0	0	1	1

Person →

5.2 Information retrieval [21]

Let us suppose that a number of pieces of information are stored on N files of length $c_j = 1, 2, \ldots, N$ and that each piece is stored on at least one file. At some time a request for M pieces of information is made and these can be obtained by searching the files in a number of ways. In order to obtain all M pieces of information with the least length of files that have to be searched a SCP will have to be solved where an element t_{ij} of the matrix T is 1 if information i is on file j and 0 otherwise.

5.3 Airline crew scheduling [1, 55, 64, 65]

Suppose that the vertices of a nondirected graph G represent the places that an airline can visit and the arcs of G represent the flight "legs" that the airline must operate at a given time. Any path in this graph (satisfying a number of constraints that may be imposed in practice) forms a feasible flight schedule which can be operated. Let there be N such feasible schedules, the cost of the jth one being calculated in some way to be c_j. The problem of finding the least-cost set of schedules so that every flight leg is operated at least once is then a covering problem where the matrix T contains an entry of unity in position (i, j) if flight leg i is contained in schedule j, and zero otherwise.

A variant to this problem (which also appears often in aircrew scheduling) is derived by insisting that a flight leg can appear once and only once in the chosen set of schedules. This leads to a SPP corresponding to the SCP defined above.

5.4 Simplification of logical (boolean) expressions [30, 50, 51, 67, 44, 43]

When simplifying a logical expression E, which is (say), given in disjunctive normal form, one need only consider the set P, of prime implicants of E. A "simplest" form for E is then obtained by finding the lowest-cardinality subset $\check{P} \subseteq P$ subject to the condition that every (conjunctive) term of E be 'covered' by (i.e. by implied by) at least one prime implicant in $\check{P}$. This is obviously a SCP, where once more the column costs $c_j = 1$ for all $j = 1, \ldots, N$.

A problem equivalent to finding a minimal representation to logical expressions occurs in the design of optimal switching circuits.

5.5 Vehicle dispatching [7, 47, 24]

In the vehicle dispatching problem a graph represents a road network with vertices representing customers except one vertex which represents the depot. A vehicle leaves the depot, supplies goods to some customers and returns to the depot thus traversing a circuit. If the "cost" of circuit j is c_j (for example c_j could be the mileage of the circuit or the time to complete it), how many vehicles should be used on which circuits so as to supply all customers (once only) in any one day and minimize cost? This problem can obviously be considered as a SPP whose columns represent all possible feasible (because of the practical constraints), circuits starting and finishing at the depot. The rows represent the customers.

An almost identical problem involves the routing of a cable to carry electricity from a substation (depot) to customers in "ring-circuits" [15]. Other practical applications of the SCP or SPP include: assembly line balancing [59, 25], political districting [29], the calculation of bounds in general integer programs [18], network planning [20] and network attack or defence [8, 10].

5.6 Other graph covering problems

A large number of problems in graph theory can be formulated as SCP's although many of these problems could be solved more efficiently by alternative graph-theoretic means discussed in other parts of this book. The purpose of the present section is to show how some of these problems are related via the SCP.

5.6.1 THE MINIMUM DOMINATING VERTEX SET. As mentioned in Section 4 of the present Chapter, the problem of finding the minimum dominating vertex set of a graph G is a SCP with the T matrix being the transpose of the adjacency matrix of G with all diagonal entries set to 1.

5.6.2 THE MAXIMUM INDEPENDENT VERTEX SET. One way of finding the maximum independent vertex set of a graph $G = (X, \Gamma)$ is to generate all maximal independent vertex sets and choose the one of largest cardinality. Another way is to consider the problem as a SCP where columns of the matrix T represent vertices of G and rows represent links, with $t_{ij} = 1$ if vertex x_j is a terminal vertex of link a_i and 0 otherwise.

It is then apparent [23] that if $\check{X} \subseteq X$ is the set of columns in the minimum solution of this SCP, the set $X - \check{X}$ is the maximum independent vertex set of G. This is no because if $\check{X}$ is a set cover of the links of G, then every link has at least one of its terminal vertices in $\check{X}$ and hence two vertices in $X - \check{X}$ cannot be adjacent, i.e. $X - \check{X}$ is independent. Moreover, if $X - \check{X}$ is independent, each link can have at most one terminal vertex in $X - \check{X}$ and hence $\check{X}$ is a set cover of the links. Hence, the complement of every vertex set cover of the links of G is an independent set, and since $\check{X}$ is the minimum cardinality such cover $X - \check{X}$ is the maximum independent set.

5.6.3 THE MINIMUM COVERING AND MAXIMUM MATCHING (see Chapter 12). What is in the literature referred to as a *minimum "covering"* is a set E of links of a graph $G = (X, A)$ so that every vertex of G is a terminal vertex of at least one link in E and $|E|$ is minimum. Thus, since E could be considered as dominating the vertices of G, that E^* with minimum $|E|$ could—in the terminology used in this Chapter—be called a *minimum dominating link set*. A problem equivalent to that of finding a minimum covering is that of finding a set M of links of a graph G so that no two links in M are adjacent. M is called a matching and that set M^* with maximum cardinality is the *maximum matching*, which in the current termonology could also be called a *maximum independent link set*. That the maximum matching and minimum covering problems are equivalent is demonstrated in Chapter 12 dealing with matchings. It is shown there that if in a minimum covering E^* the degree of a vertex x_i is $d^{E^*}(x_i)$—considering only the links in E^*—then if for every x_i with $d^{E^*}(x_i) > 1$

C

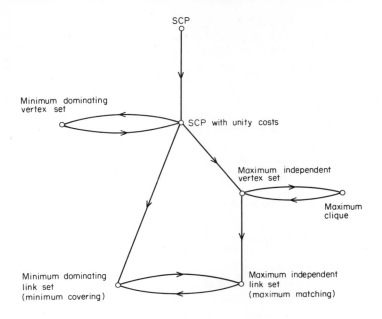

FIG. 3.5. Problem-solution implication diagram

a total of $d^{E^*}(x_i) - 1$ links incident at x_i are removed, the remaining set of links form a maximum matching. Inversely, if M^* is a maximum matching, and for every vertex x_i with $d^{M^*}(x_i) = 0$ a link incident at x_i is added, the resulting set of links is a minimum covering.

Maximum matchings and minimum coverings can be formulated as SCP's. In the case of the coverings, the columns of T represent links of the graph G, and the rows of T represent vertices, where $t_{ij} = 1$ if vertex x_i is incident with link a_j and 0 otherwise. For this case the matrix T is, therefore, the incidence matrix of G.

Figure 3.5 shows a problem-solution implication diagram, where an arc from problem α to problem β represents the fact that a solution to α implies the solution of β [34, 39]. The relation between maximum independent vertex sets and maximum cliques is via the complementary graph, and that between maximum independent vertex sets and maximum matchings is via the line graph of G. (See Chapter 10 for the definition of the line graph.)

5.6.4 COVERING A GRAPH WITH SUBGRAPHS. A whole family of problems involves the covering (or partitioning) of the vertices or links of a graph G with subgraphs or partial graphs of G having prescribed properties. In this context the columns of the T matrix of the SCP (or SPP) represent all the subgraphs or partial graphs of G having the prescribed properties, and the rows of the matrix represent the vertices or links of G. In Chapter 4, for example, the problems of finding the chromatic number of a graph G is considered as a SCP with the rows of the T matrix representing vertices and the columns representing maximal independent vertex sets of G (i.e. maximal totally-disconnected subgraphs).

Other problems involve the covering of the links of G with partial graphs having diameter less than or equal to λ [13], covering the links of G with "star" trees or covering them with simple paths and circuits [16, 42].

6. Problems P3

1. Enumerate all maximal independent sets of the graph H shown in Fig. 3.6, and hence find the independence number $\alpha[G]$.

2. Use the method of Section 2.3.2 to list all maximal independent sets of the graph of Fig. 3.7. (Note graph symmetry.)

3. Show that for a given graph $G = (X, \Gamma)$, an independent set $A \subset X$ is maximum if and only if for any arbitrary independent set $B \subset X - A$ we have:

$$B \leqslant |\Gamma(B) \cup A|$$

(See Ref. [63].)

4. In the complete nondirected graph K_n show that each link is contained in exactly $(n - 2)$ circuits of cardinality 3(i.e. triangles).

5. It is easy to show that the converse of statement 4 above is not true, i.e. if every link in a graph G is part of $n - 2$ circuits of cardinality 3, G is not necessarily K_n.

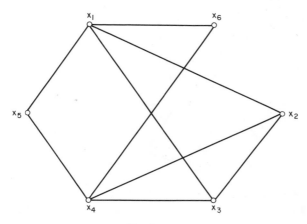

FIG. 3.6

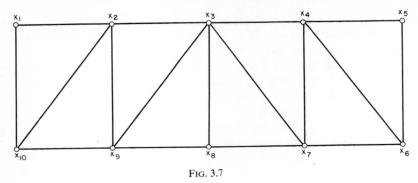

FIG. 3.7

(Figure 3.8 provides a counterexample in which each link is part of 2 such circuits but the graph does not contain K_4). However, one can use statement 4 above in order to find an upper bound on the clique number of a graph G—i.e. the largest possible K_r embedded in G—as follows:

(i)   Start with a value of r which is a lower bound on the clique number.
(ii) Remove those links not part of at least $r - 2$ circuits of cardinality 3.
(iii) Repeat (ii) above for $r = r + 1$ etc until less than $r(r - 1)/2$ links are left. The current value of r is then an upper bound on the clique number.

Show that the above algorithm is correct, and apply it to the complementary graph $\tilde{G}$ of the graph G given in Fig. 3.7. Compare the answer with the value of $\alpha[G]$ found in problem 2.

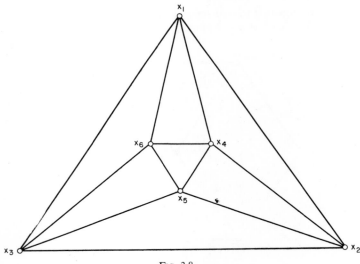

FIG. 3.8

6. A total of n symbols are to be used for transmitting information. A graph $G = (X, A)$ is given where link $(x_i, x_j) \in A$, if and only if symbol x_i could be confused with symbol x_j at the reception. The symbols are to be used in blocks of k-symbol "words". Show that the maximum number of words that could be used for transmitting information without the possibility of confusion, is the independence number $\alpha[G^k]$ of the graph $G^k = G \times G \times \dots \times G$ (k times), where the product $G' \times G''$ of two graphs $G' = (X', A')$ and $G'' = (X'', A'')$ is the graph $H = (Y, B)$ given by

$$Y = \{x_i x_j \mid x_i \in X', x_j \in X''\}$$

and

$$B = \{(x_i x_j, x_k x_l) \mid \text{either } x_i = x_k \text{ and } (x_j, x_l) \in A''$$

$$\text{or} \quad x_j = x_l \text{ and } (x_i, x_k) \in A''$$

$$\text{or } (x_i, x_k) \in A' \text{ and } (x_j, x_l) \in A''\}$$

(see Ref. [61]).

7. Given any two graphs G' and G'' show that:

$$\alpha[G' \times G''] \geqslant \alpha[G'] \, \alpha[G'']$$

8. (i) Show that in a nondirected graph G, $\alpha[G] \geqslant \beta[G]$, by showing that every maximal independent set is a dominating set.

(ii) Give an example to show that the minimum dominating set is not necessarily independent.

9. On a $k \times k$ chessboard show that it is impossible to place k queens which do not attack each other for the values $k \leqslant 4$. Find a solution for $k = 5$.

10. Moon and Moser [46] showed that the largest number of cliques ($f(n)$ say), that a graph with n vertices can have is given by:

$$3^{n/3} \qquad \text{if } n = 0 \pmod 3$$

$$4 \cdot 3^{(n-4)/3} \quad \text{if } n = 1 \pmod 3$$

$$2 \cdot 3^{(n-2)/3} \quad \text{if } n = 2 \pmod 3$$

and showed that the only graphs G which achieve $f(n)$ are the following:

(a) If $n = 0 \pmod 3$, then G consists of $n/3$ triples of vertices.
(b) If $n = 1 \pmod 3$, there are two possible classes of graphs: Either G consists of one quadruple and $(n - 4)/3$ triples, or it consists of two pairs and $(n - 4)/3$ triples.
(c) If $n = 2 \pmod 3$, then G consists of one pair and $(n - 2)/3$ triples.

In all cases a link (x_i, x_j) exists if and only if x_i and x_j belong to different groups, (pairs, triples, quadruples).
Verify the above statements for the cases of $n = 3$, $n = 4$ and $n = 5$.

11. Using the reduction tests of Section 4.2 eliminate as many rows and columns as possible from the covering problem whose $[c_j]$ and $[t_{ij}]$ matrices are given below. (Note: Apply reduction test 4 only for the case $|\mathscr{S}| = 1$.)

$$
[t_{ij}] = \begin{array}{c|cccccccc}
 & 1 & 2 & 3 & 4 & 5 & 6 & 7 & 8 \\
\hline
1 & 1 & 1 & 1 & 0 & 0 & 1 & 0 & 1 \\
2 & 1 & 0 & 1 & 0 & 0 & 1 & 0 & 1 \\
3 & 0 & 0 & 0 & 1 & 0 & 0 & 0 & 0 \\
4 & 0 & 1 & 0 & 0 & 1 & 0 & 1 & 1 \\
5 & 0 & 0 & 0 & 0 & 1 & 1 & 1 & 0 \\
6 & 1 & 1 & 0 & 0 & 0 & 0 & 1 & 0 \\
\end{array}
$$

$$[c_j] = \quad [\ 4, \quad 7, \quad 5, \quad 8, \quad 3, \quad 2, \quad 6, \quad 5\ ]$$

12. Solve the SCP remaining after the reduction in problem 11 above using the method in Section 4.4.

7. References

1. Arabeyre, J. P., Fearnley, J., Steiger, F. C. and Teather, W. (1969). The airline crew scheduling problem: A survey, *Transp. Sci.*, 3, p. 140.
2. Augustson, J. G. and Minker, J. (1970). An analysis of some graph-theoretical cluster techniques, *Jl. of ACM*, **17**, p. 571.
3. Balas, E. (1970). Project scheduling with resource constraints, *In*: "Applications of mathematical programming techniques", Beale, Ed., English Universities Press, London.
4. Balas, E. and Padberg, M. W. (1972a). On the set covering problem, *Ops. Res.* **20**, p. 1152.
5. Balas, E. and Padberg, M. W. (1972). On the set covering problem. II: An algorithm, Management Sciences Research Report No. 295, Carnegie-Mellon University.
6. Balinski, M. (1965). Integer programming: methods, uses, computation, *Man. Sci.*, **12**, p. 253.
7. Balinski, M. L. and Quandt, R. E. (1964). On an integer program for a delivery problem, *Ops. Res.*, **12**, p. 300.
8. Bellmore, M., Greenberg, H. and Jarvis, J. (1970). Multi-commodity disconnecting sets, *Man. Sci.*, **16**, p. 427.
9. Bellmore, M. and Ratliff, H. D. (1971). Set covering and involuntary bases, *Man. Sci.*, **18**, p. 194.
10. Bellmore, M. and Ratliff, H. D. (1971). Optimal defense of multi-commodity networks, *Man. Sci.*, **18**, p. 174.
11. Berge, C. (1962). "Theory of graphs", Methuen, London.
12. Bonner, R. E. (1964). On some clustering techniques, *IBM Jl. Res. and Dev.*, **8**, p. 22.
13. Bosák, J., Rosa, A. and Znám, Š. (1966). On decompositions of complete graphs into factors with given diameters, *In*: Int. Symp. on theory of graphs, Dunod, Paris, p. 37.
14. Bron, C. and Kerbosch, J. (1973). Algorithm 457—Finding all cliques of an undirected graph, *Comm. of ACM*, **16**, p. 575.
15. Burstall, R. M. (1967). Tree searching methods with an application to a network

design problem, *In*: "Machine Intelligence", Colin and Michie, Eds., Vol. 1, Oliver and Boyd, London.

16. Busacker, R. G. and Saaty, T. L. (1965). "Finite graphs and networks", McGraw-Hill, New York.

17. Chamberlin, D. D. (1971). The single assignment approach to parallel processing, *Proc. of AFIPS Conf.*, **39**, p. 263.

18. Christofides, N. (1971). Zero-one programming using non-binary tree search, *The Computer Jl.*, **14**, p. 418.

19. Christofides, N. and Korman, S. (in press). A computational survey of methods for the set covering problem, *Man. Sci.*

20. Crowston, W. B. (1968). Decision network planning models, Ph.D. Thesis, Carnegie-Mellon University.

21. Day, R. H. (1965). On optimal extracting from a multiple file data storage system: An application of integer programming, *Ops. Res.*, **13**, p. 482.

22. Desler, J. F. (1969). Degree constrained subgraphs, covers, codes and k-graphs, Ph.D. Thesis, Northwestern University, Evaston, Illinois.

23. Edmonds, J. (1962). Covers and packings in a family of sets, *Bulletin of American Mathematical Soc.*, **68**, p. 494.

24. Eilon, S., Watson-Gandy, C. D. T. and Christofides, N. (1971). "Distribution management: Mathematical models and practical analysis", Griffin, London.

25. Freeman, D. R. and Jucker, J. V. (1967). The line balancing problem, *Jl. of Industrial Engineering*, **18**, p. 361.

26. Garfinkel, R. S. (1970). Set covering; a survey, presented at XVII TIMS Conf., London.

27. Garfinkel, R. S. and Nemhauser, G. L. (1969). The set partitioning problem: set covering with equality constraints, *Ops. Res.*, **17**, p. 848.

28. Garfinkel, R. S. and Nemhauser, G. L. (1972). "Integer Programming", Wiley, New York.

29. Garfinkel, R. S. (1968). Optimal political districting, Ph.D. Thesis, The John Hopkins University.

30. Gimpel, J. F. (1965). A reduction technique for prime implicant tables, *IEEE Trans*, **EC-14**, p. 535.

31. Glover, F. (1971). A note on extreme point solutions and a paper by Lemke, Salkin and Spielberg, *Ops. Res.*, **19**, p. 1023.

32. Gomory, R. (1963). An algorithm for integer solutions to linear programs, *In*: "Recent Advances in Mathematical Programming", Graves and Wolfe, Eds., McGraw-Hill, New York.

33. Hakimi, S. L. and Frank, H. (1969). Maximum internally stable sets of a graph, *J. of Math. Anal. and Appl.*, **25**, p. 296.

34. Hakimi, S. L. (1971). Steiners' problem in graphs and its implications, *Networks*, **1**, p. 112.

35. House, R., Nelson, L. and Rado, T. (1965). Computer studies of a certain class of linear integer programs, *In*: "Recent Advances in Optimization Techniques", Lavi and Vogel, Eds., Wiley, New York.

36. Jardine, N. and Sibson, R. (1971). "Mathematical taxonomy", Wiley, London.

37. Jensen, P. A. (1971). Optimal network partitioning, *Ops. Res.*, **19**, p. 916.

38. Kapilina, R. I. and Shneyder, B. N. (1970). The problem of internal stability and methods of solving it on a universal digital computer, *In*: "Proc. of Scient. and Techn. Conf. on the results of scientific research in 1968–1969", Section on

Automation Computation and Measuring Procedures, Subsection on Graph Theory, Maskovskiy, Energticheskiy Institut Press.

39. Karp, R. M. (1972). Reducibility of combinatorial problems, *In*: "Complexity of computer computations", Miller and Thatcher, Eds., Plenum Press, New York, p. 85.

40. Lawler, E. L. (1966). Covering problems: Duality relations and a new method of solution, *Jl. of SIAM (Appl. Math.)*, **14**, p. 1115.

41. Lemke, C. E., Salkin, H. M. and Spielberg, K. (1971). Set covering by single branch enumeration with linear programming subproblems, *Ops. Res.*, **19**, p. 998.

42. Lovsaz, L. (1966). On covering of graphs, *In*: "Int. Symp. on theory of graphs", Dunod, Paris, p. 231.

43. Mayoh, B. H. (1967). Simplification of logical expressions, *Jl. of SIAM (Appl. Math.)*, **15**, p. 898.

44. McCluskey, E. J. Jr. (1956). Minimization of boolean functions, Bell. Syst. Tech. Jl., **35**, 1417.

45. Michadu, P. (1972). Exact implicit enumeration method for solving the set partitioning problem, *IBM Jl. of Res. and Dev.*, **16**, p. 573.

46. Moon, J. W. and Moser, L. (1965). On cliques in graphs, *Israel Jl. of Mathematics*, p. 23.

47. Pierce, J. F. (1967). On the truck dispatching problem—Part I, IBM Scientific Centre Technical Report 320–2018.

48. Pierce, J. F. (1968). Application of combinatorial programming to a class of all-zero-one integer programming problems, *Man. Sci.*, **15**, p. 191.

49. Pierce, J. F. and Lasky, J. S. (1973). Improved combinatorial programming algorithms for a class of all-zero-one integer programming problems, *Man. Sci.*, **19**, p. 528.

50. Pyne, J. B. and McCluskey, E. J. Jr. (1961). An essay on prime implicant tables, *Jl. of SIAM (appl. Math.)*, **9**, p. 604.

51. Quine, W. V. (1952). The problem of simplifying truth functions, *American Mathematical Monthly*, **59**, p. 521.

52. Roth, R. (1969). Computer solutions to minimum-cover problems, *Ops. Res.*, **17**, p. 455.

53. Roy, B. (1972). An algorithm for a general constrained set covering problem, *In*: "Computing and Graph Theory", Read, Ed. Academic Press, New York.

54. Roy, B. (1969, 1970). "Algebre moderne et théorie des graphes, Vol 1 and Vol 2, Dunod, Paris.

55. Rubin, J. (1971). A technique for the solution of massive set covering problems, with application to airline crew scheduling, IBM Philadelphia Scientific Center Technical Rept. No. 320-3004.

56. Rutman, R. A. (1964). An algorithm for placement of inter-connected elements based on minimum wire length, *Proc. of AFIPS Conf.*, **20**, p. 477.

57. Salkin, H. M. and Koncal, R. (1970). A pseudo dual all-integer algorithm for the set covering problem, Dept. of O.R. Technical Memo. No. 204, Case Western Reserve University.

58. Salkin, H. M. and Koncal, R. (1971). A dual algorithm for the set covering problem, Dept. of O.R. Technical Memo. No. 250, Case Western Reserve University.

59. Salveson, M. E. (1955). The assembly line balancing problem, *Jl. of Industrial Engineering*, **6**, p. 18.

60. Selby, G. (1970). The use of topological methods in computer-aided circuit layout, Ph.D. Thesis, University of London.

61. Shannon, C. E. (1956). The zero-error capacity of a noisy channel, *Trans. Inst. Elect. Eng.*, **3**, p. 3.
62. Shneyder, B. N. (1970). Algebra of sets applied to solution of graph theory problems, Engineering Cybernetics '70, English Translation by Plenum Press, p. 521.
63. Starobinets, S. M. (1973). On an algorithm for finding the greatest internally stable sets of a graph, Engineering Cybernetics '73, English Translation by Plenum Press, p. 873.
64. Steiger, F. (1965). Optimization of Swiss Air's crew scheduling by an integer linear programming model, Swissair Report O.R. SDK 3.3.911.
65 Steiger, F. and Neiderer, M. (1968). Scheduling air crews by integer programming, presented at IFIP Congress '68, Edinburgh.
66. Steinberg, L. (1961). The backboard wiring problem; a placement algorithm, *Jl. of SIAM (Review)*, **3**, p. 37.
67. Thiriez, H. (1971). The set covering problem—a group theoretic approach, Rev. Française d'Informatique et de Recherche Opérationnelle, **V-3**, p. 83.
68. Wells, M. B. (1962). Application of a finite set covering theorem to the simplification of boolean function expressions, *In*: "Information Processing '62", Popplewell, Ed., North Holland, Amsterdam.

Chapter 4

Graph Colouring

1. Introduction

A variety of problems in production scheduling, construction of examination timetables, storage or transportation of goods etc. can generally be expressed as problems in graph theory which are closely related to the so-called "colouring problem". The graphs considered in this Chapter are nondirected graphs without loops and unless otherwise stated the word "graph" should henceforth be taken to mean such a graph.

A graph G is said to be *r-chromatic* if its vertices can be coloured with r colours in such a way so that no two adjacent vertices are of the same colour. The smallest number r for which the graph is r-chromatic is called the *chromatic number* $\gamma(G)$ of the graph and finding this number is referred to as the *colouring problem*. Corresponding to this number is a colouring of the vertices which partitions them into r sets, each set containing only vertices coloured with the same colour. These sets are independent since no two vertices within a set can be adjacent.

Generally speaking the chromatic number of a graph (like the independence and dominance numbers discussed in the previous Chapter) can not be determined solely from knowledge of the number of vertices n and links m of the graph. Knowledge of the degree of each vertex is also not sufficient for calculating the chromatic number, as can be illustrated by the graphs of Figs. 4.1(a) and 4.1(b). These graphs (both of which contain $n = 12$ vertices, $m = 16$ links and the same distribution of vertex degrees d_i) have chromatic numbers of values 4 and 2 respectively. However, upper and lower bounds on the chromatic number can be calculated using the values of n, m and d_i and these bounds are discussed further in the next section.

The problem of determining the chromatic number of an arbitrary graph has been the subject of a great deal of investigation in the late 19th and during the present century, with the consequence that there is now a large number of theoretically interesting results. In the present Chapter, however, we make no attempt to discuss, or indeed even summarize these results and we intro-

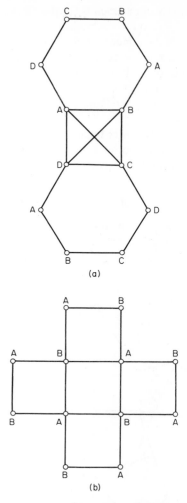

FIG. 4.1. Two graphs with same n, m and vertex degree distribution but with different chromatic numbers. (a) $\gamma(G) = 4$, (b) $\gamma(G) = 2$

duce only selected concepts which may be of significance as far as their contribution to graph colouring algorithms is concerned. This Chapter is mainly concerned with algorithms—both exact and approximate—for determining the chromatic number of an arbitrary graph and its corresponding vertex colouring.

2. Some Theorems and Bounds for the Chromatic Number

In Section 4 of Chapter 3, the concept of the clique number $\rho(G)$ of a graph G was introduced as the maximum number of vertices in a complete subgraph of G, and it was noted that since there is a one-to-one correspondence between cliques of G and maximal independent sets of the complementary graph $\tilde{G}$, it follows that $\rho(G) = \alpha(\tilde{G})$ and $\rho(\tilde{G}) = \alpha(G)$.

2.1 Lower bounds on $\gamma(G)$

Obviously, since at least $\rho(G)$ colours are needed to colour the clique of G from which the clique number is derived, it follows that $\rho(G)$ is a lower bound on the chromatic number i.e.

$$\gamma(G) \geqslant \rho(G) \qquad (4.1)$$

Zykov [26], however, has shown that there is no further relation possible between $\gamma(G)$ and $\rho(G)$ and that the difference $\gamma(G) - \rho(G)$ can be arbitrarily large. It has in fact been shown by Tutte [22] that it is possible to construct

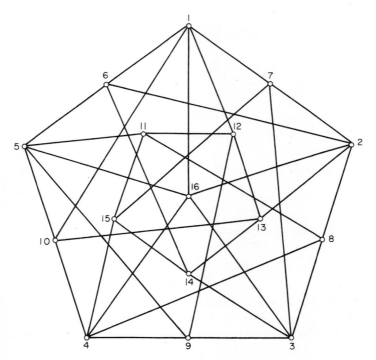

FIG. 4.2. Graph with $\rho(G) = 2$ and $\gamma(G) = 5$

a graph G containing no complete subgraph even of order 3 (i.e. for which $\rho(G) = 2$) and which will have any given, arbitrarily large, value for the chromatic number. Fig. 4.2 shows a graph with $\rho(G) = 2$ and $\gamma(G) = 5$.

Since $\alpha(G)$ is the largest number of vertices of G no two of which are adjacent, it is also the largest number of vertices which can be coloured with any one colour, and therefore:

$$\gamma(G) \geqslant \left\lceil \frac{n}{\alpha(G)} \right\rceil \tag{4.2}$$

where n is the total number of vertices of G and $\lceil x \rceil$ is the largest integer less than or equal to x.

If $\gamma(\tilde{G})$ is taken to be the chromatic number of the complementary graph $\tilde{G}$, then the following two relations have been given by Nordhaus and Gaddum [19]

$$\gamma(G) \geqslant \lfloor 2\sqrt{(n)} \rfloor - \gamma(\tilde{G}) \tag{4.3}$$

and

$$\gamma(G) \geqslant \frac{n}{\gamma(\tilde{G})} \tag{4.4}$$

where $\lfloor x \rfloor$ means the smallest integer greater than or equal to x.

Yet another lower bound on $\gamma(G)$ was proposed by Geller [8] as:

$$\gamma(G) \geqslant \frac{n^2}{(n^2 - 2m)} \tag{4.5}$$

Myers and Lin [18], however, have shown that the bound of (4.1) uniformly dominates the above bound, and hence the only merit of (4.5) is that it is simpler to calculate than (4.1).

2.2 Upper bounds on $\gamma(G)$

Lower bounds on the chromatic number are inherently more interesting than upper bounds since (if they are sufficiently tight), they could be used in a tree search procedure to determine $\gamma(G)$, whereas upper bounds could not. However, one upper bound which can easily be calculated was given by Brooks [3] as:

$$\gamma(G) \leqslant \max_{x_i \in X} [d(x_i)] + 1 \tag{4.6}$$

and other tighter upper bounds also based on the vertex degrees were given by Welsh [23], and Wilf and Szekeres [21].

Upper bounds relating $\gamma(G)$ and $\gamma(\tilde{G})$ were given by Nordhaus and Gaddum [19] as:

$$\gamma(G) \leqslant n + 1 - \gamma(\tilde{G}) \tag{4.7}$$

and

$$\gamma(G) \leqslant \left\lceil \left(\frac{n+1}{2}\right)^2 \right\rceil \bigg/ \gamma(\tilde{G}) \qquad (4.8)$$

The bounds in (4.3), (4.4), (4.7) and (4.8) are the best possible in the sense that there exist graphs for which the bounds can be attained. In most cases, however, they are too loose to be of practical significance.

2.3 The four-colour conjecture

Graphs which can be drawn in the plane without any two of its links crossing each other, are called *planar* graphs. These graphs are of both theoretical and practical importance and possess some colouring properties which ought to be mentioned in this chapter.

The five colour theorem [13]
Every planar graph can be coloured with five colours so that no two adjacent vertices are of the same colour, i.e. if G is planar then $\gamma(G) \leqslant 5$.

The four colour "theorem" (Unproven)
Every planar graph can be coloured with four colours so that no two adjacent vertices are of the same colour, i.e. $\gamma(G) \leqslant 4$.

The first appearance of the four colour conjecture is believed to have occurred in a conversation between Augustus de Morgan and his pupil F. Guthrie in about 1850, and in printed form in a letter (dated 23rd October 1852) from de Morgan to Sir William Rowan Hamilton. Since then "proofs", "counterproofs" and other related conjectures and theorems have appeared with increasing frequency with the result that a very voluminous literature on the subject has amassed and the "theorem" has become perhaps the most celebrated unsolved problem in mathematics. The problem will not be discussed here further and the interested reader is referred to an excellent book by Ore [20].

3. Exact Colouring Algorithms

In this section we describe some algorithms which are exact in the sense that they guarantee an optimal colouring and the correct value of the chromatic number for any arbitrary graph.

3.1 A dynamic programming method

3.1.1 MAXIMAL r-SUBGRAPHS. An r-subgraph of a graph $G = (X, \Gamma)$ is a subgraph $\langle S_r[G] \rangle$ where $S_r[G] \subseteq X$, and which is r-chromatic. If no set of vertices $H \supset S_r[G]$ exists so that the subgraph $\langle H \rangle$ is r-chromatic; then the graph $\langle S_r[G] \rangle$ is called a *maximal r-subgraph* of G. Obviously, for a specified value of r there is, in general, a large number of maximal r-subgraphs of a given graph G. Thus, for example, a subgraph having a maximal independent set of G as its vertices (this is a totally disconnected subgraph having no links) can be called a maximal 1-subgraph since there can be no subgraph having a larger set of vertices which is 1-chromatic.

The chromatic number of a graph G is then given by the smallest value of r so that $S_r[G] = X$ for at least one of the maximal r-subgraphs.

THEOREM 1. *If a graph is r-chromatic it can be coloured with r (or fewer) colours colouring first with one colour a maximal independent set $S_1[G]$, next colouring with another colour a set $S_1[\langle X - S_1[G] \rangle]$ and so on, until all the vertices are coloured.*

Proof. The fact that such a colouring using only r colours always exists can be shown as follows. Say a colouring with r colours exists so that one or more sets coloured with the same colour are not maximal independent sets in the above-mentioned sense. Numbering the colours in an arbitrary way we can always colour with colour 1 those vertices $\overline{V}_1$ which are not coloured with colour 1 and which form a maximal independent set together with the vertices V_1 that are already of colour 1. This new colouring is possible because the set of vertices $\overline{V}_1$ is not adjacent to the set of vertices V_1, and hence those vertices which are adjacent to $\overline{V}_1$ have colours different from 1 and are therefore unaffected by the change of the colour of the vertices in $\overline{V}_1$. Considering now the subgraph $\langle X - V_1 \cup \overline{V}_1 \rangle$ the same procedure will lead to a recolouring containing a maximal independent set of colour 2 and so on.

Colourings of the type indicated in the theorem will be called *optimal independent colourings*.

3.1.2 RECURRENCE RELATIONS. The above theorem can be used to establish a recurrence relation so that the maximal r-subgraphs of a graph can be obtained from the maximal $(r - 1)$-subgraphs.

Let $Q_{r-1}[G]$ be the family of maximal $(r - 1)$-subgraphs of G and let $S_{r-1}^j[G]$ be the vertex set of the jth one of these subgraphs. The set $S_1^k[G^j]$ is then the vertex set of the kth maximal 1-subgraph (independent set), of the graph $G^j \equiv \langle X - S_{r-1}^j[G] \rangle$, formed by those vertices of G not included in the $(r - 1)$-subgraph $\langle S_{r-1}^j[G] \rangle$.

The family $Q_r[G]$ of all maximal r-subgraphs of G can then be calculated as follows:

Form all the sets H^i from the expression

$$H^i = S_{r-1}^j[G] \cup S_1^k[G^j] \tag{4.9}$$

for all $j = 1, 2, \ldots, q_{r-1}$, and $k = 1, 2, \ldots, q_1^j$; where q_{r-1} and q_1^j are the numbers of $(r-1)$-subgraphs and 1-subgraphs respectively.

The family of maximal r-subgraphs is then contained in the family Θ of the sets H^i ($i = 1, 2, \ldots, q_1^j q_{r-1}$), and can be obtained from it by excluding those sets of Θ which are contained in other sets.

Thus $Q_r[G]$ is given by:

$$Q_r[G] = \{H^i \,|\, H^i \in \Theta, \text{ and } H^i \not\subseteq H^j \text{ for any } H^j \in \Theta, \ j \neq i\} \qquad (4.10)$$

3.1.3 AN ALGORITHM BASED ON MAXIMAL r-SUBGRAPHS. It was mentioned in the above section that the chromatic number of a graph G is the smallest value of r so that the set of vertices $S_r[G]$ of some maximal r-subgraph is the complete set X of vertices of G. The recurrence relations (4.9) and (4.10) can, therefore, be used to generate progressively maximal 1-subgraphs, 2-subgraphs, etc., checking at each stage whether any subgraph that has been generated includes all the vertices of G.[†]

The following is a description of an algorithm to find the chromatic number of a graph G and the colouring realizing this number.

Step 1. Set $r = 1$. Find the vertex sets $S_r^j[G]$, $j = 1, \ldots, q_r$, of the maximal r-subgraphs of G. (Say there are q_r such sets.) Let $Q = \{S_r^j[G] \,|\, j = 1, \ldots, q_r\}$. Set $j = 1$.

Step 2. Find a maximal independent set $S_1[G^j]$ of the graph $G^j = \langle X - S_r^j[G] \rangle$. If one exists go to step 3; if all of them have already been found go to step 6.

Step 3. Calculate $S = S_r^j[G] \cup S_1[G^j]$.

Step 4. If $S = X$, stop. The number $(r + 1)$ is the chromatic number. The subsets that were introduced into the set S give the required colouring. (These subsets can be kept separate with markers as they are introduced.) If $S \neq X$, go to step 5.

Step 5. (i) If $S \subseteq S'$ for any $S' \in Q$ go to step 2.
     (ii) If $S \supset S'$ for any $S' \in Q$, set $Q = Q - \{S'\}$ for all such S'. Set $Q = Q \cup \{S\}$ and go to step 2.
     (iii) If neither of cases (i) and (ii) above apply, set $Q = Q \cup \{S\}$ and go to step 2.

Step 6. If $j < q_r$ set $j = j + 1$ and go to step 2.
If $j = q_r$ set $j = 1$, $r = r + 1$, $q_r =$ number of sets in Q, and go to step 2.

If the algorithm is not stopped when the first set $S = X$ at step 4, it will continue to produce an alternative colouring with $r + 1$ colours, if such a

[†] The algorithm described in this section can be considered as dynamic programming or breadth-first tree search. A depth-first version having considerable advantages has recently been described by Wang, (*J. ACM*, **21**, 1974, p. 385).

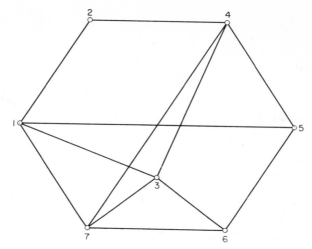

FIG. 4.3. Graph for example 3.1.4

colouring exists. One should note, however, that the algorithm will not give a complete enumeration of all possible colourings with $r + 1$ colours but will only produce the optimal independent colourings. Such colourings may be only a small fraction of the total possible number of colourings using $r + 1$ colours.

3.1.4 EXAMPLE. Consider the 7-vertex graph G shown in Fig. 4.3.
Step 1. The vertex sets of the maximal 1-subgraph are:

$S_1^1[G] = \{1, 4, 6\}$; $S_1^2[G] = \{2, 3, 5\}$; $S_1^3[G] = \{2, 5, 7\}$; $S_1^4[G] = \{2, 6\}$.
Hence $q_1 = 4$. $Q = \{S_1^1[G], S_1^2[G], S_1^3[G], S_1^4[G]\}$.

Iterative applications of steps 2–5 produces:

for $S_1^1[G]$
Step 2. $G^1 = \langle X - S_1^1[G]\rangle = \langle\{2, 3, 5, 7\}\rangle$; $S_1[G^1] = \{2, 3, 5\}$.
Step 3. $S = \{1, 4, 6 \uparrow 2, 3, 5\}$.
Step 5. $Q = [\{1, 4, 6 \uparrow 2, 3, 5\}]$.
Step 2. $S_1[G^1] = \{2, 5, 7\}$.
Step 3. $S = \{1, 4, 6 \uparrow 2, 5, 7\}$.
Step 5. $Q = [\{1, 4, 6 \uparrow 2, 3, 5\}, \{1, 4, 6 \uparrow 2, 5, 7\}]$.
for $S_1^2[G]$
Step 2. $G^2 = \langle\{1, 4, 6, 7\}\rangle$; $S_1[G^2] = \{1, 4, 6\}$.

Step 3. $S = \{2, 3, 5 \uparrow 1, 4, 6\}$, eliminated by step 5 (i).

Step 2. $S_1[G^2] = \{7\}$.

Step 3. $S = \{2, 3, 5 \uparrow 7\}$.

*Step 5. $Q = [\{1, 4, 6 \uparrow 2, 3, 5\}, \{1, 4, 6 \uparrow 2, 5, 7\}, \{2, 3, 5 \uparrow 7\}]$.

for $S_1^3[G]$

Step 2. $G^3 = \langle\{1, 3, 4, 6\}\rangle; S_1[G^3] = \{1, 4, 6\}$.

Step 3. $S = \{2, 5, 7 \uparrow 1, 4, 6\}$, eliminated by step 5(i).

Step 2. $S_1[G^3] = \{3\}$.

Step 3. $S = \{2, 5, 7 \uparrow 3\}$, eliminated by step 5(i).

for $S_1^4[G]$

Step 2. $G^4 = \{1, 3, 4, 5, 7\}, S_1[G^4] = \{1, 4\}$.

Step 3. $S = \{2, 6 \uparrow 1, 4\}$, eliminated by step 5(i).

Step 2. $S_1[G^4] = \{3, 5\}$.

Step 3. $S = \{2, 6 \uparrow 3, 5\}$, eliminated by step 5(i).

Step 2. $S_1[G^4] = \{5, 7\}$.

Step 3. $S = \{2, 6 \uparrow 5, 7\}$, eliminated by step 5(i).

Thus, at the end of the first iteration through steps 2 to 5 the family Q of sets $S_2^j[G]$ corresponding to the maximal 2-subgraphs is as shown at the step marked with an asterisk. It should be noted here that 5 out of the eight generated sets S were eliminated by the dominance tests of step 5.

Continuing in the same way we can calculate the maximal 3-subgraphs etc. In fact the next set to be generated $S_3^1[G] = \{1, 4, 6 \uparrow 2, 5, 3 \uparrow 7\}$ satisfies $S_3^1[G] = X$ at step 4 and hence the chromatic number of the graph is 3 with the optimum colouring being: $\{1, 4, 6\}, \{2, 5, 3\}, \{7\}$. If one decides to continue, another possible colouring is found as:

$$S_3^2[G] = \{1, 4, 6 \uparrow 2, 5, 7 \uparrow 3\} = X$$

all other $S_r^j[G]$ are either contained in the previous two sets or do not contain all the vertices in X.

Notice that colourings such as $\{5, 3\}, \{1, 4, 6\}, \{2, 7\}$ which are possible but are not optimal independent are not produced.

3.2 Formulation as a zero-one programming problem

Let q be an upper bound to the chromatic number of a graph G. This upper bound may be any one of the bounds given in Section 4 of this Chapter, or the number of colours resulting from one of the heuristic methods of solution to the colouring problem and which are described in detail later on.

Let $\Xi = [\xi_{ij}]$ be a matrix allocating vertices to colours so that

$$\xi_{ij} = 1 \quad \text{if vertex } x_i \text{ is of colour } j$$

$$= 0 \quad \text{otherwise.}$$

If $\mathbf{A} = [a_{ik}]$ is the adjacency matrix of the graph G with all diagonal elements set to zero, then the following two conditions guarantee a feasible colouring of the vertices of G (i.e. a feasible allocation matrix $[\xi_{ij}]$).

$$\sum_{j=1}^{q} \xi_{ij} = 1 \quad (\text{for all } i = 1, 2, \dots, n) \tag{4.11}$$

$$L(1 - \xi_{ij}) - \sum_{k=1}^{n} a_{ik}\xi_{kj} \geqslant 0 \begin{cases} \text{for all } i = 1, 2, \dots, n \\ \text{and } j = 1, 2, \dots, q \end{cases} \tag{4.12}$$

Condition (4.11) ensures that a vertex can be coloured with one and only one colour.

In condition (4.12) L is a very large positive number (any integer greater than n will do). If vertex x_i is coloured with colour j (i.e. $\xi_{ij} = 1$), then the first term of (4.12) is zero. The second term of (4.12) must also then be zero in order to satisfy the inequality; since both a_{ik} and ξ_{kj} are non-negative. Thus the condition (4.12) guarantees that if vertex x_i is of colour j then no adjacent vertex is of the same colour. If vertex x_i is not of colour j ($\xi_{ij} = 0$), then the first term of (4.12) becomes L and since the second term of (4.12) cannot possibly attain the value L (the largest value it can attain is in fact $(n - 1)$), any number of vertices x_k adjacent to vertex x_i can be coloured with colour j and the inequality would still be satisfied. Notice that if x'_k and x''_k are both adjacent to vertex x_i and also adjacent to each other, condition (4.12) written in terms of x_i will not prevent x'_k and x''_k being coloured with the same colour j. However condition (4.12) written in terms of x'_k (or x''_k) will ensure that the colour of these two vertices is not the same.

Let us now associate with each colour j a penalty p_j, the penalties being chosen so that:

$$p_{j+1} > h \cdot p_j \quad (\text{where } p_1 \text{ is taken as unity}) \tag{4.13}$$

and where h is an upper bound on the largest number of vertices that can be coloured with any one colour, i.e. h is any number greater than the independence number $\alpha(G)$ of the graph. In the absence of a better bound, h may be set equal to n at no extra computational cost.

The problem of colouring the vertices of a graph with the minimum number of colours can then be expressed as

Minimize

$$z = \sum_{j=1}^{q} \sum_{i=1}^{n} p_j \xi_{ij} \tag{4.14}$$

subject to constraints (4.11) and (4.12).

The minimization of expression (4.14) ensures that a colour $j + 1$ is never used for colouring vertices, if colours 1 to j are sufficient for a feasible colouring. The vertex-to-colours allocation matrix $[\xi_{ij}^*]$ which is a solution to the above zero-one linear programming problem is then the optimal colouring, and the number of colours used is the chromatic number of the graph.

An alternative condition to (4.12) is given by Berge [1] as:

$$\sum_{i=1}^{n} b_{ik}\xi_{ij} \leqslant 1 \quad \begin{cases} \text{for all } k = 1, 2, \ldots, m \\ \text{and } j = 1, 2, \ldots, q \end{cases} \tag{4.15}$$

where $[b_{ik}]$ is the incidence matrix so that $b_{ik} = 1$ if vertex x_i is a terminal vertex to link a_k, and $b_{ik} = 0$ otherwise. Condition (4.15) then expresses the fact that only one, or none, of the two end vertices of any link can be coloured with a colour j.

Although this condition is more direct than condition (4.12) it requires a total of mq constraints whereas (4.12) requires only nq constraints. Since the number of links (m) in a connected graph is usually much larger than the number of vertices (n), condition (4.12) may be preferable from the computational point of view. An appreciation of the computational saving involved can be obtained by noting that for a 100-vertex graph which contains only 20% of the links of the corresponding complete graph, condition (4.15) would require 1000 constraints for each colour, whereas condition (4.12) would require only 100 constraints for each colour.

3.3 Formulation as a set covering problem (SCP)

Since, for any feasible colouring of a graph G, the set of vertices coloured with the same colour must form an independent set, any feasible colouring can then be considered as a *partitioning* of the vertices of G into such sets. Moreover, if every independent set is expanded by adding to it other vertices of G until it becomes maximal, then the colouring may also be considered as a *covering* of the vertices of G by maximal independent sets. Obviously, after such an expansion some vertices of G will belong to more than one maximal independent set. This suggests alternative feasible colourings (using the same total number of colours) since it is possible to re-colour such a vertex with any of the colours corresponding to the maximal independent sets that it belongs to.

Thus, let us suppose that all t maximal independent sets of G have been generated—for example using the algorithm given in Section 2.3 of the previous Chapter—and let us form an $n \times t$ matrix $\mathbf{M} = [m_{ij}]$ so that $m_{ij} = 1$ if vertex x_i is in the jth maximal independent set and $m_{ij} = 0$ otherwise. If we now associate unity costs c_j with each maximal independent set j, the colouring problem simply becomes the problem of finding the least number of columns of **M** which cover all the rows. Each column in the solution to this SCP then corresponds to a colour which can be used for all the vertices in the maximal independent set represented by the column.

Although the number t of columns of the **M** matrix may be quite large even for moderate graphs, this formulation of the colouring problem as an SCP is preferable to the direct formulation as a general zero-one programming problem because SCP's can be solved for sizes (number of variables) at least one order of magnitude larger than those of general zero-one programming problems. (See Chapter 3.)

3.3.1 EXAMPLE. For the graph of Fig. 4.4 there is a total of 15 maximal independent sets, and these are shown as the columns of matrix **M** given below, where 0 entries are left blank.

Maximal independent sets

	1	2	3	4	5	6	7	8	9	10	11	12	13	14	15
x_1	1	1	1	1											
x_2					1	1	1	1							
x_3	1	1							1	1	1	1			
x_4			1		1	1							1		
x_5	1			1			1		1						
x_6					1			1		1	1		1	1	1
x_7							1	1			1	1	1		
x_8				1											1
x_9		1						1			1	1		1	
x_{10}					1										

$$\mathbf{M} =$$

(↑ under columns 2, 4, 6, 12)

The solution to this covering problem involves 4 columns although there is more than one such solution. For example, the column sets {4, 6, 10, 14}, {4, 6, 10, 12}, {4, 6, 10, 11}, {4, 6, 10, 8} and {4, 6, 10, 2} are all solutions to the covering problem. If we choose the last of these solutions (shown arrowed), and associate colours a, b, c, d with columns 4, 6, 10 and 12 respectively, then vertex x_1 (belonging to both columns 2 and 4) can be coloured with either colour a or d and vertex x_3 (belonging to both columns 2 and 10) can be coloured with either colour c or d, so that alternative optimal colourings may be produced.

It may be interesting to note here that since the independence number of the graph is $\alpha[G[= 3$ and $n = 10$ the lower bound on $\gamma[G]$ is 4, according to eqn 4.2. This bound is achieved for the present graph.

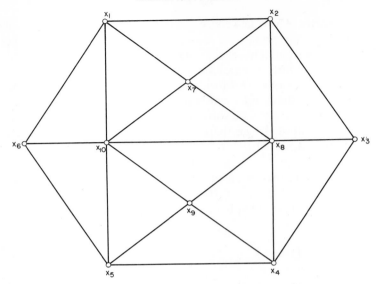

FIG. 4.4. Graph for example 3.3.1

3.4 A direct implicit enumeration (tree search) algorithm

A very simple implicit enumeration method could be used to find the chromatic number of a graph with surprising effectiveness considering the lack of sophistication [4], [9]. The method proceeds as follows.

Suppose that the vertices are ordered in some way and are renumbered so that x_i is the ith vertex in this ordering. An initial feasible colouring can then be obtained as follows:

    (i)   Colour x_i with colour 1.

    (ii)   Colour each of the remaining vertices sequentially so that a vertex x_i is coloured with the lowest-numbered colour that is "feasible" (i.e. which has not been used so far to colour any vertices adjacent to x_i).

Let q be the number of colours required by the above colouring. If a colouring using only $(q - 1)$ colours exists, then all vertices coloured with q must be recoloured with $j < q$. If x_{i*} is the first vertex in the vertex ordering which has been coloured q, and since (from (ii) above) it has been so coloured because it could not be coloured with any of the colours $j < q$, this vertex can only be recoloured with $j < q$ if at least one of its adjacent vertices is also recoloured. Thus, a backtracking step from x_{i*} can be taken as follows.

Of these vertices $x_1, \ldots, x_{i*-1}$ which are adjacent to x_{i*} find the last one in the vertex ordering (i.e. the one with the largest index) and let this be x_k. If x_k is coloured with colour j_k, recolour x_k with the lowest-numbered feasible alternative colour j'_k, $j'_k \geqslant j_k + 1$.

If $j_k^i < q$ continue by recolouring sequentially all the vertices x_{k+1} to x_n using method (ii) above, and provided that colour q is not needed. If this is possible a new better colouring using less than q colours has been found, otherwise if a vertex is encountered which requires colour q, then backtracking can again take place from such a vertex.

If $j_k' = q$, or no alternative feasible colour j_k' exists (see (a) below), then backtracking can take place immediately from vertex x_k. The algorithm terminates when backtracking reaches vertex x_1.

The following observations can help to speed up the above implicit enumeration procedure.

(a) Whatever the vertex ordering, the feasible colours j for vertex x_i are $j \leqslant i$ (provided $i < q$). This is apparent since only $i - 1$ vertices precede x_i in the vertex ordering and hence colours $j > i$ will never be needed. Thus for vertex x_1 the only feasible colour is 1, for vertex x_2 the feasible colours are 1 and 2 (unless x_2 is adjacent to x_1 in which case only colour 2 is feasible) etc.

(b) From (a) above it is apparent that it would be computationally beneficial to order the variables in such a way so that the first ρ (say) vertices form the largest clique of G. This would imply that each vertex x_i ($1 \leqslant i \leqslant \rho$) has only one feasible colour, i.e. colour i, and the algorithm can stop earlier when backtracking reaches vertex x_ρ.

Although the implicit enumeration procedure described in this section is a very primitive tree search—in which no bounds are calculated to limit the search—it has nevertheless been found to be at least as good an algorithm for graph colouring as any of the other available methods. Thus, a 30-vertex graph with about $\frac{1}{3}$ the number of links of the equivalent complete graph could be optimally coloured in approximately 30 seconds on an IBM 1130 computer.

4. Approximate Colouring Algorithms

There are many heuristic graph colouring procedures which can find good approximations to the chromatic number of those graphs that are too large for the colouring to be determined optimally using the exact methods mentioned earlier. In the present section we briefly describe one such procedure and a number of its variations.

4.1 Sequential methods based on vertex ordering

In this, simplest of all methods, the vertices are initially arranged in descending order of their degrees.

Colour the first vertex with colour 1 and scan the list of vertices downwards colouring with 1 any vertex which is not adjacent to another vertex that has already been coloured with 1. Starting from the top of the list, colour the first uncoloured vertex by 2 and again scan the list downwards colouring with 2 any uncoloured vertex which is not connected by a link to another vertex that has already been coloured with 2. Proceed in the same way with colours 3, 4, etc., until all the vertices have been coloured. The number of colours used is then taken as an approximation to the chromatic number of the graph [23, 25].

A simple modification to the above heuristic procedure would be to reorder—at each stage after a vertex is coloured—the remaining uncoloured vertices in descending order of their degrees, excluding from the calculation of the degrees those links which have an already coloured vertex as a terminal.

In the above description of the procedure it was tacitly assumed that if two vertices have the same degrees, then the order in which they are placed in the list is random. In such cases, the dilemma could be resolved by calculating the 2-step degrees $d_i^{(2)}$ of those vertices x_i with the same degrees (1-step degrees); where $d_i^{(2)}$ is defined as the number of paths of cardinality 2 emanating from vertex x_i.† These vertices can then be ordered according to $d_i^{(2)}$. If there are still vertices with the same d_i and $d_i^{(2)}$ one could calculate their 3-step degree $d_i^{(3)}$ (defined in a similar way), and order these according to $d_i^{(3)}$ etc.

Alternatively, one could order all the vertices directly according to their $d_i^{(2)}$ or their $d_i^{(3)}$, and apply the same sequential method of colouring [24]. Thus, the colouring method described above is not a single method but a class of sequential methods all of which depend on a vertex ordering, either static i.e. once and for all, or dynamic i.e. changing during the colouring process. The ordering itself can be based on a number of possible criteria depending on the vertex degrees or some other related measure. Computational results and comparisons between sequential colouring methods when applied to random graphs are given by Matula, Marble and Isaacson [16], and by Williams [24]. The limitations of these heuristic methods were demonstrated by Mitchem [17]., who showed that it is possible to construct graphs for which any of these heuristics yield arbitrarily bad estimates for the chromatic number.

5. Generalizations and Applications

The colouring problem, in its pure form mentioned earlier in this Chapter, seldom appears in practical problems. However, generalizations and slight

---

† If A is the adjacency matrix of the graph G with all diagonal elements set to 0, then $d_i = \sum\limits_{j=1}^{n} a_{ij}$. The 2-step degrees are then given by $d_i^{(2)} = \sum\limits_{j=1}^{n} a_{ij}d_j$ and, in general, the k-step degrees are given recursively by

$$d_i^{(k)} = \sum_{j=1}^{n} a_{ij}d_j^{(k-1)}.$$

variants of the problem appear very often in a large number of diverse application areas. The purpose of this section is to indicate the various generalizations that are most often encountered, and is by no means an exhaustive list of applications.

5.1 The simple loading problem [6]

Consider the problem of n items to be loaded into boxes. Let each item form the vertex of a graph G. Whenever two items x_i and x_j cannot be placed in the same box (for example if item x_i can contaminate item x_j), then a link (x_i, x_j) is introduced in G. If the boxes are of infinite capacity so that they can each hold any number of items, the problem of finding the smallest number of boxes to accommodate the items is that of finding the chromatic number of of G; where each "colour" corresponds to a box and items coloured similarly are loaded into the same box.

(i) BOXES OF THE SAME CAPACITY. In reality boxes have a capacity and suppose that the capacity Q is the same for all boxes. This is equivalent to saying that no more than Q vertices can be coloured with the same colour. In terms of the 0–1 programming formulation given in Section 3 of this Chapter for the pure colouring problem, the present capacitated problem can be written as:

> Minimize expression (4.14)
> subject to constraints (4.11) and (4.12)
> and also to

$$\sum_{i=1}^{n} \xi_{ij} \leqslant Q \text{ (for all } j = 1, 2, \ldots, q) \tag{4.16}$$

where the solution $[\xi_{ij}^*]$ is once more the optimal item-to-colour (or box) allocation matrix, and q is an upper bound on the number of boxes that may be required.

(ii) BOXES OF DIFFERENT CAPACITIES AND WITH COSTS. If, on the other hand, the boxes have all different "weight" capacities and different costs, (say Q_j and v_j respectively, for box j); the vertices have different "weights" (say w_i for item i), and what is required is to load the items into boxes so as to satisfy:

(a) The conditions imposed by the graph G,
(b) The capacity restrictions imposed by the boxes, and
(c) Minimize the total cost of boxes used,

then the problem becomes:

$$\text{Minimize } z = \sum_{j=1}^{q} \psi_j v_j \tag{4.17}$$

subject to constraints (4.11), (4.12) and also to:

$$\sum_{i=1}^{n} \xi_{ij} w_i \leq Q_j \tag{4.18}$$

$$\left.\begin{array}{l} \\ \end{array}\right\} \text{for all } j = 1, \ldots, q$$

and

$$\sum_{i=1}^{n} \xi_{ij} \leq L\psi_j \tag{4.19}$$

where variable ψ_j takes the value 1 if box j is loaded with any item, and 0 otherwise. L is once more any positive integer greater than n; and q is now the total number of available boxes. Constraint (4–18) ensures that no box is overloaded and constraint (4.19) ensures that if $\psi_{j_0} = 0$ (i.e. if box j_0 is not used), then all ξ_{ij_0} $(i = 1, \ldots, n)$ are also zero.

In this most general case, it has become necessary to introduce extra variables ψ_j since the boxes (colours) can no longer be penalized in the way shown in expression (4.14). It should be noted that the more the problem is generalized the less significant becomes its "colouring" aspect. Thus, in the last-mentioned problem, two subproblems can be recognized with: (a) The "colouring" aspect represented by constraint (4.12), and (b) The "knapsack"† aspect represented by constraint (4.18). The other two constraints, i.e. (4.11) and (4.19), being structural in nature.

Because of these two interrelated aspects, the general problem is much more difficult to solve than the pure colouring problem.

Practical problems which can be formulated as graph colouring problems (or generalizations thereof), are:

5.2 Scheduling examination timetables [25] [24]

In the timetable problem, examinations are to be scheduled into time slots. Each examination can be represented by a vertex of a graph and a link between any two vertices is added if there is any candidate who is taking both examinations. The problem is to schedule the examinations into the smallest number of sitting periods, so that no two examinations take place simultaneously if any one candidate must take both of them. This problem is precisely equivalent to the problem of colouring the vertices of the related graph with the smallest number of colours. The chromatic number of the graph is then the smallest number of sittings so that all the examinations take place.

If the number of examinations that can take place in any one period is limited (say because of room availability), then the problem becomes one of type (i) of Section 5.1 above where Q would now be the number of rooms available for the examinations.

---

† For the *knapsack problem* see [10] and the Appendix.

5.3 Resource allocation

Consider n elementary jobs which are to be executed by allocating m available resources. Let us suppose that each of these elementary jobs can be executed in a unit of time but that some set S_i of resources must be available for job i to be performed. We can now form a graph G on n vertices each of which corresponds to a job. An arc (x_i, x_j) is added between two vertices of G whenever jobs i and j have at least one resource requirement in common, i.e. when $S_i \cap S_j \neq \phi$. This implies that jobs i and j cannot be performed at the same time because of the availability of resources. A colouring of G then implies an allocation of resources to jobs, so that those jobs corresponding to vertices with the same colour are performed simultaneously. The greatest resource utilization (i.e. shortest time required to perform all n jobs) is then achieved by the optimal colouring of the vertices of G.

6. Problems P4

1. For the graph shown in Fig. 4.5 calculate upper and lower bounds for the chromatic number using the results given in Section 2.1 and 2.2.

2. For the above graph use the algorithms of Sections 3.1 and 3.4 to calculate the chromatic number and its associated colouring. Compare the computational effort involved.

3. For the graph shown in Fig. 3.6 of Chapter 3 calculate the chromatic number and associated colouring by formulating the problem as an SCP.

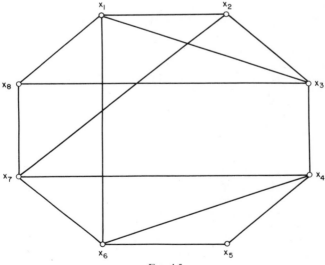

FIG. 4.5

4. For the graph shown in Fig. 4.5 use the sequential methods based on d_i, $d_i^{(2)}$ and $d_i^{(3)}$ to find the "optimal" colouring.

5. Prove that every tree on n vertices ($n \geqslant 2$) has chromatic number 2.

6. Show that a graph is 2-chromatic if and only if all its circuits are of even cardinality.

7. Show that two colours are sufficient for colouring the regions generated by the intersection of straight lines in the plane, so that no two adjacent regions are of the same colour.

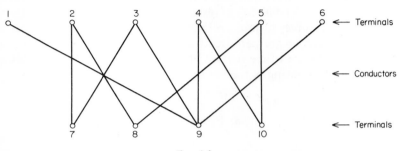

FIG. 4.6

8. The graph of Fig. 4.6 represents an electrical connection diagram with vertices corresponding to terminals and links corresponding to straight conducting plated metal strips. For the connections to be physically possible they must not cross each other and it is therefore necessary to distribute the links onto a number of parallel boards so that on any one board the conductors do not cross. (The terminals are available on all boards).

Find the least number of boards necessary for realizing these connections. (See References [14] and [7]).

9. Eleven types of products A, B, ..., K are to be manufactured at 3 locations, with each type being manufactured in only one of the 3 plants. The percentage commonality of parts and materials used in the products is given by the matrix:

	A	B	C	D	E	F	G	H	I	J	K
A											
B	55	—									
C	60	80	—								
D	85	65	40	—							
E	45	90	35	80	—						
F	80	75	30	90	90	—					
G	50	30	70	35	65	30	—				
H	50	60	90	40	35	70	30	—			
I	60	45	40	80	30	70	20	25	—		
J	45	40	45	35	80	20	90	25	85	—	
K	80	70	85	30	35	40	25	75	75	60	—

It is considered desirable to have products with large percentage commonalities manufactured at the same plant. The problem is to allocate the products to plants in such a way so that the minimum commonality between any two products allocated to the same plant is as large as possible.

7. References

1. Berge, C. (1962). "Theory of Graphs", Methuen, London.
2. Bondy, J. A. (1969). Bounds for the chromatic number of a graph, *Jl. of Combinatorial Theory*, **7**, p. 96.
3. Brooks, R. L. (1941). On colouring the nodes of a network, *Proc. Cambridge Philosophical Soc.*, **37**, p. 194.
4. Brown, J. R. (1972). Chromatic scheduling and the chromatic number problem, *Man. Sci.*, **19**, p. 456.
5. Christofides, N. (1971). An algorithm for the chromatic number of a graph, *The Computer Jl.*, **14**, p. 38.
6. Eilon, S. and Christofides, N. (1971). The loading problem, *Man. Sci.*, **17**, p. 259.
7. Even, S. (1973). "Algorithmic combinatorics", Macmillan, New York.
8. Geller, D. P. (1970). Problem 5713, *American Mathematical Monthly*, **77**, p. 85.
9. Gillian, R. J. (1970). The chromatic number of a graph, M.Sc. Thesis, Imperial College, London University.
10. Gilmore, P. C. and Gomory, R. E. (1967). The theory and computation of knapsack functions, *Ops. Res.*, **15**, p. 1045.
11. Graver, J. E. and Yackel, J. K. (1968). Some graph theoretic results associated with Ramsey's theorem, *Jl. of Combinatorial Theory*, **4**, p. 125.
12. Harary, F. (1969). "Graph Theory", Addison-Wessley, Reading, Massachusetts.
13. Heawood, P. J. (1890). Map-colour theorems, *Quart. Jl. of Mathematics, Oxford Series*, **24**, p. 322.
14. Lin, C. L. (1968). "Introduction to combinatorial mathematics", McGraw-Hill, New York.
15. Matula, D. W. (1968). A min–max theorem for graphs with application to colouring, *Jl. of SIAM (Review)*, **10**, p. 481.
16. Matula, D. W., Marble, G. and Isaacson, J. D. (1972). Graph colouring algorithms, *In*: "Graph Theory and Computing", Read, Ed., Academic Press, New York, p. 109.
17. Mitchem, J. (to appear). On various algorithms for estimating the chromatic number.
18. Myers, B. R. and Lin, R. (1972). A lower bound on the chromatic number of a graph, *Networks*, **1**, p. 273.
19. Nordhous, E. A. and Gaddum, J. W. (1956). On complementary graphs, *American Mathematical Monthly*, **63**, p. 175.
20. Ore, O. (1967). "The four colour problem", Academic Press, New York.
21. Szekeres, G. and Wilf, H. S. (1968). An inequality for the chromatic number of a graph, *Jl. of Combinatorial Theory*, **4**, p. 1.
22. Tutte, W. T. (B. Descartes). (1954). Solution of advanced problem No. 4526, American Mathematical Monthly, **61**, p. 352.
23. Welsh, D. J. A. and Powell, M. B. (1967). An upper bound on the chromatic number of a graph and its application to timetabling problems, *The Computer Jl.*, **10**, p. 85.

24. Williams, M. R. (1968). A graph theory model for the solution of timetables, Ph.D. Thesis, University of Glasgow.
25. Wood, D. C. (1969). A technique for colouring a graph applicable to large scale timetabling problems, *The Computer Jl.,* **12**, p. 317.
26. Zykov, A. A. (1952). On some properties of linear complexes (in Russian), *Mat. Sbornik, 24/26*, p. 163, American Math. Soc. Translation, No. 79.

Chapter 5

The Location of Centres

1. Introduction

Problems of finding the "best" location of facilities in networks or graphs, abound in practical situations. In particular, if a graph represents a road network with its vertices representing communities, one may have the problem of locating optimally a hospital, police station, fire station, or any other "emergency" service facility. In such cases, the criterion of optimality may justifiably be taken to be the minimization of the distance (or travel time) from the facility to the most remote vertex of the graph, i.e. the optimization of the "worst-case". In a more general problem, a large number (and not just one) of such facilities may be required to be located. In this case the furthest vertex of the graph must be reachable from at least one of the facilities within a minimum distance. Such problems, involving the location of emergency facilities and whose objective is to minimize the largest travel distance to any vertex from its nearest facility, are, for obvious reasons, called *minimax location problems*. The resulting facility locations are then called the *centres* of a graph.

In some location problems, however, a more appropriate objective would be to minimize the total sum of the distances from vertices of the graph to the central facility (assuming that only one such facility is to be located). Such objectives are, for example, better suited to the problem of locating a depot in a road network where the vertices represent customers to be supplied by the depot, or to the problem of locating switching centres in a telephone network where the vertices represent subscribers. Problems of this type are generally referred to as *minisum location problems*, although the objective function is often not simply the sum of distances but the sum of various functions of distance (See Chapter 6). The facility locations resulting from the solution to a minisum problem are called the *medians* of a graph.

The purpose of this Chapter is to discuss the minimax location problem for general weighted graphs with a set of weights c_{ij} (representing lengths), associated with its arcs, and another set of weights v_j (representing, say, size or importance), associated with its vertices. Algorithms for the optimum location of centres in such graphs are given together with computational results for medium sized grphs.

The minisum location problem is discussed separately in the next Chapter and although the two location problems are obviously related, these Chapters can be read independently since the methods used for solving the two problems are different.

2. Separations

For any vertex x_i of a graph $G = (X, \Gamma)$, let $R^0_\lambda(x_i)$ be the set of those vertices x_j of G which are reachable from vertex x_i by a path of weighted length $v_j d(x_i, x_j)$ less than or equal to λ, and let $R^t_\lambda(x_i)$ be the set of those vertices x_j of G from which vertex x_i can be reached by a path of weighted length $v_j d(x_j, x_i) \leqslant \lambda$.

Thus

$$R^0_\lambda(x_i) = \{x_j | v_j d(x_i, x_j) \leqslant \lambda, x_j \in X\}$$

and

$$R^t_\lambda(x_i) = \{x_j | v_j d(x_j, x_i) \leqslant \lambda, x_j \in X\}$$

(5.1)

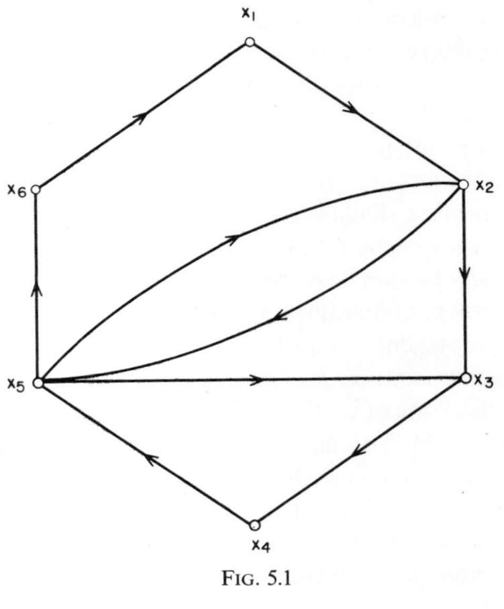

FIG. 5.1

We now define two *separation* numbers for every vertex x_i as follows:

$$s_o(x_i) = \max_{x_j \in X} \left[v_j d(x_i, x_j) \right]$$

and

$$s_t(x_i) = \max_{x_j \in X} \left[v_j d(x_j, x_i) \right]$$

(5.2)

The numbers $s_o(x_i)$ and $s_t(x_i)$ are called the *outseparation* and *inseparation* of the vertex x_i respectively. One should note here that $s_o(x_i)$ is the largest number in row x_i of the matrix $D'(G)$ derived by multiplying every column j of the distance matrix $D(G) = [d(x_i, x_j)]$ by v_j; and $s_t(x_i)$ is the largest number in column x_i of the matrix $D''(G)$ derived by multiplying every row j of the distance matrix $D(G)$, by v_j.

Consider for example the directed graph of Fig. 5.1 and suppose that the graph has vertex and arc weights of value unity. The distance matrix of the graph is:

$$D(G) =$$

	x_1	x_2	x_3	x_4	x_5	x_6	$s_o(x_i)$
x_1	0	1	2	3	2	3	3
x_2	3	0	1	2	1	2	3
x_3	4	3	0	1	2	3	4
x_4	3	2	2	0	1	2	3
x_5	2	1	1	2	0	1	2*
x_6	1	2	3	4	3	0	4
$s_t(x_i)$	4	3*	3*	4	3*	3*	

The outseparations and inseparations are shown in adjoining columns and rows respectively.

If λ_o is the smallest length λ so that for a vertex x_i

$$R_\lambda^o(x_i) = X$$

(i.e. all vertices of G are reachable from x_i within a minimum weighted distance λ_o), then, it follows from eqns (5.1) and (5.2) that:

$$s_o(x_i) = \lambda_o$$

(5.3)

Similarly, if λ_t is the smallest length λ so that

$$R_\lambda^t(x_i) = X$$

D

then $s_t(x_i) = \lambda_t$.

It is quite apparent that for a graph G to have finite outseparation and inseparation numbers for all the vertices it must be strongly connected, i.e. every vertex must be able to reach every other vertex along a path of some finite length.

3. Centre and Radius

A vertex x_o^* for which:

$$s_o(x_o^*) = \min_{x_i \in X} [s_o(x_i)] \tag{5.4)(a)}$$

is called an *outcentre* of the graph G; and similarly a vertex x_t^* for which:

$$s_t(x_t^*) = \min_{x_i \in X} [s_t(x_i)] \tag{5.4)(b)}$$

is called an *incentre* of G.

For a given graph there may be more than one outcentre or incentre thus forming outcentral or incentral sets.

The outseparation of the outcentre is called the *outradius* $\rho_0 = s_o(x_o^*)$, and the inseparation of the incentre is called the *inradius* $\rho_t = s_t(x_t^*)$.

Thus for the example of Fig. 5.1, and referring to the distance matrix above, we see that there is only one outcentre (that is vertex x_5), giving an outradius of 2 units; whereas there are four incentres forming the incentral set $\{x_2, x_3, x_5, x_6\}$, giving an inradius of 3 units.

3.1. The location of an emergency centre

Consider the problem of a number of communities (interlinked by a road network), which has to be served by a single hospital, police station or fire station. It is required for some reason (i.e. availability of manpower, or other amenities) that these facilities must be located in one of these communities and not just any arbitrary point along the road.

Let us now assume that the arc "lengths" c_{ij} of the graph G whose vertices represent the communities and whose arcs represent the road, form a matrix corresponding to the travel times between these communities. This matrix is not necessarily symmetrical, i.e. $c_{ij} \neq c_{ji}$ since the travelling times may be affected by slopes in the road, one-way streets, etc.

In the case of locating a police station or a fire station, what is of interest is the time that is required to reach the most distant of these communities and the problem is, therefore, to locate the police (or fire-station) so as to minimize this time. The problem can be made more realistic if weights v_j are assigned to the vertices of a graph representing the probabilities of the police or fire-brigade being required by the various communities; these weights may, for example, be taken as proportional to the population of each community. The vertex which minimizes the time to reach the most distant community is then the outcentre of the graph.

In the case of locating a hospital, what may be of interest is the time that an ambulance takes to reach the most distant community and return back to the hospital. If we define the combined in–outseparation of a vertex x_i as:

$$s_{ot}(x_i) = \max_{x_j \in X} \{v_j[d(x_i, x_j) + d(x_j, x_i)]\}$$

then that vertex x_{ot}^* which produces a minimum of the expression:

$$\min_{x_i \in X} [s_{ot}(x_i)]$$

may be called the combined in-outcentre.

For the graph of Fig. 5.1 having the distance matrix $D(G)$ given earlier, the combined in-outcentre is x_5 having an in–outradius of 5 units.

4. The Absolute Centre

Equations (5.3) define the separation numbers for any $x_i \in X$. We can now generalize the definition to include artificial points that may be placed on the

FIG. 5.2. Location on an arc

arcs. Thus, if $a = (x_i, x_j)$—see Fig. 5.2—represents an arc of a graph whose weight (length) is c_{ij}, a point y placed on the arc can be defined by specifying the length $l(x_i, y)$ of the section (x_i, y) and assuming that

$$l(x_i, y) + l(y, x_j) = c_{ij} \tag{5.5}$$

The separations $s_o(y)$ and $s_t(y)$ of a point y, whether it be a vertex of G or an artificially placed point on an arc of G, are then given by:

$$s_o(y) = \max_{x_i \in X} [v_i d(y, x_i)] \tag{5.6}$$

and similarly from eqn (5.2) for $s_t(y)$.

A point y_o^* for which:

$$s_o(y_o^*) = \min_{y \text{ on } G} [s_o(y)] \tag{5.7}$$

is called the *absolute outcentre* of the graph; and similarly for y_t^* the *absolute incentre*.

The outseparation of the absolute outcentre is called the *absolute outradius* $r_o = s_o(y_o^*)$, and the inseparation of the absolute incentre is called the *absolute inradius* $r_t = s_t(y_t^*)$.

EXAMPLE. Consider the nondirected graph G of Fig. 5.3, where all the vertex weights and link lengths are unity. Since the graph is nondirected, the outseparations and inseparations are the same.

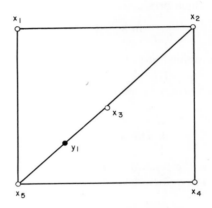

FIG.5.3. Absolute centre

The distance matrix of the graph is:

	x_1	x_2	x_3	$x_.$	x_5
x_1	0	1	2	2	1
x_2	1	0	1	1	2
$D(G) = x_3$	2	1	0	2	1
x_4	2	1	2	0	1
x_5	1	2	1	1	0

from which it is seen that any vertex of G is a centre and the radius is 2 units.

If we now choose a point y_1 on link (x_3, x_5) so that $l(x_5, y_1) = \frac{1}{2}$ (and therefore $l(y_1, x_3) = \frac{1}{2}$), the distances of this point from the vertices of G are given by the array:

	x_1	x_2	x_3	x_4	x_5
y_1	$1\frac{1}{2}$	$1\frac{1}{2}$	$\frac{1}{2}$	$1\frac{1}{2}$	$\frac{1}{2}$

producing a maximum separation of $1\frac{1}{2}$ units. Thus point y_1 is more "central" than any of the vertices of G. In fact point y_1 is an absolute centre of G and so are all the points in the middle of links (x_1, x_2), (x_2, x_3), (x_5, x_4), (x_4, x_2) and (x_5, x_1). None of the vertices of G are absolute centres. Thus, in general, there may be one or more absolute centres and these may be located either at the vertices or on the arcs.

5. Algorithms for the Absolute Centre

The centre and radius of a graph can be found directly from the weighted distance matrix as was shown in Sections 2 and 3. We now give two methods for finding the absolute centre of a graph, and will illustrate the methods by an example.

5.1. The method of Hakimi [7]

This method is very simple and for a nondirected graph is as follows: (For a directed graph the method is unchanged and one could substitute each nondirected concept with its directed counterpart.)

(i) For each link a_k of the graph find that point (or points) y_k^* on a_k which has the minimum separation.

(ii) Of all the y_k^* ($k = 1, 2, \ldots, m$) choose the smallest as the absolute centre of G.

The first step in the method is done as follows:

Consider the link a_k of the graph G as shown in Fig. 5.4. We then have: (from eqn (5.6))

$$s(y_k) = \max_{x_i \in X} [v_i d(y_k, x_i)]$$

$$= \max_{x_i \in X} [v_i \min \{l(y_k, x_\beta) + d(x_\beta, x_i), l(y_k, x_\alpha) + d(x_\alpha, x_i)\}] \qquad (5.8)$$

since the distance $d(y_k, x_i)$ may be either the length of a path through x_α or that of a path through x_β. The lengths $l(y_k, x_\alpha)$ and $l(y_k, x_\beta)$ are the lengths of the respective parts of link a_k.

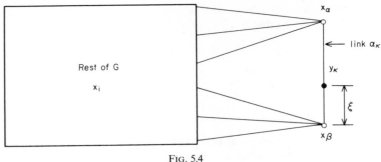

FIG. 5.4

Now let $l(y_k, x_\beta) = \xi$, and since $l(y_k, x_\alpha) = c_{\alpha\beta} - l(y_k, x_\beta) = c_{\alpha\beta} - \xi$; eqn (5.8) becomes:

$$s(y_k) = \max_{x_i \in X} \min \left[v_i\{\xi + d(x_\beta, x_i)\}, v_i\{c_{\alpha\beta} + d(x_\alpha, x_i) - \xi\} \right] \qquad (5.9)$$

For a given x_i, we can find the smallest of the two terms in the square parentheses of eqn (5.9) for every value of ξ, $(0 \leqslant \xi \leqslant c_{\alpha\beta})$, by plotting the two terms

$$\left. \begin{aligned} T_i &= v_i\{\xi + d(x_\beta, x_i)\} \\[2mm] T_i' &= v_i\{c_{\alpha\beta} + d(x_\alpha, x_i) - \xi\} \end{aligned} \right\} \qquad (5.10)$$

and

separately against ξ and taking the lower envelope of the two resulting straight lines.

We repeat the process for all the other $x_i \in X$ and obtain all other lower envelopes on the same plot. We now draw the overall upper envelope to all the previously obtained lower envelopes and this final envelope—according to eqn (5.9)—gives the separation $s(y_k)$ for all values of ξ i.e. for all points y_k on the link a_k. The envelope (which is made up of piecewise linear sections) may have a number of minima. The position of the point y_k which produces the lowest minimum is, according to eqn (5.7), the absolute centre y_k^* subject to the initial restriction that y_k must lie on link a_k. Thus, the absolute centre of the graph is found as the minimum of the y_k^* for all the links a_k ($k = 1, 2, \ldots, m$).

5.2. The location of an emergency centre (general case)

Consider once more the problem of a number of communities which have to be served by a single hospital, fire station or police station. If the restriction that these facilities must be located in one of the communities (see Section 3.1) is relaxed, then the optimum location of the facilities is at the absolute centre(s) of the corresponding graph.

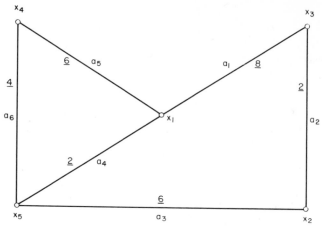

FIG. 5.5. Graph for example in section 5.2

Consider, for example, the nondirected graph of Fig. 5.5 in which each community is represented by a vertex and the length of a link (x_i, x_j) is the time, in minutes, taken to travel from community x_i to community x_j. The vertices of the graph have unit "weights" and the "distance" (time) matrix of this graph is:

		x_1	x_2	x_3	x_4	x_5
	x_1	0	8	8	6	2
	x_2	8	0	2	10	6
$D(G) =$	x_3	8	2	0	12	8
	x_4	6	10	12	0	4
	x_5	2	6	8	4	0

The centre of the graph is obviously either x_1 or x_5 and the radius is 8.

Let us consider link a_1 (Fig. 5.5) first and measure distance ξ from x_3 towards x_1.

The two terms T_i and T_i' of eqn (5.10) (for $i = 1, 2, \ldots, 5$) become:

$$T_1 = \xi + d(x_3, x_1) = \xi + 8$$
$$T_1' = c_{3,1} + d(x_1, x_1) - \xi = 8 - \xi$$

similarly:

$$T_2 = \xi + 2 \qquad T'_2 = 16 - \xi$$
$$T_3 = \xi \qquad T'_3 = 16 - \xi$$
$$T_4 = \xi + 12 \qquad T'_4 = 14 - \xi$$
$$T_5 = \xi + 8 \qquad T'_5 = 10 - \xi$$

These terms are shown plotted in Fig. 5.6(a) against $(0 \leqslant \xi \leqslant 8)$. There are two local absolute centres y_1^* and these are seen to be at distances of 6 and 8 minutes from x_3. The separation of both these centres is 8 minutes.

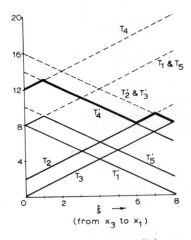

FIG. 5.6 (a). Location on link a_1

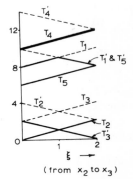

FIG. 5.6 (b). Location on link a_2

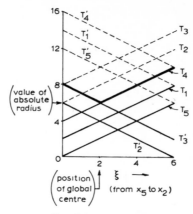

FIG. 5.6 (c). Location on link a_3

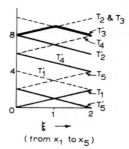

FIG. 5.6 (d). Location on link a_4

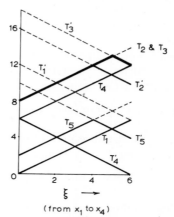

FIG. 5.6 (e). Location on link a_5

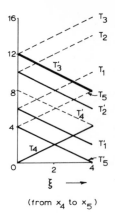

FIG. 5.6 (f). Location on link a_6

Similarly one obtains plots for the other links a_2, a_3, a_4, ... as shown in Figs. 5.6(b) to (f). The absolute centre (y^*) of the graph is that local centre which has the minimum separation. This is seen to be on link a_3 (Fig. 5.6(c)) at a distance of 2 minutes from x_5. The absolute radius r (separation of y^*) is then seen from Fig. 5.6(c) to be 6 minutes.

5.3. Modified Hakimi's method

The previously described method of Hakimi requires a search for a local centre along every existing link in the graph G. If the number of links in G is large then the computing time required for the search may be quite excessive. The present modification to the method determines upper and lower bounds on the absolute local radii associated with the local centres on each link, and these bounds may therefore enable one to reduce the number of links that may have to be searched.

Any local centre located on link (x_i, x_j) will, (as can be seen from eqn (5.8) with the l terms set to zero), have associated with it an absolute local radius (r_{ij} say), which must be at least as large as p_{ij}; where

$$p_{ij} = \max_{x_s \in X} \left[v_s \min \{d(x_s, x_i), d(x_s, x_j)\} \right] \tag{5.11}$$

Thus p_{ij} is a *lower bound* on the absolute radius of the graph, provided that the absolute centre lies on link (x_i, x_j). Hence the quantity

$$P = \min_{(x_i, x_j) \in A} [p_{ij}] \tag{5.12}$$

(where A is the set of links in the graph), is a valid lower bound on the absolute radius.

Suppose the absolute centre is in the centre of link (x_i, x_j); then, according to eqn (5.8), the absolute radius is $p_{ij} + \frac{1}{2}v_{s^*}c_{ij}$ where v_{s^*} is the value of that x_s which produces the maximum p_{ij} according to eqn (5.11). Hence the quantity

$$H = \min_{(x_i, x_j) \in A} [p_{ij} + \tfrac{1}{2}v_{s^*}c_{ij}] \tag{5.13}$$

is a valid *upper bound* on the absolute radius. Thus any link (x_{i_0}, x_{j_0}) for which $p_{i_0 j_0} \geq H$ may be excluded from the search for the absolute centre.

Considering the previous example of the graph of Fig. 5.5 where $v_i = 1$ for all $x_i \in X$; we calculate from eqn (5.11):

Centre on $a_1 = (x_3, x_1)$ gives $p_{3,1} = \max (2, 6, 2) = 6$

Centre on $a_2 = (x_2, x_3)$ gives $p_{2,3} = \max (8, 10, 6) = 10$

Centre on $a_3 = (x_5, x_2)$ gives $p_{5,2} = \max (2, 2, 4) = 4$

Centre on $a_4 = (x_1, x_5)$ gives $p_{1,5} = \max (6, 8, 4) = 8$

Centre on $a_5 = (x_1, x_4)$ gives $p_{1,4} = \max (8, 8, 2) = 8$

Centre on $a_6 = (x_4, x_5)$ gives $p_{4,5} = \max (2, 6, 8) = 8$

The upper bound H is, therefore, (according to eqn (5.13))

$$H = \min (6 + 4, 10 + 1, 4 + 3, 8 + 1, 8 + 3, 8 + 2) = 7$$

and hence links a_2, a_4, a_5 and a_6 whose $p_{ij} > 7$ can be excluded as candidates for the location of the absolute centre. Only the remaining two links a_1 and a_3 need be searched as explained previously and shown in Figs. 5.6(a) and 5.6(c). By the use of upper and lower bounds, the search is therefore reduced to $\frac{1}{3}$ of its previous level. The overall lower bound P as calculated by eqn (5.12) is

$$P = \min (6, 10, 4, 8, 8, 8) = 4$$

a value which is not realized in practice since the actual value of the absolute radius (r) of the graph was found in Section 5.2 to be 6.

5.4. An iterative method
Once more, in order to simplify the explanation, the method will be described in terms of a nondirected graph although the replacement of every nondirected concept by its directed counterpart will leave the method unchanged.

Let $Q_\lambda(x_i)$ represent the set of all points y on the graph G from which vertex x_i is reachable within a weighted distance λ. Thus

$$Q_\lambda(x_i) = \{y \,|\, v_i d(y, x_i) \leq \lambda, y \text{ on } G\} \tag{5.14}$$

which is analogous to eqn (5.1)—giving the sets $R_\lambda^0(x_i)$ and $R_\lambda^t(x_i)$—except that now y is any point on G and need not be a vertex. The absolute radius r is obviously the smallest value of λ so that from some point y on G all the vertices of the graph can be reached within a weighted distance less than or equal to λ. Hence r is the smallest value of λ (say λ_{min}) so that

$$\bigcap_{x_i \in X} [Q_\lambda(x_i)] \equiv Q_\lambda(x_1) \cap Q_\lambda(x_2) \cap \ldots \cap Q_\lambda(x_n) \neq \varnothing \qquad (5.15)$$

Therefore, we can start with any small value of λ, calculate the sets $Q_\lambda(x_i)$ for all $i = 1, 2, \ldots, n$, and check eqn (5.15). If the equation is not satisfied increase λ by a small increment; recalculate the $Q_\lambda(x_i)$ with the new value of λ and again check eqn (5.15). We can repeat the process until eqn (5.15) is satisfied and the value of λ on that iteration would then be the absolute radius r of the graph G. Moreover, since the increments to λ are small the intersection set

$$\bigcap_{x_i \in X} [Q_\lambda(x_i)] \qquad (5.16)$$

will, for all practical purposes, contain on the final iteration not a field but just a single point which is the absolute centre y^*. (It may contain more than one point if there is more than one absolute centre.)

Since $d(x_i, x_j)$ is the shortest distance between any two vertices x_i and x_j then it is quite apparent that if λ is smaller than one half the weighted distance between x_i and x_j i.e. if

$$\lambda \leqslant \frac{v_i v_j}{v_i + v_j} d(x_i, x_j), \quad \text{then:} \quad Q_\lambda(x_i) \cap Q_\lambda(x_j) = \varnothing$$

and the total intersection set in expression (5.16) would then be empty.

Therefore the iterative method for locating the absolute centre may be started with a value of λ given by

$$\lambda_{\text{initial}} = \max_{x_i \neq x_j \in X} \left[\frac{v_i v_j}{v_i + v_j} d(x_i, x_j) \right] \qquad (5.17)$$

since r must be $\geqslant \lambda_{\text{initial}}$. In view of this the value $\delta = 2\lambda_{\text{initial}}$ may be called the *diameter* of the graph, but it should be noted that this is not necessarily twice the value of the absolute radius as defined in Section 4. (See problem P5.11).

A detailed description of a more general algorithm, of which the present algorithm is only a part, is given later in Section 8.

6. Multiple Centres (p-Centres)

The concept of a centre of a graph can be generalized to more than a single centre so that a set of p points forms a multi-centre (p-centre).

Let X_p be a subset (containing p vertices) of the set X of vertices of the graph $G = (X, \Gamma)$. We will write $d(X_p, x_i)$ to mean the shortest distance between any one of the vertices in the set X_p and vertex x_i; i.e.

$$d(X_p, x_i) = \min_{x_j \in X_p} [d(x_j, x_i)]$$

similarly we will write $d(x_i, X_p)$ to mean $\min_{x_j \in X_p} [d(x_i, x_j)]$

In a way analogous to the definition of the separation number of a vertex (see Section 3), we can define separation numbers for the set of vertices X_p as follows:

$$s_o(X_p) = \max_{x_j \in X} [v_j d(X_p, x_j)]$$

and

$$s_t(X_p) = \max_{x_j \in X} [v_j d(x_j, X_p)]$$

(5.18)

where $s_o(X_p)$ and $s_t(X_p)$ are the outseparation and inseparation numbers of the set X_p.

A set X_{po}^* for which:

$$s_o(X_{po}^*) = \min_{X_p \subseteq X} [s_o(X_p)]$$

(5.19)

is called the *p-outcentre* of the graph G, and similarly for the *p-incentre* X_{pt}^*.

In Sections 2 and 3 it was pointed out that the centre of a graph could be formed very simply from the weighted distance matrix. The *p*-centre, however, could be calculated in the same way (by complete enumeration) only for small graphs and for small values of *p*. Thus one could generate all possible sets of vertices $X_p \subseteq X$ containing *p* vertices and then use eqns (5.18) and (5.19) directly to find the sets X_{po}^* and X_{pt}^* forming the *p*-centre. Assuming that the distance matrix already exists, this direct use of the equations requires a total of $p(n - p)\binom{n}{p}$ comparisons which for (say) $n = 100$ and $p = 5$ becomes $3 \cdot 58 \times 10^{10}$, a number of comparisons that is well above the limit that present day computers could perform in a reasonable time.

6.1. The location of a number of emergency centres

In Section 3.1 we have discussed the problem of locating a single hospital, police station or fire station in a graph representing a practical road network. However, the situation often arises where a single facility will not be able to meet various requirements and what is then needed is to locate a number of these facilities in the best possible way. The following problem may then arise.

Find the smallest number and location of fire stations (say) so that no community is further away than a prespecified distance from a station, and, with that number of stations given, the distance of the furthest community from a station is a minimum.

If we assume that the fire stations must be located at the vertices of the corresponding graph G, then the problem is one of finding the p-centres of G for $p = 1, 2, 3, \ldots$ until the separation of the p-centre becomes less than or equal to the prespecified distance. The last value of p will then be the smallest number of stations, and the p-centre the optimum location of the fire stations meeting the requirements.

The algorithm for the calculation of p-centres is a special case of an algorithm for the solution of the more general problem of finding p-centres without the restriction that they must lie at the vertices of the graph. The discussion of the algorithm for finding p-centres will therefore be postponed until the more general problem is discussed. (See Section 8.5.)

7. Absolute p-centres

If the restriction that the points forming the p-centre must lie at the vertices of a graph is lifted so that points lying on the arcs are also admissible, then this more general set of p points is called the *absolute p-centre*. Thus, what has in Section 5 been called the absolute centre, could, with the present terminology also be called the absolute 1-centre.

The problem of finding the absolute p-centre can now generally be stated as follows:

(a) Find the optimal location anywhere on the graph of a given number (say p) of centres so that the distance (time) required to reach the most remote vertex from its nearest centre is a minimum.

A second problem, which is very closely related to problem (a) above, and which—as shown later—can be solved by the same method which solves problem (a), is:

(b) For a given "critical" distance, find the smallest number (and location) of centres so that all the vertices of the graph lie within this critical distance from at least one of the centres.

It is this general problem of finding absolute p-centres which occurs most often in practice, but this is a considerably more difficult problem to solve than any of its restricted versions. The method of Hakimi [7, 8], described in Section 5 for solving the single absolute centre problem, cannot be generalized to find absolute p-centres, whereas Singer [12] has proposed a heuristic method for finding such centres.

In this section we present an iterative algorithm for the absolute p-centre of a graph. This algorithm converges to the optimal solution very rapidly as can be seen from the computational results given later on. The method has two further advantages:

(i) It can be terminated when the required tolerance (accuracy) for the position of the centres has been reached.
and

(ii) The method can easily be modified to generate suboptimal solutions and hence enables the analysis of the sensitivity or robustness of solutions.

For simplicity in notation we will only consider graphs which are non-directed, since any extension to directed graphs is straightforward.

Let Y_p be a set of p points anywhere on the graph G.

The separation $s(Y_p)$ of the set Y_p is then

$$s(Y_p) = \min_{x_j \in X} \left[v_j d(Y_p, x_j) \right] \tag{5.20}$$

where once more

$$d(Y_p, x_j) = \min_{y_i \in Y_p} \left[d(y_i, x_j) \right]$$

The absolute p-centre of G is then defined as the set of points Y_p^* so that:

$$s(Y_p^*) = \min_{Y_p \text{ on } G} \left[s(Y_p) \right] \tag{5.21}$$

8. An Algorithm for finding Absolute p-Centres

Consider each vertex x_i in turn and "penetrate" along all possible routes leading from it, up to a distance $\delta_i = \lambda/v_i$; where λ is a predefined constant which we shall call the *penetration*.

Let $Q_\lambda(x_i)$ be the set of all points y on G from which vertex x_i is reachable within a distance δ_i for a given value of λ. The sets $Q_\lambda(x_i)$ are given by eqn (5.14) Section 5.2. These sets can be calculated quite easily by an algorithm similar to Dijkstra's algorithm [4] for finding the shortest paths (see Chapter 8).

Let us now define a *region* Φ_λ as a set of points y on G all of which can reach exactly the same set of vertices of G within the distances δ_i for a given λ. A region may, for example, be a section of a link or may contain just one point. Thus, for the graph shown in Fig. 5.7, where the vertex weights are all unity and the link lengths are given by $c_{1,2} = 4$, $c_{2,3} = 8$ and $c_{3,1} = 6$ units; taking $\lambda = 3$ units (say) gives $\delta_i = 3$ for all i and produces the following different six regions:

Region 1: Point a. (The only point that can reach both x_1 and x_3 within a distance of 3 units)

Region 2: The section of link (x_1, x_2) from b to c. (Any point in this region can reach both x_1 and x_2)

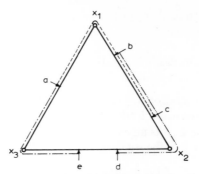

FIG. 5.7. Regions formed by taking $\lambda = 3$
$Q_\lambda(1)$: — — — — — — —
$Q_\lambda(2)$: — · · — · · —
$Q_\lambda(3)$: — — — — —

Region 3: Composed of link sections (a, x_1) and (x_1, b). (This region can reach x_1 only)

Region 4: Composed of link sections (a, x_3) and (x_3, e). (This region can reach x_3 only)

Region 5: Composed of link sections (c, x_2) and (x_2, d). (This region can reach x_2 only)

Region 6: The section of link (x_2, x_3) from e to d. (This region cannot reach any vertex)

In general the regions Φ_λ can be calculated from the reachable sets Q_λ as follows:

The regions that do not reach any vertex are given by:

$$\Phi_\lambda(0) = \{y \mid y \text{ on } G\} - \bigcup_i Q_\lambda(x_i) \qquad (5.22)$$

where the second term excludes all regions of G that can reach *some* vertex x_i.

The regions that can reach the t vertices $x_{i_1}, x_{i_2}, \ldots, x_{i_t}$ (for any $t = 1, 2, \ldots, n$) and no other vertex are given by:

$$\Phi_\lambda(x_{i_1}, x_{i_2}, \ldots, x_{i_t}) = \bigcap_{q=1,\ldots,t} Q_\lambda(x_{i_q}) - [\bigcap_{q=1,\ldots,t} Q_\lambda(x_{i_q})] \cap [\bigcup_{q=t+1,\ldots,n} Q_\lambda(x_{i_q})] \qquad (5.23)$$

where, once more, the second term excludes regions that can reach other vertices in addition to $x_{i_1}, x_{i_2}, \ldots, x_{i_t}$.

Although the number of regions may—at first sight—be expected (from eqn 5.23) to be very large, the set intersections soon become empty in practice, and the number of regions is not excessive. Computationally efficient methods for calculating the regions and further reducing their number are given later on in this section.

8.1. Description of the algorithm

The skeleton of an algorithm for finding the absolute p-centre of a graph $G = (X, A)$ for a given p can now be outlined as follows.

Step 1. Put $\lambda = 0$.

Step 2. Increase λ by a small amount $\Delta\lambda$.

Step 3. Find the sets $Q_\lambda(x_i)$ for all $x_i \in X_n$ and calculate the regions Φ_λ.

Step 4. Form the bipartite graph $G' = (X' \cup X, A')$ where X' is a set of vertices each representing a region, and A' is a set of arcs so that an arc between a region-vertex and vertex x_i exists if and only if x_i can be reached from that region.

Step 5. Find the minimum dominating set of G' (See Chapter 3).

Step 6. If the number of regions in the above set is greater than p return to step 2; otherwise stop. The regions in this set then form the absolute p-centre of the original graph G.

One should note (see Chapter 3) that the number of regions in the minimum dominating set represents (by definition) the smallest number of points on G from which all the vertices of G can be reached within the penetration distance λ for that iteration. It should also be noted that in the process of deriving the absolute p-centre by the above algorithm one would also derive the absolute $(n - 1)$, $(n - 2)$ etc. centres. Thus, at the end of any iteration (going from step 2 to step 6), during which the number of regions in the minimum dominating set decreases from some level (say l) to $l - 1$, the regions in that set will be the absolute $(l - 1)$-centre, and the value of λ for that iteration will be the "absolute $(l - 1)$-radius" i.e. it is the "critical" value of λ below which no $(l - 1)$-centre exists, which can reach every vertex of the graph within the weighted distance λ.

If what is required is the smallest value of p so that every vertex is reachable from some centre within a given critical distance (problem (b), in Section 8), then steps 3 to 6 of the above algorithm should be performed just once with λ set to this critical value. The corresponding number of regions in the minimum dominating set is then the required value of p, and the regions in this set form the required p-centre.

8.2. Computational aspects

Let us assume that λ has a fixed constant value, from which the distances $\delta_i = \lambda/v_i$ are calculated. Any link of the graph may be reachable either in whole or in part—or not reachable at all—within a distance δ_i from a vertex x_i. If only a portion of a link (from one end to some limit point along the link) is reachable, a "marker" is placed at the limit point. These markers contain all the information necessary to describe the sets $Q_\lambda(x_i)$. Thus $Q_\lambda(x_i)$ consists of all the links (or parts of links) on the shortest paths between any marker and vertex x_i.

When markers have been placed for all vertices the totality of all the markers will divide each link into a number of sections, a section being characterized by the vertices which can be reached from it. Thus, each section can be represented by a binary (zero–one) vector $\{j_1, j_2, \ldots, j_n\}$ of length n; where $j_k = 1$ if vertex x_k is reachable from the section and zero otherwise.

The totality of all link sections with the same binary vector, form a region and hence all the regions Φ_λ given by eqn (5.23) can be calculated from the markers. Since a region is reachable from the set of vertices for which $j_k = 1$ (in the binary vector representing this region), and from no other vertices, these binary vectors will henceforth be referred to as *strict intersections* (SI).

The representation of each region by a binary vector does not contain any information as to where this region lies in the graph. However, this representation is computationally very advantageous both from the computing memory and execution times points of view, particularly as far as step 5 of the algorithm is concerned. Once the SI's forming the absolute p-centre are found by the algorithm it is quite a simple, and computationally insignificant step, to recalculate to which sections of links these SI's correspond.

Step 4 of the algorithm requires the construction of a bipartite graph G' with a vertex representing every region. This may lead to quite large graphs which will greatly affect the computation times for step 5. However, the following theorem reduces the size of the graph G' by excluding those regions which cannot affect the optimal answer, although if more than one optimum answer exists they may also exclude some of these. (Other reduction tests are given in Section 4.2 Chapter 3.)

THEOREM 1. *For a given λ, a minimal dominating set of G' can be found by excluding from the set of vertices X' those vertices which correspond to SI's dominated by other SI's in X'. We say that $(SI)_1$ dominates $(SI)_2$ if $(SI)_1 \otimes (SI)_2 = (SI)_2$, where $\otimes$ signifies a Boolean product.* The proof of the above theorem is immediate.

In the outline of the algorithm given in the last section, λ is shown (for reasons of clarity) to be increased by a small amount at each iteration. Many more efficient ways of varying λ exist and the computational results shown in Section 8.4 are based on a program using a biased binary search on λ.

8.3. Example

Consider the graph G shown in Fig. 5.8 where the link lengths are shown next to the links and the vertex weights are all taken to be unity. What is

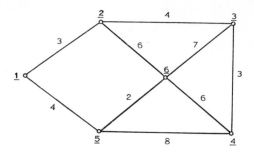

FIG. 5.8. Graph for example 8.3
Underlined numbers are vertex numbers
All other numbers are link costs (lengths)

required is to find the absolute p-centre for the smallest value of p such that every vertex is within $3\frac{1}{2}$ units from at least one of the p centres.

Setting $\lambda = 3\frac{1}{2}$, the markers representing the sets $Q_\lambda(x_i)$ are placed immediately by inspection and are shown in Fig. 5.9. The numbers in the circles attached to each marker are the nodes to which the marker corresponds. Thus, marker $(1, 3)$ on link $(2, 3)$ means that the marked point is just reached by vertices 1 and 3 within a distance $\lambda = 3\frac{1}{3}$. As an example the reachable set $Q_\lambda(5)$ for vertex 5 is shown marked in Fig. 5.9. There are 33 link sections in all, including a blank section (from which no vertex can be reached within

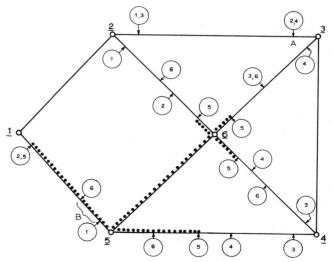

FIG. 5.9. Markers for reachable sets. $Q_\lambda(5)$ shown dotted

a distance of $3\frac{1}{2}$ units), between markers 4 and 5 on link (4, 5). Some of these 33 sections have the same SI's and they combine to form a total of 18 regions. The SI's of those regions are as follows:

(1) 000000	(7) 000011	(13) 000101
(2) 010000	(8) 110000	(14) 001001
(3) 001000	(9) 001100	(15) 111000
(4) 000100	(10) 100010	(16) 011100
(5) 000010	(11) 011000	(17) 110010
(6) 000001	(12) 010001	(18) 100011

Thus, for example, the section between markers 1 and 6 on link (1, 5) belongs to a region whose SI is 100011, since any point of the graph between these two markers can reach vertices 1, 5 and 6 (and no other) within a distance of $\lambda = 3\frac{1}{2}$ units.

After elimination of the dominated SI's only 7 regions (SI numbers 12 to 18 above) are left and the graph G' at the end of step 4 of the algorithm is then as shown in Fig. 5.10. Finding the minimum dominating set of this graph is a problem that could be solved using the set covering algorithm described earlier in Chapter 3. However, in this simple case the minimum set is obvious by inspection and is the set composed of SI's 16 and 18 in the above list. The

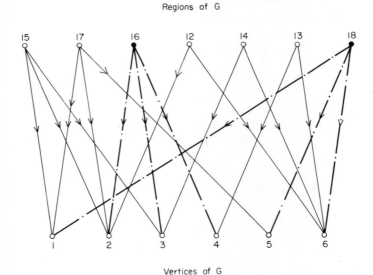

Regions of G

Vertices of G

FIG. 5.10. Graph G' for example 8.3
● Regions in minimum dominating set

region corresponding to SI 16 is made up of just one point at the position of marker $(2, 4)$ on link $(2, 3)$ and is shown as point A in Fig. 5.9. The region corresponding to SI 18 is the section of link $(1, 5)$ between markers 6 and 1 and is shown as region B in Fig. 5.9.

Thus, for $\lambda = 3\frac{1}{2}$ two centres are required, one at point A and the other at any point in region B. The fact that region A is a point means that these two centres also form the absolute 2-centre and $\lambda = 3\frac{1}{2}$ is the minimum possible critical distance for the existence of a 2-centre. Thus if $\lambda < 3\frac{1}{2}$ region A would disappear altogether and a 3-centre would then be required.

8.4. Computational results

The algorithm for finding the absolute p-centre has been tested on the CDC 6600 computer. Each binary vector for the SI's is stored in one word so that set operations can be done most conveniently using the Boolean functions. Since the word length of the CDC 6600 is composed of 60 bits, graphs of up to 60 vertices could be handled directly.

In certain problems, particularly when the number of regions in the absolute p-centre being sought (i.e. the value of p) is large, step 5 of the algorithm given in Section 8.1 takes a significant percentage of the total time taken to find the solution. The times taken by this step are therefore shown separately in Table 5.1 together with the total solution times. Table 5.1 shows the computation times (in seconds) and the number of iterations necessary to find the absolute 1, 2, 3, 4 and 5 centres of a number of graphs. An iteration is here taken to mean the number of times that steps 3 to 6 of the algorithm had to be performed (for different values of λ) in order to obtain the absolute p-centre within the required tolerance. The graphs in Table 5.1 were randomly generated connected nondirected graphs and the tolerance was in all cases taken to be 1 % of the average of the link lengths.

It can be seen from Table 5.1 that the algorithm given in Section 8.1 can determine absolute p-centres efficiently for quite large graphs.

8.5. Use of the general algorithm for finding p-centres

The algorithm of Section 8.1 can obviously be used for the more restricted case of Section 6 i.e. for finding the p-centre. All that is necessary in this case is to alter step 6 of the present algorithm (see Section 8.1), so that what is checked is not the *number* of regions in the minimum dominating set found at step 5, but the *number of vertices of G contained* in these regions. If this number of vertices is greater than p the return to step 2 is made. Otherwise, the algorithm ends and the set of vertices contained in the set of regions forming the dominating set is the p-centre of the graph.

Since in this case only the vertices contained in the regions [given by

TABLE 5-1. Computational results for sample graphs

Number of centres in absolute p-centre (value of p)

Graph		1			2			3			4			5		
n†	m‡	A	B	C	A	B	C	A	B	C	A	B	C	A	B	C
10	20	0·34	—	7	0·30	0·04	6	0·15	0·04	3	0·33	0·06	7	0·44	0·08	7
10	30	0·51	—	7	0·55	0·08	7	0·46	0·09	7	0·35	0·05	6	0·46	0·06	6
20	30	0·86	—	6	1·13	0·05	8	1·16	0·11	8	1·36	0·45	7	14·00	13·00	8
20	40	1·05	—	5	1·73	0·06	6	1·49	0·43	7	0·37	0·22	1	3·51	1·94	7
20	60	1·75	—	7	1·63	0·05	6	1·29	0·18	5	2·58	1·06	7	6·02	4·50	7
20	80	2·90	—	8	3·20	0·10	8	2·70	0·41	6	3·60	1·70	8	16·10	12·00	9
20	100	2·20	—	4	5·41	0·19	9	2·85	0·37	6	4·25	1·53	6	—	—	—
20	120	3·32	—	4	4·35	0·09	6	4·21	0·58	7	6·38	2·86	6	—	—	—
30	40	5·00	—	12	3·83	0·05	10	3·82	0·45	10	6·00	2·10	11	—	—	—
30	60	2·22	—	4	3·16	0·06	6	2·66	0·12	5	18·10	14·50	7	—	—	—
30	80	3·55	—	5	6·90	0·17	10	7·20	1·56	9	8·40	3·73	6	—	—	—
30	100	6·60	—	8	7·10	0·25	8	6·50	1·20	6	23·20	18·80	6	—	—	—
40	60	12·40	—	14	11·20	0·22	13	11·80	2·53	10	—	—	—	—	—	—
40	80	7·40	—	6	10·00	0·51	8	27·00	13·50	11	—	—	—	—	—	—
50	80	17·80	—	12	8·11	0·11	6	17·70	0·95	12	24·55	16·60	13	31·34	25·88	12

† Number of vertices.
‡ Number of links.
A Total computing times (seconds on CDC 6600).
B Computing times for step 5 of the algorithm.
C Number of iterations.

eqns (5.22) and (5.23)], are required, the algorithm of Section 8.1 can be used unchanged in order to find p-centres, if the regions at step 3 are calculated from eqns (5.22) and (5.23) using the sets $R_\lambda(x_i)$ instead of the sets $Q_\lambda(x_i)$. Moreover, if all the $n(n-1)$ distance in the distance matrix are initially arranged in ascending order as a linear array $[f_1, f_2, \ldots, f_{n(n-1)}]$ then at step 2 of the algorithm, λ need only take these values and not merely be increased by arbitrary small amounts at each iteration. An obvious search for λ is then to start with $\lambda = f_1$ and continue with $\lambda = f_2$, f_3, etc. in a sequential fashion, although it is quite apparent that a binary search of the set $\{f_1, f_2, \ldots, f_{n(n-1)}\}$ would be a better alternative.

9. Problems

1. Find the centre of the graph shown in Fig. 5.11, where arc costs are shown next to the arcs and the vertex weights v_j are given by the vector $[5, 3, 1, 7, 4, 6]$.

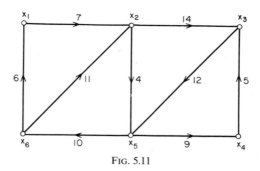

FIG. 5.11

2. Find the absolute centre and radius of the nondirected graph shown in Fig. 5.12 assuming that all vertex weights are unity. Use both Hakimi's method and the iterative method of Section 5.4. Compare the computational effort involved in the two methods.

3. Show that for any nondirected graph G with unity vertex weights, there exists a path between some pair of vertices x_i and x_j such that the absolute centre y^* of G lies half-way between x_i and x_j on this path.

4. Use the property given in problem 3 above to find the absolute centre of the graph in Fig. 5.12.

5. Redefine what is meant by "half-way" and show that the property given in (3) above can be generalized to nondirected graphs with arbitrary vertex weights $[v_j]$.

6. Use the algorithm of Section 8 to find the absolute 3-centre of the graph in Fig. 5.12 and to show that all vertices of G can be reached from at least one of the regions in

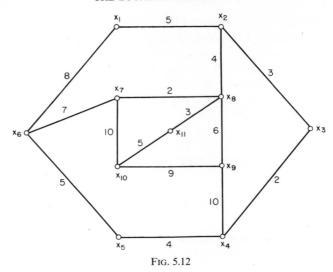

FIG. 5.12

the absolute 3-centre within a distance of $\lambda = 6$ units. What is the minimum value of p for the absolute p-centre to be able to reach all vertices of G within a distance of $\lambda = 5\frac{1}{2}$ units?

7. Use the algorithm of Section 8.5 to find the 3-centre of the graph in Fig. 5.12 and show that $\lambda = 8$ is the least possible value of λ for a 3-centre to exist.

8. Prove that if in a nondirected graph G, X_p^* is a p-centre, then we can define a direction of the links of G so that X_p^* remains a p-(out)centre for the resulting directed graph and the access-distance $d(X_p^*, x_i)$ of all vertices x_i remain unchanged.

9. Show that if G is a connected nondirected graph with unity link costs, then the smallest value of p for a p-centre to exist when reachabilities are restricted to λ, is bounded by:

$$p \leqslant \frac{n}{[\lambda] + 1}$$

where $[\lambda]$ is the largest integer less than or equal to λ. (See Ref. [5]). Assume $n > \lambda$.

10. Using the result in problem (9) above, show that for any positive integer δ, all n vertices of a connected nondirected graph G can be covered by

$$\frac{n}{[\delta/2] + 1}$$

of its subgraphs having diameters $\leqslant \delta$. (See Section 5.4 and Ref. [5]).

11. Show that for a connected nondirected graph G the radius ρ and diameter δ satisfy the relationships $\delta/2 \leqslant \rho \leqslant \delta$. Give examples in which the equalities in the two bounds are achieved.

12. Show that for a directed graph G the equivalent relationships to those of problem (11) above, do not apply.

13. Redefine the outseparation $s_o(y)$ and the absolute outcentre y_o^* of a graph G as follows:

$$s_o(y) = \max_{y' \text{ on } G} [d(y, y')]$$

and y_o^* is a point y for which:

$$s_o(y_o^*) = \min_{y \text{ on } G} [s_o(y)].$$

Generalize the methods given in Section 5 to find these redefined absolute outcentres.

10. References

1. Bollobas, B. (1968). Graphs of given diameter, *In*: "Theory of Graphs", Erdos and Katona, Eds., Academic Press, New York.
2. Bosak, J., Rosa, A. and Znam, Š. (1968). On decomposition of complete graphs into factors with given diameters, *In*: "Theory of Graphs", Erdos and Katona, Eds., Academic Press, New York.
3. Christofides, N. and Viola, P. (1971). The optimum location of multi-centres on a graph, *Opl. Res. Quart.*, **22**, p. 45.
4. Dijkstra, D. (1959). A note on two problems in connection with graphs, *Numerische Mathematik*, **1**, p. 269.
5. Erdos, Gerencsér, L. and Máté, A. (1969). Problems of graph theory concerning optimal design, Colloquia Mathematica Societatis Janos Bolyai, No. 4, Combinatorial theory and its applications, Balatonfured (Hungary), p. 317.
6. Frank, H. and Frisch, I. T. (1971). "Communication transmission and transportation networks", Addison Wesley, Reading, Massachusetts.
7. Hakimi, S. L. (1964). Optimum location of switching centres and the absolute centres and medians of a graph, *Ops. Res.*, **12**, p. 450.
8. Hakimi, S. L. (1965). Optimum distribution of switching centres in a communication network and some related graph theoretic problems, *Ops. Res.* **13**, p. 462.
9. Minieka, E. (1970). The m-centre problem, *J. of SIAM (Review)*, **12**, p. 138.
10. Rosenthal, M. R. and Smith, S. B. (1967). The m-centre problem, Presented at the 31st National ORSA Meeting, New York.
11. Sakowicz, A. F. (1970). A planning model for the location of communication centres, Presented at the 38th National ORSA Meeting, Detroit, Michigan.
12. Singer, S. (1968). Multi-centres and multi-medians of a graph with an application to optimal warehouse location, Presented at the 16th TIMS Conf.
13. Toregas, C., Swain, R., Revelle, C. and Bergman, L. (1971). The location of emergency service facilities, *Ops. Res.*, **19**, p. 1363.

Chapter 6

The Location of Medians

1. Introduction

In certain problems associated with the location of facilities on a graph, what is required is to locate a facility in such a way so as to minimize the sum of all shortest distances from the facility to the vertices of the graph. The optimum location of the facility is called the *median* of the graph, and because of the nature of the objective function this class of problems is referred to as the *minisum location problem*. The problem appears often in practice in a variety of forms: The location of switching centres in telephone networks, substations in electric power networks, supply depots in a road distribution network (where the vertices represent customers), and the location of sorting offices for letter post are some of the areas where minisum location problems occur.

In contrast to this type of problem is the *minimax location problem* which occurs often in the location of emergency facilities such as fire, police or ambulance stations, and which was the subject matter of the previous Chapter. In the present Chapter we discuss the minisum location problem. In particular, we discuss the problem of finding the *p-median* of a given graph G; that is the problem of locating a given number (p say) of facilities optimally so that the sum of the shortest distances to the vertices of G from their nearest facility is minimized. The problem of finding the p-median can be made slightly more general by associating with each vertex x_j a weight v_j (representing, say, size or importance), so that the objective to be minimized becomes the sum of the "weighted" distances.

2. The Median

For a given graph $G = (X, \Gamma)$ let us define two *transmission* numbers for every vertex $x_i \in X$, as follows:

106

$$\left.\begin{array}{l} \sigma_o(x_i) = \sum_{x_j \in X} v_j d(x_i, x_j) \\[2mm] \sigma_t(x_i) = \sum_{x_j \in X} v_j d(x_j, x_i) \end{array}\right\} \tag{6.1}$$

and

where $d(x_i, x_j)$ is the shortest distance from vertex x_i to vertex x_j. The numbers $\sigma_o(x_i)$ and $\sigma_t(x_i)$ are called the *outtransmission* and *intransmission* of the vertex x_i respectively. The number $\sigma_o(x_i)$ is the sum of the entries of row x_i of a matrix obtained by multiplying every column j of the distance matrix $D(G) = [d(x_i, x_j)]$ by v_j; and $\sigma_t(x_i)$ is the sum of the entries of column x_i of a matrix obtained by multiplying every row j of the distance matrix $D(G)$ by v_j.

A vertex $\bar{x}_o$ for which:

$$\sigma_o(\bar{x}_o) = \min_{x_i \in X} [\sigma_o(x_i)] \tag{6.2}$$

is called an *outmedian* of the graph G, and a vertex $\bar{x}_t$ for which:

$$\sigma_t(\bar{x}_t) = \min_{x_i \in X} [\sigma_t(x_i)] \tag{6.3}$$

is called an *inmedian* of G.

Referring once more to the graph shown in Fig. 5.1 (where all v_j and c_{ij} are taken to be unity), we find the outtransmission and intransmission numbers as shown in the rows and columns adjoining the distance matrix:

		x_1	x_2	x_3	x_4	x_5	x_6	$\sigma_o(x_i)$
	x_1	0	1	2	3	2	3	11
	x_2	3	0	1	2	1	2	9
$D(G) =$	x_3	4	3	0	1	2	3	13
	x_4	3	2	2	0	1	2	10
	x_5	2	1	1	2	0	1	7*
	x_6	1	2	3	4	3	0	13
$\sigma_t(x_i) \rightarrow$		13	9*	9*	12	9*	11	

From the transmission numbers the outmedian is seen to be x_5 with $\sigma_o(x_5) = 7$, whereas there are three inmedians (x_2, x_3 and x_5) all with intransmission 9.

2.1 The location of a depot

Consider the problem of supplying a number of customers from a single depot. The customers may be grouped together in neighbourhoods so that each group of customers will require *an integral number of vehicle loads*. Thus, a vehicle leaves the depot, supplies a group of customers and returns to the depot. We can represent the customer groups by the vertices of a graph and the road network by the arcs. In practice each customer group will be weighted by a number v_j representing the "importance" of that customer (i.e. v_j may be proportional to the total annual demand, or the frequency·with which a vehicle has to visit that customer group in order to satisfy its demand).

The objective here is to find a location for the depot so that the total mileage covered by the vehicle fleet is minimized. If the distance matrix $D(G)$ represents actual mileages, the required location is that vertex $\bar{x}_{o,t}$ for which the sum of the inspanning and outspanning numbers is a minimum. The vertex $\bar{x}_{o,t}$ may then be called the in–outmedian, and may be calculated from an equation similar to (6.1). It is shown in the next section, that there is no point (vertex or other arbitrary point) along any road at which the depot can be located to produce a smaller total mileage than that incurred by the location at the vertex $\bar{x}_{o,t}$.

3. Multiple Medians (p-Medians)

In the same way that the single centre was generalized to the p-centre in the previous chapter, the single median can be generalized to the p-median as follows.

Let X_p be a subset of the set X of vertices of the graph $G = (X, \Gamma)$ and let X_p contain p vertices. Once more we will write

$$d(X_p, x_j) = \min_{x_i \in X_p} \left[d(x_i, x_j) \right] \qquad (6.4(\text{a}))$$

and

$$d(x_j, X_p) = \min_{x_i \in X_p} \left[d(x_j, x_i) \right] \qquad (6.4(\text{b}))$$

If x_i' is the vertex of X_p which produces the minimum in eqns 6.4(a) or (b) we will say that vertex x_j is *allocated* to x_i'. The transmission numbers for the set X_p of vertices are then defined in ways analogous to those for a single vertex i.e.

$$\left. \begin{aligned} \sigma_o(X_p) &= \sum_{x_j \in X} v_j d(X_p, x_j) \\[2mm] \sigma_t(X_p) &= \sum_{x_j \in X} v_j d(x_j, X_p) \end{aligned} \right\} \qquad (6.5)$$

and

where $\sigma_o(X_p)$ and $\sigma_t(X_p)$ are the outtransmission and intransmission of the set X_p of vertices.

A set $\overline{X}_{po}$ for which

$$\sigma_o(\overline{X}_{po}) = \min_{X_p \subseteq X} [\sigma_o(X_p)] \tag{6.6}$$

is now called the *p-outmedian* of the graph G, and similarly for the *p-inmedian*.

Just as for the case of the *p*-centres discussed in the previous chapter, it is not computationally practical to use eqns (6.4), (6.5) and (6.6) directly to find the *p*-medians of graphs of even moderate size. Algorithms for the computation of *p*-medians will be given in Section 5.

3.1 Absolute p-medians

In order to simplify the discussion, let us consider a nondirected graph G, drop the suffices *o* and *t* and take the case of the median (1-median) first. The question arises as to whether there exists a point y on some link (not necessarily a vertex) of G so that the transmission number:

$$\sigma(y) = \sum_{x_j \in X} v_j d(y, x_j)$$

is less than that of the median of G. We would then call that point $\bar{y}$ with the minimum $\sigma(y)$ the *absolute median* of G.

In this section we will show [5, 6] that, contrary to the case of the centre of a graph, there is no point $\bar{y}$ with $\sigma(\bar{y}) < \sigma(\bar{x})$, i.e.

THEOREM 1. *There exists at least one vertex x of $G = (X, A)$ for which $\sigma(x) \leqslant \sigma(y)$ for any arbitrary point y on G.*

Proof. Let y be a point on link (x_a, x_b) distance ξ from x_a. Then:

$$d(y, x_j) = \min [\xi + d(x_a, x_j), c_{ab} - \xi + d(x_b, x_j)] \tag{6.7}$$

where c_{ab} is the length of link (x_a, x_b).

Let X_a be the set of those vertices x_j for which the first term in eqn (6.7) is smallest and let X_b be the set of vertices for which the second term is smallest. We can then write:

$$\sigma(y) = \sum_{x_j \in X} v_j d(y, x_j) = \sum_{x_j \in X_a} v_j [\xi + d(x_a, x_j)]$$

$$+ \sum_{x_j \in X_b} v_j [c_{ab} - \xi + d(x_b, x_j)] \tag{6.8}$$

Since, from the triangle inequality for distances, we have:

$$d(x_a, x_j) \leqslant c_{ab} + d(x_b, x_j) \tag{6.9}$$

substitution of $d(x_a, x_j)$ for $c_{ab} + d(x_b, x_j)$ in the second term of eqn (6.8) yields:

$$\sigma(y) \geq \sum_{x_j \in X_a} v_j [\xi + d(x_a, x_j)] + \sum_{x_j \in X_b} v_j [d(x_a, x_j) - \xi] \qquad (6.10)$$

Noting that $X_a \cup X_b = X$ and rearranging (6.10) we get:

$$\sigma(y) \geq \sum_{x_j \in X} v_j d(x_a, x_j) + \xi \Big[\sum_{x_j \in X_a} v_j - \sum_{x_j \in X_b} v_j \Big] \qquad (6.11)$$

Since by a suitable choice of which end vertex of link (x_a, x_b) to call x_a and which one to call x_b we can always have:

$$\sum_{x_j \in X_a} v_j \geq \sum_{x_j \in X_b} v_j,$$

and noting that the first term on the right-hand side of (6.11) is $\sigma(x_a)$, inequality (6.11) becomes:

$$\sigma(y) \geq \sigma(x_a)$$

Thus, vertex x_a has $\sigma(x_a)$ at least as low as $\sigma(y)$ and the theorem follows.

Theorem 1 can be generalized quite simply [16] to the case of the absolute p-median and thus obtain:

THEOREM 2. *There exists at least one subset $X_p \subset X$ containing p vertices, such that $\sigma(X_p) \leq \sigma(Y_p)$ for any arbitrary set Y_p of p points on the links or vertices of the graph $G = (X, A)$.*

Theorems 1 and 2 above apply when the transmissions $\sigma(x)$ and $\sigma(x_p)$ are defined by expressions of the form (6.1) and (6.5). Levy [20], Goldman [12], and Goldman and Mayers [14] later showed that both these theorems remain valid in cases where the transmissions are defined as the sum of arbitrary concave functions of the weighted distances.

In view of Theorems 1 and 2, the concept of absolute medians is not a useful one (contrary to the case for absolute centres discussed in Chapter 5), and therefore for the rest of this chapter we will concentrate attention on the p-median problem.

4. The Generalized p-Median

The problem of finding the p-median of a graph is the central problem in a general class studied in the literature under the names of "facility location and allocation" [5, 6, 9] or "depot location" [8, 29, 30, 10, 15, 7, 28]. These problems are somewhat more general than that of the p-median, in that fixed costs f_i are associated with the vertices x_i. We then define the *generalized p-median* problem as follows.

Given a graph $G = (X, A)$, with shortest distance matrix $[d(x_i, x_j)]$, vertex weights v_i and vertex fixed costs f_i the problem is to find a subset $\overline{X}_p$ containing p vertices so that

$$z = \sum_{x_i \in \overline{X}_p} f_i + \sigma(\overline{X}_p) \tag{6.12}$$

is minimized.

Thus, the objective is now to minimize not just the transmission $\sigma(\overline{X}_p)$ of $\overline{X}_p$ but the total function z which includes a fixed cost f_i for every vertex x_i in $\overline{X}_p$. In practice f_i represents the fixed cost associated with building a facility (depot, factory, switching centre), at the location represented by vertex x_i. The p-median problem then corresponds to the case when all f_i are equal (to say f) so that the first term of eqn (6.12) becomes a constant pf regardless of the choice of the set X_p.

A problem very similar to that of the generalized p-median defined above is one in which $|\overline{X}_p|$ is not required to be exactly p but any number less than or equal to p. The problem of minimizing expression (6.12) subject to $|\overline{X}_p| \leqslant p$ is a version of the generalized p-median problem that is often encountered in practice.

Practical depot location problems invariably involve constraints which do not exist in the pure p-median problem. One usual such constraint is a restriction on the maximum and minimum values that the number:

$$\sum_{x_j \text{ allocated to } x_i} v_j \tag{6.13}$$

can take, for any median vertex $x_i \in \overline{X}_p$. Expression (6.13) is a measure of the amount of material transmitted from x_i (throughput), and is therefore also a measure of the physical size of depot x_i [8, 15, 7, 28].

Obviously, the same kind of generalization introduced by the restrictions on the value of expression (6.13), is also applicable to the problems of finding p-centres and absolute p-centres in graphs, as discussed in the previous chapter. However, this generalization is of very much smaller significance in the practical energency facility location problems which translate into p-centre problems and, in any case, these restrictions can be introduced into the algorithm given in Section 8 of the previous chapter with negligible effort.

The difficulties of solving practical depot location problems, are not due to the variations and additional restrictions discussed above, but are inherent in the pure median problem itself. In view of this we will now discuss some of the available methods for finding the p-median of a graph. In order to simplify the nomenclature we will drop the suffices o and t for "out" and "in" medians respectively, since the methods apply equally to both.

5. Methods for the p-Median Problem

5.1 Formulation as an integer programme

Let $[\xi_{ij}]$ be an allocation matrix so that:

$$\xi_{ij} = 1 \text{ implies that vertex } x_j \text{ is allocated to vertex } x_i$$

$$= 0 \text{ otherwise.}$$

Further, we will take $\xi_{ii} = 1$ to imply that vertex x_i is a median vertex and $\xi_{ii} = 0$ otherwise. The p-median problem can then be stated as follows:
Minimize:

$$z = \sum_{i=1}^{n} \sum_{j=1}^{n} d_{ij}\xi_{ij} \qquad (6.14)$$

subject to:

$$\sum_{i=1}^{n} \xi_{ij} = 1 \text{ for } j = 1, \ldots, n \qquad (6.15)$$

$$\sum_{i=1}^{n} \xi_{ii} = p \qquad (6.16)$$

$$\xi_{ij} \leqslant \xi_{ii} \qquad \text{for all } i, j = 1, \ldots, n \qquad (6.17)$$

and

$$\xi_{ij} = 0 \text{ or } 1 \qquad (6.18)$$

where $[d_{ij}]$ is assumed to be the weighted distance matrix of the graph, i.e. it is the distance matrix with every column j multiplied by v_j. Equation (6.15) ensures that any given x_j is allocated to one and only one median vertex x_i. Equation (6.16) ensures that there are exactly p median vertices, and constraint (6.17) guarantees that $\xi_{ij} = 1$ only if $\xi_{ii} = 1$, i.e. allocations are made only to vertices that are in the median set. If $[\bar{\xi}_{ij}]$ is the optimal answer to the problem defined by eqns (6.14) to (6.18) above, then the p-median is:

$$\overline{X}_p = \{x_i | \bar{\xi}_{ii} = 1\}$$

If constraint (6.18) of the above problem is replaced by:

$$\xi_{ij} \geqslant 0 \qquad (6.19)$$

then the resulting problem is a linear program (LP) whose solution can easily be obtained. (Note that an upper bound of value 1 on ξ_{ij} is not needed since $\xi_{ij} \leqslant 1$ is already implied by constraint (6.15).)

The solution to the LP is not necessarily all integer and fractional values of ξ_{ij} can occur Revelle and Swain [26] report that such fractional values occur rarely, and therefore one could use the LP formulation to obtain the p-median for most problems. In any case, if fractional values of some ξ_{ij} occur, then a resolution of these uncertainties can be obtained by a tree-search procedure [11] in which one branch fixes some fractional variable ξ_{ij} to 0 and another the same ξ_{ij} to 1. One can then proceed to resolve the LP's for each of the two resulting subproblems and so on until all ξ_{ij} become either 0 or 1.

An alternative approach via linear programming is given by Marsten [23], who shows that the solution $[\bar{\xi}_{ij}]$ corresponding to the p-median of a graph as described by eqn (6.14) to (6.18), is an extreme point of a certain polyhedron H, and that all other p-medians for $1 \leqslant p \leqslant n$ are also extreme points of H. By using Lagrange multipliers and parametric linear programming, Marsten gives a method of traversing a path among a few of the extreme points of H. This path successively generates the p-medians of the graph G in descending order of p, although certain p-medians (for some values of p) may be missed and never generated, or, conversely, extreme points of H may be generated which do not correspond to p-medians of G, i.e. contain fractional values of ξ_{ij}. Thus, although this method is both theoretically and computationally attractive, it may fail to produce the p-median of a graph for the specific value of p that may be required. Reference [23] reports the case of a complete 33-vertex graph all of whose p-medians have been successfully generated for $p = 33, 32, \ldots, 10$, but whose 9-median and 8-median could not be obtained by this method.

5.2 A direct tree search algorithm

Instead of an explicit formulation of the p-median problem as an integer programme one could use a direct tree search approach which is better suited to exploiting the structure of the problem. In the present section we describe one such tree search approach in which each subproblem generated by the branching from a tree node is produced by setting—for a given vertex x_j—a variable ξ_{ij} to 1 for some vertex x_i. The setting of $\xi_{ij} = 1$ implies that vertex x_j is allocated to vertex x_i, and which, obviously, also implies that x_i is a median vertex.

This tree search can be conveniently carried out as follows.

Set up a matrix $M = [m_{kj}]$ the jth column of which contains all the vertices of the graph G arranged in ascending order of their distance from vertex x_j. Thus, if $m_{kj} = x_i$, then vertex x_i is the kth nearest vertex to x_j. Obviously, the nearest vertex to vertex x_j is itself, i.e. $m_{1j} = x_j$.

The search proceeds by allocating sequentially—starting from x_1 and finishing with x_n—all the vertices x_j of the graph. A vertex x_j is initially allocated to vertex m_{1j} then m_{2j}, m_{3j} and so on until all possibilities are implicitly enumerated. The following observations can now be made.

1. Since, in the optimal solution there are p median vertices, each allocation of a vertex in this solution must be the best of the p possible allocations to the median vertices, i.e. there are at least p-1 more costly ways of allocating

any one vertex. Hence, the last p-1 rows of matrix M can be permanently removed without any possibility of the optimal p-median solution being affected.

2. Suppose that an allocation of vertex x_j, to vertex $m_{k'j'}$ ($=x_i$ say) has been made. For an unallocated vertex x_j ($j' < j \leqslant n$) corresponding to column j of the M matrix, let x_i be the kth nearest vertex to x_j i.e. let $m_{kj} = x_i$. Then, obviously, all entries m_{lj} with $l > k$ could be neglected (marked), since the allocation of $x_{j'}$ to x_i has already implied that x_i is a median vertex, and x_j can be allocated with lower cost to x_i rather than to any of the vertices m_{lj}, $l = k + 1, k + 2, \ldots$ Clearly, if at some backtracking step during the search the allocation of vertex $x_{j'}$ to x_i is altered, then the entries m_{lj} have to be considered again (unmarked).

3. Let vertex $x_{j'}$ be allocated to vertex $m_{k'j'}$. Then all vertices $m_{1j'}, m_{2j'}, \ldots, m_{(k'-1)j'}$ can be assumed not to be median vertices, since if they were, $x_{j'}$ could have been allocated to any one of them—rather than to $m_{k'j'}$—with lower resultant cost. These vertices could, therefore, be marked (neglected) in all columns $j > j'$. Once more, the marking of these vertices is temporary and must be removed whenever the allocation of $x_{j'}$ to $m_{k'j'}$ is changed.

4. If the top t, say, entries in a column j corresponding to an unallocated vertex x_j are marked, and the $(t + 1)$-st entry i.e. $m_{(t+1)j}$ is a median vertex, then x_j must be allocated to vertex $m_{(t+1)j}$ and no other alternative need be considered until some of the top t entries are again unmarked (due to a change in the previous allocations that resulted in some of these markings). This is a direct consequence of observations 2 and 3 above.

5. If at some stage q of the tree search, a total of p median vertices are implied by the allocations already made, then the remaining unallocated vertices can now be allocated to their nearest median vertex. This is obviously the optimal completion of the partial solution corresponding to the allocations up to that stage, and the next backtracking step must therefore involve a change in the allocation at stage q.

5.2.1 CALCULATION OF A LOWER BOUND. Observations 1 to 5 above can be used to limit the size of the tree search by reducing the number of alternative possible allocations of a vertex x_j at any one stage. In addition, at some stage when the last vertex to have been allocated is $x_{j'}$, a lower bound on the value of the overall optimal solution (given the allocations already made), can be used to limit the search further.

Thus, suppose that the allocations to date (up to and including the allocation of vertex $x_{j'}$), imply a total of p' ($p' < p$) median vertices. Then, the remaining allocations must imply a total of another $p-p'$ median vertices. Let J be the set of indices of the as yet unallocated vertices. In general, J is the set of indices j where $j' < j \leqslant n$, but excluding the indices of those vertices whose allocation might have already been forced by the allocations of the first j' vertices $x_1, x_2, \ldots, x_{j'}$—according to observation 4 above.

Let $m_{\alpha jj}$ and $m_{\beta jj}$ be the topmost and second topmost entry in column j which are unmarked. The best possible allocation of vertex x_j is then to vertex $m_{\alpha jj}$. If the number of distinct vertices $m_{\alpha j}$ for $j \in J$ is h and $h = p - p'$, then all these best possible allocations for the as yet unallocated vertices are feasible (i.e. produce a total of p median vertices). These allocations then constitute the optimal completion of the partial solution implied by the current allocations of the vertices $x_1, x_2, \ldots, x_{j'}$. In such an event, the result is noted and backtracking can occur from the current partial solution.

If, however, $h > p - p'$ then at least $(h - p + p')$ of the best assignments must be changed to second best or worse in order to produce a total of p-median vertices. The minimum additional cost of allocation is then the sum of the $(h - p + p')$ smallest differences:

$$v_j[d(x_j, m_{\beta jj}) - d(x_j, m_{\alpha jj})] \tag{6.20}$$

over all vertices $x_j, j \in J$.

A lower bound on the cost of the optimal solution given the current partial solution is then the sum of the costs of the allocations already made, plus the sum

$$\sum_{j \in J} v_j d(x_j, m_{\alpha jj})$$

plus the sum of the $(h - p + p')$ smallest differences, given by expression (6.20).

It is also possible that h is less than $p - p'$, in which case the best completion of the current partial solution leads to less than p median vertices. However, since it is quite apparent that the transmission $\sigma(\overline{X}_p)$ of the optimal p-median $\overline{X}_p$ monotonically decreases as p increases, it follows that when $h < p - p'$ occurs the current partial solution is certainly not part of the optimal p-median solution and backtracking can then occur.

5.3 An alternative direct search algorithm

The tree search procedure described in Section 5.2 generated subproblems by considering alternative allocations, i.e. by specifying values for the allocation variables ξ_{ij}. A different search procedure [18] is to generate subproblems in a binary tree search by specifying whether or not a given vertex is to be a median vertex. In this case, the state of the search can be represented by sets S^+, S^- and F corresponding, respectively, to those vertices currently fixed as median vertices, those vertices fixed as non-median vertices and those vertices about which no decision has as yet been made.

A lower bound on the cost of the optimal solution, given the decisions already made about the vertices in S^+ and S^-, can then be calculated [18] in a way similar to that explained in the previous section.

Thus, a vertex $x_j \in S^-$ must be allocated to some other vertex in the set $S^+ \cup F$ and the best such allocation has cost:

$$u'_j = \min_{x_i \in S^+ \cup F} [v_j d(x_j, x_i)] \tag{6.21}$$

On the other hand, a vertex $x_j \in F$ that is not going to become a median vertex, must incur a cost of at least:

$$u''_j = \min_{\substack{i \neq j \\ x_i \in S^+ \cup F}} [v_j d(x_j, x_i)] \tag{6.22}$$

The lower bound is then calculated as:

$$\sum_{x_j \in S^-} u'_j + U'' \tag{6.23}$$

where U'' is the sum of the $n - p - |S^-|$ smallest numbers u''_j. It should be noted that $n - p - |S^-|$ is the number of vertices which are as yet free but which in the final solution cannot be median vertices and must therefore be allocated to others.

Jarvinen et al. [18] use the bound of expression (6.23) in an isocost type of tree search where branching always takes place from the tree node having the least lower bound. Depth first searches of the type discussed earlier in this section and employing bounds similar to that of (6.23), (slightly modified to deal with the generalized p-median problems encountered in practical depot location studies), can also be found in the literature [29, 30, 15, 7, 28].

5.4 An approximate algorithm

A heuristic method based on vertex substitution is described by Teitz and Bart [32]. The method proceeds by choosing any p vertices at random to form the initial set S which is assumed to be an approximation to the p-median set $\overline{X}_p$. The method then tests if any vertex $x_j \in X - S$ could replace a vertex $x_i \in S$ as a median vertex and so produce a new set $S' = S \cup \{x_j\} - \{x_i\}$ whose transmission $\sigma(S')$ is less than $\sigma(S)$. If so, the substitution of vertex x_i by x_j is performed thus obtaining a set S' which is a better approximation to the p-median set $\overline{X}_p$. The same tests are now performed on the new set S' and so on, until a set $\overline{S}$ is finally obtained for which no substitution of a vertex in $\overline{S}$ by another vertex in $X - \overline{S}$ produces a set with transmission less than $\sigma(\overline{S})$. This final set $\overline{S}$ is then taken to be the required approximation to $\overline{X}_p$.

5.4.1 DESCRIPTION OF THE ALGORITHM

Step 1. Select a set S of p vertices to form the initial approximation to the p-median. Call all vertices $x_j \notin S$ "untried".

Step 2. Select some "untried" vertex $x_j \notin S$ and for each vertex $x_i \in S$, compute the "reduction" Δ_{ij} in the set transmission if x_j is substituted for x_i, i.e. compute:

$$\Delta_{ij} = \sigma(S) - \sigma(S \cup \{x_j\} - \{x_i\}) \qquad (6.24)$$

Step 3. Find $\Delta_{i_o j} = \max_{x_i \in S} [\Delta_{ij}]$.

(i) If $\Delta_{i_o j} \leqslant 0$ call x_j "tried" and go to step 2.
(ii) If $\Delta_{i_o j} > 0$ set $S \leftarrow S \cup \{x_j\} - \{x_i\}$ call x_j "tried" and go to step 2.

Step 4. Repeat steps 2 and 3 until all vertices in X–S have been tried. This is referred to as a cycle. If, during the last cycle no vertex substitution at all has been made at step 3(i), go to step 5. Otherwise, if some vertex substitution has been made, call all vertices "untried" and return to step 2.

Step 5. Stop. The current set S is the estimated p-median set $\overline{X}_p$.

The above algorithm will quite obviously, not produce an optimal answer in all cases [18]. A counter example is provided by the non-directed graph of Fig. 6.1 where link costs are shown next to the links and vertex weights are taken to be all unity. If the 2-median is required and the initial choice of the set S is $\{x_3, x_6\}$ with transmission $\sigma(S) = 8$, then no substitution of a single vertex can lead to a set with lower transmission. However, the set $\{x_3, x_6\}$ is not the 2-median of the graph and there are two 2-median sets $\{x_1, x_4\}$ and $\{x_2, x_5\}$ both with transmission $\sigma(\overline{X}_2) = 7$.

The algorithm described in this section is in fact only one of a family of algorithms based on local optimization and on the idea of *λ-optimality* first

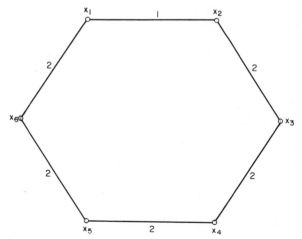

FIG. 6.1. Counterexample to approximate algorithm of section 5.4

introduced by Lin [22] for the travelling salesman problem, and subsequently extended and used by others [3, 4, 19] for a variety of combinatorial problems.

In the current problem a set S of p vertices is called λ-optimal ($\lambda \leqslant p$), if the substitution of any λ vertices of S by any other λ vertices not in S, cannot produce a new set with transmission less than that of S. Obviously, if S is the p-median set $\overline{X}_p$ of a graph, then S is p-optimal. It is also quite obvious that if S' is a λ'-optimal set and S'' is a λ''-optimal set, then, if $\lambda'' > \lambda'$, $\sigma(S'') \leqslant \sigma(S')$.

According to the above definitions, the answer produced by the algorithm of Section 5.4.1 could be called 1-optimal, and similar algorithms producing 2-optimal, 3-optimal etc. answers could be described. In order to ensure that a given set S is λ-optimal, a total of:

$$\binom{p}{\lambda}\binom{n-p}{\lambda}$$

potential substitutions (and hence calculations of transmissions σ) must be performed. This number increases rapidly with λ and hence practical algorithms cannot use values of λ much above 2 or 3.

6. Problems P6

1. Prove that for any arbitrary set Y_p of p points on the links or vertices of a graph G, there exists at least one subset $X_p \subset X$ with $|X_p| = p$ such that $\sigma(X_p) \leqslant \sigma(Y_p)$.

2. For the graph shown in Fig. 6.2 find the medians, 2-medians, 3-medians and 4-medians.

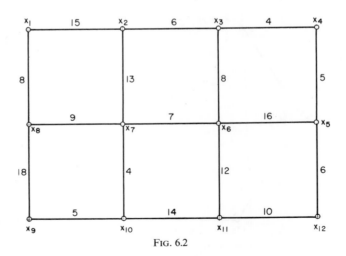

FIG. 6.2

3. An iterative technique for solving the 1-median problem for the case of an infinite graph embedded in a Euclidean plane is as follows: [8].

Let G be the infinite complete graph representing the Euclidean plane and let S be a specified subset of vertices consisting of the n points with coordinates $\{(x_i, y_i) \mid i = 1, \ldots, n\}$. The problem is to locate one median vertex at a point (x_m, y_m) in the plane so that the expression:

$$\sigma_m = \sum_{i=1}^{n} v_i d_{im}$$

is minimized; where v_i is the "weight" of the point with coordinates (x_i, y_i) and d_{im} is the Euclidean distance:

$$d_{im} = \sqrt{[(x_i - x_m)^2 + (y_i - y_m)^2]}$$

At the minimum we have:

$$\frac{\partial \sigma_m}{\partial x_m} \sum_{i=1}^{n} \left[\frac{v_i(x_i - x_m)}{d_{im}} \right] = 0 \tag{P7.1}$$

and

$$\frac{\partial \sigma_m}{\partial y_m} = \sum_{i=1}^{n} \left[\frac{v_i(y_i - y_m)}{d_{im}} \right] = 0 \tag{P7.2}$$

Equations (P7.1) and (P7.2) can be solved iteratively for x_m and y_m by rewriting the expressions as follows:

$$x_m = \sum_{i=1}^{n} \left[\frac{v_i x_i}{d_{im}} \right] \Big/ \sum_{i=1}^{n} \left[\frac{v_i}{d_{im}} \right] \tag{P7.3}$$

$$y_m = \sum_{i=1}^{n} \left[\frac{v_i y_i}{d_{im}} \right] \Big/ \sum_{i=1}^{n} \left[\frac{v_i}{d_{im}} \right] \tag{P7.4}$$

Now starting from a given point (x_m, y_m) we can calculate the distances d_{im} then use expressions (P7.3) and (P7.4) to recalculate a new better point (x_m, y_m). We can now recalculate the distances d_{im} recompute x_m and y_m and so on until no further change in x_m or y_m takes place.

Using the above iterative scheme find the 1-median for the case where the set S is made up of points at coordinates: (2, 2), (2, 6), (4, 8), (5, 5), (6, 1), (8, 6) and (9, 3) with all weights $v_i = 1$.

Show that the iterations result in the global optimum for (x_m, y_m) regardless of the initial starting location of this point.

4. Extend the iterative scheme of problem 3 above to the general case of the p-median. Does the extended method provide the global optimum for the p-median?

5. If a graph $G = (X, \Gamma)$ is a tree and the removal of a link (x_i, x_j) separates G into two connected components $G_i = (X_i, \Gamma)$ and $G_j = (X_j, \Gamma)$ with $x_i \in X_i$, $x_j \in X_j$ and $X_i \cup X_j = X$, and if we write:

$$v(S) = \sum_{x_i \in S}^{i} v_i,$$

then show that a median vertex $\bar{x}$ of G satisfies:

(i)   If $v(X_i) \geqslant v(X_j)$, then $\bar{x} \in X_i$

(ii) If $v(X_i) \leqslant v(X_j)$, then $\bar{x}$ remains unchanged if G is replaced by G_j with the weight of x_j changed to $v_j + v(X_i)$.

Hence give an algorithm for finding a median vertex of a tree and which is more efficient than the direct use of equations (7.2) or (7.3). (See Ref. [13].)

7. References

1. Beale, E. M. L. (1970). Selecting an optimum subset, *In*: "Integer and nonlinear programming", Abadie, Ed., North Holland, Amsterdam.
2. Cabot, A. V., Francis, R. L. and Stary, M. A. (1970). A network flow solution to a rectilinear distance facility location problem, *AIIE Trans.*, **2**, p. 132.
3. Christofides, N. and Eilon, S. (1969). An algorithm for the vehicle dispatching problem, *Opl. Res. Quart.*, **20**, p. 309.
4. Christofides, N. and Eilon, S. (1972). Algorithms for large scale travelling salesman problems, *Opl. Res. Quart.*, **23**, p. 511.
5. Cooper, L. (1963). Location–allocation problems, *Ops. Res.*, **11**, p. 331.
6. Curry, G. R. and Skeith, R. W. (1969). A dynamic programming algorithm for facility location and allocation, *AIIE Trans.*, **1**, p. 133.
7. Efroymson, E. and Ray, T. L. (1966). A branch and bound algorithm for plant location, *Ops. Res.*, **14**, p. 361.
8. Eilon, S., Watson-Gandy, C. D. T. and Christofides, N. (1971). "Distribution management: mathematical modelling and practical analysis", Griffin, London.
9. Ellwein, L. B. and Gray, P. (1971). Solving fixed charge location–allocation problems with capacity and configuration constraints, *AIIE Trans.*, **3**, p. 290.
10. Elson, D. G. (1972). Site location via mixed-integer programming, *Opl. Res. Quart.*, **23**, p. 31.
11. Garfinkel, R. S. and Nemhauser, G. L. (1972). "Integer programming", Wiley, New York.
12. Goldman, A. J. (1969). Optimal locations for centres in a network, *Transp. Sci.*, **3**, p. 352.
13. Goldman, A. J. (1970). Optimal centre location in simple networks, Paper FP 6.1, presented at the 38th National Meeting of the Operations Research Soc. of America, Detroit.
14 Goldman, A. J. and Mayers, P. R. (1965). A domination theorem for optimal locations, *Ops. Res.*, **13**, 13–147 (Abstract).
15 Gray, P. (1967). Mixed integer programming algorithm for site selection and other fixed charge problems having capacity constraints, Department of Operations Research, Report No. 6, Stanford University.
16. Hakimi, S. L. (1965). Optimum distribution of switching centres in a communications network and some related graph theoretical problems.
17. Hakimi, S. L. (1964). Optimum locations of switching centres and the absolute centres and medians of a graph, *Ops. Res.*, **12**, p. 450.
18. Jarvinen, P., Rajala, J. and Sinervo, H. (1972). A branch and bound algorithm for seeking the *p*-median, *Ops. Res.*, **20**, p. 173.
19. Kerningan, B. W. and Lin, S. (1970). An efficient heuristic procedure for partioning graphs, *Bell Syst. Tech. Jl.*, **49**, p. 291.

20. Levy, J. (1967). An extended theorem for location on a network, *Opl. Res. Quart.*, **18**, p. 433.
21. Lin, S. (1965). Computer solutions of the travelling salesman problem, *Bell. Syst. Tech. Jl.*, **44**, p. 2245.
22. Maranzana, F. E. (1964). On the location of supply points to minimize transport costs, *Opl. Res. Quart.*, **15**, p. 261.
23. Marsten, R. E. (1972). An algorithm for finding almost all the *p*-medians of a network, Paper 23, Northwestern University, Evaston, Illinois.
24. Rao, M. R. (1972). The rectilinear facilitities location problem, Working paper No. F7215, The graduate school of management, University of Rochester, Rochester, New York.
25. Revelle, C. S., Marks, D. and Liebman, J. C. (1970). An analysis of private and public sector location models, *Man. Sci.*, **16**, p. 692.
26. Revelle, C. S. and Swain, R. W. (1970). Central facilities location, *Geographical Analysis*, **2**, p. 30.
27. Rojeski, P. and Revelle, C. S. (1970). Central facilities location under an investment constraint, *Geographical Analysis*, **2**.
28. Sá, G. (1964). Branch and bound and approximate solutions to the capacitated plant-location problem, *Ops. Res.*, **17**, p. 1005.
29. Spielberg, K. (1969). An algorithm for the simple plant location problem with some side conditions, *Ops. Res.*, **17**, p. 85.
30. Spielberg, K. (1969). Plant location with generalized search origin, *Man. Sci.*, **16**, p. 165.
31. Surkis, J. (1967). Optimal warehouse location, XIV Int. TIMS Conf., Mexico City, Mexico.
32. Teitz, M. B. and Bart, P. (1968). Heuristic methods for estimating the generalized vertex median of a weighted graph, *Ops. Res.*, **16**, p. 955.

Chapter 7

Trees

1. Introduction

One of the most important concepts of graph theory, and one which appears often in areas superficially unrelated to graphs, is that of a tree. The present chapter introduces directed and nondirected trees and gives a brief mention of the areas of application. Some of these applications are discussed in greater detail in other chapters of this book, where concepts dual to that of the tree are also introduced.

Definitions. The following definitions of a nondirected tree can be quite easily shown to be all equivalent to each other. (See problem P7.1).

A nondirected tree is:

(i) A connected graph of n vertices and $(n - 1)$ links,

or (ii) A connected graph without a circuit,

or (iii) A graph in which every pair of vertices is connected with one and only one elementary path.

If $G = (X, A)$ is a nondirected graph of n vertices, then a *spanning tree of G* is defined as a partial graph of G which forms a tree according to the above definition. Thus, if G is the graph of Fig. 7.1(a), then the graph of Fig. 7.1(b) is a spanning tree of G, and so is the graph of Fig. 7.1(c). From the above definition it can be seen that a spanning tree of G could also be considered as a minimal connected partial graph of G, where "minimal" is used in the sense of Chapter 2—i.e. no subset of the links of the tree exists which forms a connected partial graph of G.

The concept of the tree as a mathematical entity was first proposed by Kirchhoff [36] in connection with the definition of fundamental circuits used in the analysis of electrical networks. About 10 years later Cayley [5] rediscovered trees independently and produced most of the early works in this area.

A directed tree (also called an *arborescence*), is defined in a similar way as follows.

122

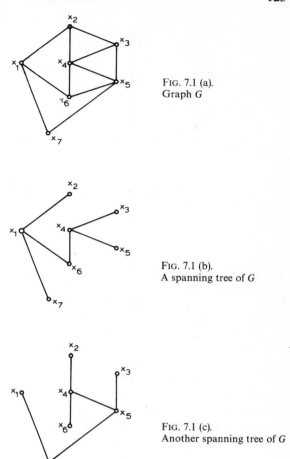

FIG. 7.1 (a).
Graph G

FIG. 7.1 (b).
A spanning tree of G

FIG. 7.1 (c).
Another spanning tree of G

Definition. A directed tree is a directed graph without a circuit, for which the indegree of every vertex, except one (say vertex x_1), is unity: the indegree of x_1 (called the root of the tree) being zero.

Figure 7.2 shows a graph which is a directed tree with vertex x_1 as the root. The above definition immediately implies that a directed tree of n vertices contains $(n-1)$ arcs and is connected. It is also quite obvious that not every directed graph has a spanning directed tree, as can, for example, be seen from the graph of Fig. 7.3.

It should be noted that a nondirected tree can be converted into a directed one by arbitrarily picking any vertex for the root and choosing directions for the links so that there is a path (there can only be one path) between the

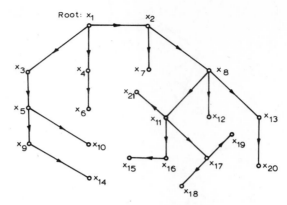

FIG. 7.2. A directed tree

root and every other vertex. Conversely, if $T = (X, B)$ is a directed tree, $\bar{T} = (X, \bar{B})$—where $\bar{B}$ is the set of arcs B with their directions disregarded—is a nondirected one.

The "family tree" in which male persons are represented by vertices and arcs are drawn from the parents to their children, is one familiar example of a directed tree, the root being the earliest person in the family that can be identified.

In the present chapter we give an algorithm for generating all spanning trees of an arbitrary nondirected graph G, and also methods for calculating directly the shortest of the spanning trees of G when costs c_{ij} are associated with its arcs. The shortest spanning tree (SST) of a graph has obvious applications in cases where roads (gas pipelines, electric power lines, etc.) are to be

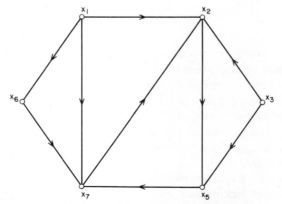

FIG. 7.3. Graph with no directed spanning tree

used to connect n points together in such a way so as to minimize the total length of road that has to be constructed. If the n points to be connected are on a Euclidean plane, they can be represented as vertices of a complete graph G with arc costs being the straight line distances between the corresponding end points. The SST of G is then (provided no road junctions outside the given n points are allowed) the required minimum-cost road network. If junctions outside the given n points are allowed, then an even shorter road network may be possible, and finding it is a problem known as Steiner's problem. This latter problem is briefly discussed in the last section of this chapter.

2. The Generation of all Spanning Trees of a Graph

In some situations it is necessary to be able to generate a complete list of all the spanning tree of a graph G. This may, for example, be the case when the "best" tree needs to be chosen, but the criterion to be used for deciding what tree is "best" is very complex (or even subjective in part) so that direct optimization (not involving complete enumeration) is impractical. In other situations, for example in the computation of the block transfer function of a system [42, 6], or in the computation of the determinant of certain important matrices in macroeconomic theory [2], the manipulations could be simplified by first generating all spanning trees of an associated graph.

The number of distinct spanning trees of a completely connected non-directed graph of n vertices was first calculated by Cayley [4] to be n^{n-2}. A bibliography of over 25 papers proving this formula is given by Moon [45]. (See also problem 6). Formulae giving the number of trees in more

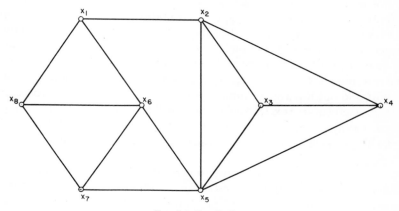

FIG. 7.4. Graph G.

general graphs are given in Riordan [49]. Although these formulae tend to be very complicated and their derivation is outside the scope of this book it is perhaps worthwhile to note the following result.

THEOREM 1 *Given a graph G with no loops, let B_o be its incidence matrix with one row removed, (i.e. with $n - 1$ independent rows), and B_o^t be the transpose of B_o. Then the determinant $|B_o . B_o^t|$ gives the number of distinct spanning trees of G.*

The proof of the above theorem can be found in [53] and is reproduced in [1]. (See also problem 5).

2.1 Elementary tree transformations

Consider two (directed or nondirected) spanning trees $T_1 = (X, A_1)$ and $T_2 = (X, A_2)$ of a graph $G = (X, A)$. The "distance" between these two trees is denoted by $d(T_1, T_2)$ and is defined to be the number of arcs in T_1 not in T_2, (or equivalently the number of arcs in T_2 not in T_1—since both T_1 and T_2 have $n - 1$ arcs). If the distance $d(T_1, T_2) = 1$, i.e. if

$$A_1 \cup A_2 - A_1 \cap A_2 = \{a_1, a_2\} \text{ where } a_1 \in A_1 \text{ and } a_2 \in A_2$$

then T_2 could be derived from T_1 be removing arc a_1 from T_1 and introducing arc a_2. Such a transformation of T_1 into T_2 is called an *elementary tree transformation*.

THEOREM 2 *If T_o and T_k are spanning trees of a graph and $d(T_o, T_k) = k$ then T_k can be obtained from T_o by a series of k elementary tree transformations.*

Proof: Let $a_o^1, a_o^2, \ldots, a_o^k$ be k arcs of T_o not in T_k and $a_k^1, a_k^2, \ldots, a_k^k$ be k arcs of T_k not in T_o. If arc a_k^1 is added to the tree T_o a circuit would result by the definition of a tree. (Fig. 7.5(a) shows a non-directed tree T_o, shown in heavy lines, with an arc $a_k^1 = (x_5, x_6)$ of the tree T_4 given in Fig. 7.5(e) shown in dotted lines.) At least one of the arcs on this circuit is not in the set of arcs of T_k, since T_k is a tree, and hence could be removed thus breaking the circuit and forming a new tree T_1. Since T_1 has one more arc in common with T_k (than T_o did), and one less arc in common with T_o: $d(T_1, T_k) = k - 1$. Proceeding in this way elementary tree transformations will produce trees $T_2, T_3 \ldots T_{k-1}, T_k$ where for any two consecutive trees T_i and $T_{i+1}, d(T_i, T_{i+1}) = 1$. Figures 7.5(b) to (e) shows 4 elementary tree transformations to derive T_4 from T_0.

2.2 A search procedure to generate all trees of a graph

From what has already been said at the beginning of this section, it is quite obvious that since the number of spanning trees of a graph increases so rapidly with the number of arcs of G, what is needed is an efficient method

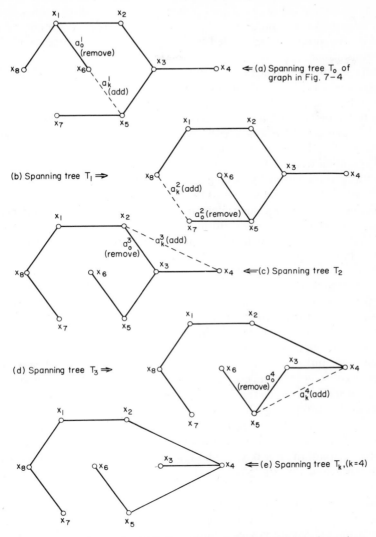

FIG. 7.5. Generation of T_4 from T_0 by elementary tree transformations

for generating the list of spanning trees exhaustively and without duplication. One obvious way of generating this list is to start with an initial spanning tree T_o and use the elementary tree transformations of the previous section to generate subsequent trees. Methods based on this principle have been given by Mayeda [42], Paul [47], Chen [8], and others [20]. However, there is an inherent disadvantage in using the elementary tree transformations, in that access to previously generated trees is required in order to generate a new tree. With the number of generated trees being so large it is necessary—from the computational point of view—to store trees that have just been generated in auxiliary store, since the fast computer store will inevitably be inadequate. Thus, since access times to the auxiliary store are very much longer than those to the fast store, any method based on elementary tree transformations is bound to be inefficient, unless a transformation is used, which when applied to the last generated tree repeatedly, eventually produces all spanning trees. (See Section 2.4). It is, therefore, apparent that a better algorithm for generating all spanning trees of a graph, would be one which produces a list of such trees without duplication, by placing them in auxiliary store as they are generated but without requiring access to them again in order to continue.

In the next section one such algorithm is described based on a decision-tree search principle. Other algorithms, based on generating all major square submatices of the reduced incidence matrix B_0 referred to in Theorem 1, are given in [10, 9, 43].

2.2.1 THE BASIC METHOD. The algorithm is described below for a nondirected graph G although its extention to the problem of enumerating all spanning arborescences of a directed graph is quite obvious.

The method starts by indexing the links of G from 1 to m. At each step (forward branch of the decision–tree search), a link is chosen which—together with the other links already chosen at previous steps—will form part of the tree being constructed. Thus, before picking such a link a check is made to ensure that its addition to the partially formed tree (which at this stage is a collection of subtrees) does not close any circuit. If a circuit is formed the link is rejected and a higher-numbered link is considered next. If no circuit results the link is added to those already chosen and a new forward step is taken until a spanning tree is formed. Links are considered in ascending order of their index and this leads to an exhaustive decision–tree search without duplication.

To facilitate the checking for the formation of a circuit when a link is added, each vertex x_j carries a double label of the form (r_j, p_j), and to facilitate the manipulation of the subtrees, each subtree is arbitrarily "rooted" at one of its vertices and considered as an arborescence. Similar forms of vertex labelling for circuit detection are given by Johnson [32], Srinivasan and Thompson [52] and by Glover and Klingman [15].

The first label r_j is the "root" of the subtree to which x_j belongs. Initially $r_j = x_j$ for all vertices x_j. At some stage, two subtrees T_1 and T_2 will be merged by the addition of a link $a_l = (x_\alpha, x_\beta)$, with x_α in T_1 and x_β in T_2. If at this stage r_1 is the root label of the vertices in T_1 and r_2 the root label of the vertices in

T_2 and if $r_1 < r_2$ (say), then all the vertices in T_2 have their root labels updated from r_2 to r_1 so that the two subtrees T_1 and T_2 become a single new tree T_1.

The second label p_j on vertex x_j is the predecessor vertex of x_j, i.e. if (x_k, x_j) is an arc in the subtree then $p_j = x_k$. The predecessor label for the root vertex is set to zero. In the example of the tree shown in Fig. 7.6(a), $p_5 = x_6$ $p_{13} = x_9$ and $p_9 = 0$.

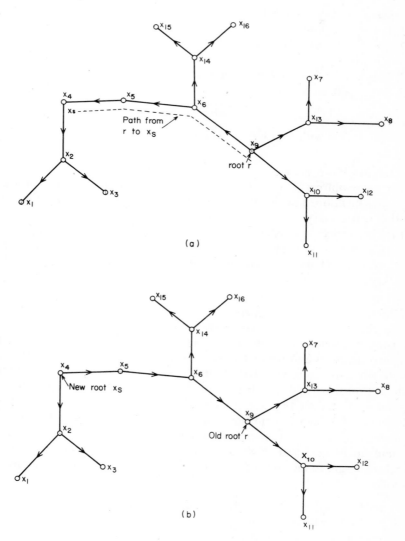

FIG. 7.6. Change of root from x_9 in (a) to x_4 in (b)

A. *Change in the root of a tree*

If the root of a tree T is r, a change of the root of T from r to x_s can be immediately achieved by simply reversing the direction of the arcs on the path from r to x_s and leaving all other arcs unchanged as shown in Fig. 7.6. Thus the necessary label changes for a change in root from r to x_s are:

To change predecessor labels
    1. Let $x_j = x_s$ and $z = p_j$.
    2. Set $x_i = z$ and re-set $z = p_i$.
    3. Update $p_i = x_j$.
    4. If $x_i = r$ go to step 5, otherwise set $x_j = x_i$ and return to step 2.
    5. Set $p_s = 0$, stop.

To change root labels
    1. All vertices with root labels r change their labels to x_s.

B. *Merging of two subtrees*

When two subtrees T_1 and T_2 are to be merged by the addition of a link (x_α, x_β) as mentioned earlier, the following changes to the labels are, therefore, all that is necessary:
    (i) The vertices with root labels r_2 have their labels changed to r_1.
    (ii) Change the root of T_2 from r_2 to x_β using the method explained in A above, and then change the predecessor label of x_β from $p_\beta = 0$ to $p_\beta = x_\alpha$.

C. *Division of a tree in two parts*

Since the method of tree generation that is being described below is a decision-tree search, the need arises (in backtracking steps) to remove some link in order to try another one. In such a situation the removal of a link would divide some tree into two parts, T_1 and T_2 say, and the labels of one of these subtrees must be updated. If the link to be removed is (x_α, x_β) with x_α in T_1 and x_β in T_2, then if $p_\beta = x_\alpha$ (i.e. the arc in the tree is directed from x_α to x_β) the labels of T_1 can remain unchanged and those of T_2 must be changed, whereas if $p_\alpha = x_\beta$ (i.e. the arc is directed from x_β to x_α) the labels of T_2 can remain unchanged and those of T_1 must be changed. Assuming that the labels of T_2 are to be changed we proceed to re-establish a root for this tree as follows:

    1. Set $S = \{x_\beta\}$ and $p_\beta = 0$ (x_β will be the root of T_2).
    2. Find all vertices x_j with $p_j \in S$ and change their root labels to $r_j = x_\beta$. If no such vertices exist stop.
    3. Update $S = S \cup \{x_j | p_j \in S\}$ and return to step 2.

One should note that with the exception of the newly formed root vertex x_β, the predecessor labels of all other vertices do not require any changing. One should also note that the number of times step 2 and 3 above need to be executed, is equal to the cardinality of the largest cardinality path in T_2 starting from vertex x_β.

2.2.2 DESCRIPTION OF THE ALGORITHM. Choose any arbitrary vertex x^* of the graph and let its degree be d^*. Number the d^* links incident on x^* sequentially as $a_1, a_2, \ldots, a_{d^*}$ and number the remaining links of G arbitrarily from a_{d^*+1} to a_m. The links will be considered for inclusion into the tree being formed in ascending order of their allocated suffixes.

Step 1: Set all labels (r_i, p_i) with $r_i = x_i$ and $p_i = 0$, $\forall x_i \in X$. Set $k = 1$.

Step 2: Choose link $a_k = (x_i, x_j)$, say, to investigate. If $k \leqslant m$ (the number of links in the graph) goto 2(i); if $k = m + 1$, no more links remain, goto step 5.

   (i) If $r_i = r_j$ it implies that x_i and x_j are in the same subtree and the addition of a_k would close a circuit. Reject link a_k i.e. set $k = k + 1$ and return to step 2.

   (ii) If $r_i \neq r_j$ the link a_k can be added to the links of the subtrees. Goto step 3.

Step 3: Merge the two subtrees whose vertices have root labels r_i and r_j respectively, using the method described in B above.

Step 4: If $n - 1$ links have been added a tree has been obtained. Store this tree and goto step 5. If less than $n - 1$ links have been added set $k = k + 1$ and return to step 2.

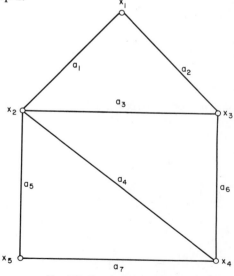

FIG. 7.7. Graph for example 2.3

Step 5: Backtrack: Remove the last added link, say a_l. If a_l was the only remaining added link and $l = d^*$, *stop*. All trees have been generated. (Any further decision-tree branching will always leave vertex x^* isolated.)

Otherwise, update labels as described in C above, set $k = l + 1$ and return to step 2.

2.3 Example

We want to generate all spanning trees of the graph shown in Fig. 7.7. Choosing arbitrarily $x^* = x_1$, we have $d^* = 2$ and we number links (x_1, x_2) and (x_1, x_3), a_1 and a_2 respectively. All other links are numbered randomly a_3 to a_7 as shown in Fig. 7.7.

The decision-tree search generated by the algorithm of Section 2.2.2 is shown in Fig. 7.8. The links (shown in circles), on any path from the top node of the search-tree to a node on the bottom level, form a tree. These trees are numbered 1 to 21 and they are shown in Fig. 7.9.

It is quite easy to verify that the total number of spanning trees of the graph shown in Fig. 7.7 is 21. Thus, from Theorem 1, Section 2, the incidence matrix B of this graph, with the links given an arbitrary orientation from the smallest indexed vertex to the largest, is:

$$
\begin{array}{c}
 & \begin{array}{ccccccc} a_1 & a_2 & a_3 & a_4 & a_5 & a_6 & a_7 \end{array} \\
\begin{array}{c} x_1 \\ x_2 \\ B = x_3 \\ x_4 \\ x_5 \end{array} &
\left[\begin{array}{ccccccc}
1 & 1 & 0 & 0 & 0 & 0 & 0 \\
-1 & 0 & 1 & 1 & 1 & 0 & 0 \\
0 & -1 & -1 & 0 & 0 & 1 & 0 \\
0 & 0 & 0 & -1 & 0 & -1 & 1 \\
0 & 0 & 0 & 0 & -1 & 0 & -1
\end{array}\right]
\end{array}
$$

Removing row x_2 (say) to form matrix B_o, the product matrix $B_o.B_o^t$ becomes:

$$
\begin{array}{c}
 & \begin{array}{cccc} 1 & 2 & 3 & 4 \end{array} \\
B_o.B_o^t = \begin{array}{c} 1 \\ 2 \\ 3 \\ 4 \end{array} &
\left[\begin{array}{cccc}
2 & -1 & 0 & 0 \\
-1 & 3 & -1 & 0 \\
0 & -1 & 3 & -1 \\
0 & 0 & -1 & 2
\end{array}\right]
\end{array}
$$

and the determinant $|B_o.B_o^t| = 21$ which according to Theorem 1 is the total number of spanning trees of the graph.

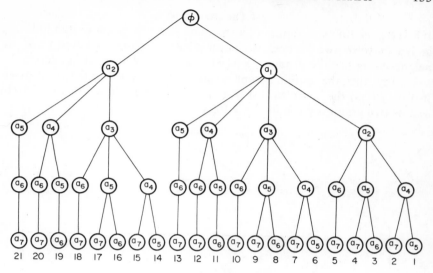

FIG. 7.8. Complete search-tree for example 2.3

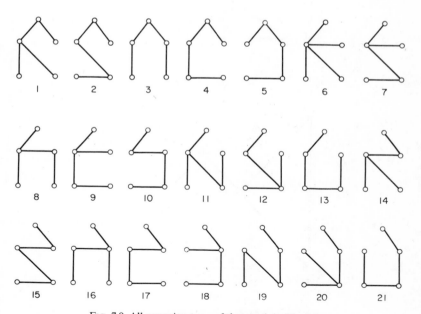

FIG. 7.9. All spanning trees of the graph in Fig. 7.7

2.4 The tree graph

If a vertex is drawn to represent a spanning tree of a graph G, and links are added between two vertices whenever their two corresponding trees are neighbouring (i.e. the distance between the trees—as defined in Section 2.1 —is unity), then the resulting graph is called a *tree graph*. For the graph given in Fig. 7.10(a) the complete list of spanning trees is shown in Fig. 7.10(b) and its tree graph is given in Fig. 7.10(c).

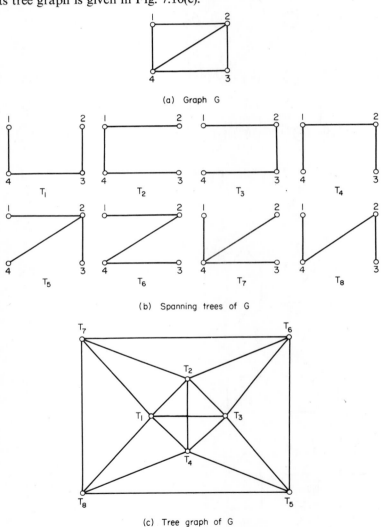

(a)   Graph G

(b)   Spanning trees of G

(c) Tree graph of G

FIG. 7.10. A graph G and its tree graph

It has been shown by Cummings [15] and Shank [51] that the tree graph of any connected graph G is Hamiltonian. Methods for finding a Hamiltonian circuit in the tree graph—and hence of listing all spanning trees of G—were later given by Kishi and Kajitani [37, 38] and by Kamae [33]. These methods are primarily of theoretical interest and do not lead to computationally efficient procedures.

3. The Shortest Spanning Tree (SST) of a Graph

Consider a connected nondirected graph $G = (X, A)$ with costs c_{ij} associated with its links (x_i, x_j). Of the many spanning trees of G that may be possible, we want to find the one for which the sum of the costs of its links is a minimum. This problem appears, for example, in the case where the vertices are terminals of an electric network which have to be connected together and one wants (in order to minimize stray effects) to use as short a length of wire as possible. Alternatively, if the vertices represent towns to be joined with a pipeline network, the shortest length of pipe that can be used (provided that out-of-town pipe junctions are not allowed) is given by the shortest spanning tree of the corresponding graph. A less direct application of the shortest spanning tree is as an intermediate step in the solution of the travelling salesman problem, itself a problem which often appears in practice and which is discussed in detail in Chapter 10.

One should perhaps note here that the SST of a graph bears no relation to the tree giving all shortest paths from a chosen vertex. Thus, for the graph shown in Fig. 7.11(a), where the numbers next to the links represent link costs, the tree giving all shortest paths from vertex x_1 (say) is shown in Fig. 7.11(b) whereas the SST is shown in Fig. 7.11(c).

The calculation of the shortest spanning tree (SST) of a graph is one of the few problems in graph theory which can be considered completely solved. Thus, let T_i and T_j be any two subtrees produced by the addition of links during the construction of the SST. If T_i is also used to represent the set of vertices of the subtree, then Δ_{ij} can be defined to be the shortest distance from a vertex in T_i to a vertex in T_j i.e.

$$\Delta_{ij} = \min_{x_i \in T_i} \left[\min_{x_j \in T_j} \{c(x_i, x_j)\} \right], \qquad i \neq j \tag{7.1}$$

It can then be quite easily shown that a repeated application of the following operation will produce the SST of a graph.

Operation I: For some subtree T_s, find that subtree T_{j*} for which $\Delta_{sj*} = \min_{T_j} [\Delta_{sj}]$, and let $(\overline{x_s, x_{j*}})$ be the link whose cost produced Δ_{sj*} in

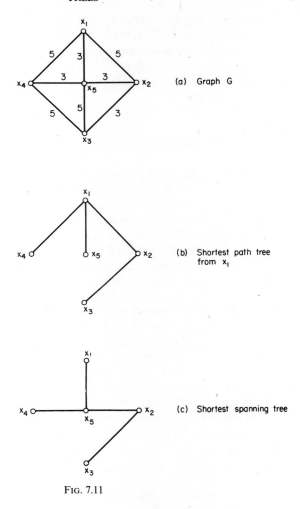

FIG. 7.11

equation (7.1). Link $(\overline{x_s, x_{j*}})$ is then in the SST and can be added to the other links of the partially formed SST.

Proof: Assuming that the links in the subtrees at some stage, k say, are in the final SST, let us suppose that link $(\overline{x_s, x_{j*}})$ as chosen above is not in the SST. Since subtree T_s must, by definition, be finally connected to some other subtree, some link (x_i, x_j) with $x_i \in T_s$ and $x_j \notin T_s$ must be in the SST. The removal of link (x_i, x_j) from this tree will divide it in two connected components, and its replacement by link $(\overline{x_s, x_{j*}})$ will form a new tree shorter than the SST, which is a contradiction. Hence, subject to the above assumption, it is possible to add $(\overline{x_s, x_{j*}})$ to the links of the partially formed SST at stage k and proceed to stage $k + 1$. One should note here that this is regardless of the subtree T_s chosen. Also, since at the initial stage, i.e. before any links are chosen, the assumption is nonexistent (and hence true), it follows that repeated application of the above operation will finally produce the SST of a graph.

The various methods [40, 48, 41, 46, 17, 35, 26] available for finding the SST are all based on particular cases of the above operation. The first of these is a method due to Kruskal [40] and is as follows.

3.1 The Kruskal algorithm

Step 1: Start with a completely disconnected graph T of n vertices.

Step 2: Order the links of G in ascending order of cost.

Step 3: Starting from the top of this list add links into T provided that this addition does not close a circuit in T.

Step 4: Repeat step 3 until $(n - 1)$ links have been added. T is then the SST of graph G.

This algorithm chooses to add into the partially formed tree T the absolute shortest link which is feasible, rather than simply the shortest link between one subtree of T, T_s say, and any other subtree (as suggested by Operation I). Since the link chosen is obviously the shortest between *some* subtree and any other subtree, the choice rule of this algorithm is a special case of Operation I. However, the situation may arise in this algorithm in which the next shortest link chosen from the list of step 2 may be between two vertices of the same subtree which would then make this link infeasible since its addition would close a circuit. Thus, during step 3, links must be tested for feasibility before they are added to T. This test can be performed most effectively—by a single comparison using the labelling procedure described in Section 2.2.1—in an exactly analogous manner to that shown in the second step of the algorithm of Section 2.2.2.

A computationally more expensive step is step 2, which for a graph with m links would require of the order of $m \log_2 m$ operations to produce a complete list of links in ascending order of cost. In general, however, one would not need the complete list, since it is quite likely that the $(n - 1)$ feasible links forming the SST may be found after examining only the top $r < m$ of the links in the list. This immediately suggests that the sorting procedure used at step 2 should be a multipass routine in which at the end of the pth pass the top p links are correctly placed. With such a procedure [34] one could then perform a single pass at step 2 (thus producing the shortest link at the top of the list), then test the top link in step 3; return to step 2 for a second pass, test the second-from-the-top link in step 3 etc., until after some number r of such tries $(n - 1)$ links have been added to T thus forming the SST. At the end of this process only r links have been effectively sorted in $r \log_2 m$ operations, the remaining $(m - r)$ links not having been required.

From what has been said above, it is quite obvious that despite the refinements, the Kruskal algorithm is more suitable for relatively sparse graphs rather than for complete ones. For these latter graphs $m = n(n-1)/2$ and for such cases Prim [48] and Dijkstra [17] have described other algorithms based on a more efficient particularization of Operation I.

3.2 The algorithm of Prim [48]

This algorithm produces in SST by growing only one subtree T_s (say) containing more than a single vertex and considering the remaining single vertices to form one subtree each. Subtree T_s is then grown continuously by adjoining that link $(x_i x_j), x_i \in T_s, x_j \notin T_s$ with the minimum cost c_{ij} until $(n - 1)$ links are added and T_s becomes the required SST. This particular form of Operation I was first suggested by Prim [48] and an efficient technique for its implementation was given by Dijkstra [17], and by Kevin and Whitney [35].

The algorithm proceeds by labelling each vertex $x_j \notin T_s$ with $[\alpha_j, \beta_j]$ where at any step α_j is the vertex of T_s nearest to vertex x_j and β_j is the length of this link (α_j, x_j). At any one step during the algorithm that vertex—say x_{j*} —with the smallest β_j is appended to T_s by the addition of link (α_{j*}, x_{j*}). Since T_s has now acquired a new vertex x_{j*}, the labels $[\alpha_j, \beta_j]$ for those vertices $x_j \notin T_s$ may now need updating (if, for example, $c(x_j, x_{j*})$ is less than the existing label β_j), and the process continued. This labelling procedure can be seen to be very similar to that used for the shortest path problem using the Dijkstra algorithm. (Chapter 8, Section 2.1.)

The algorithm is as follows:

Step 1: Let $T_s = \{x_s\}$, where x_s is any arbitrarily chosen vertex, and $A_s = \emptyset$. (A_s will be the set of links forming the SST).

Step 2: For all $x_j \notin T_s$ find a vertex $\alpha_j \in T_s$ so that:

$$c(\alpha_j, x_j) = \min_{x_i \in T_s} [c(x_i, x_j)] = \beta_j$$

and set the label of x_j as $[\alpha_j, \beta_j]$.
If no such vertex α_j can be found, i.e. if $\Gamma(x_j) \cap T_s = \emptyset$, set the label of x_j as $(0, \infty)$.

Step 3: Choose that vertex x_{j*} so that

$$\beta_{j*} = \min_{x_j \notin T_s} [\beta_j].$$

Update $T_s = T_s \cup \{x_{j*}\}$, $A_s = A_s \cup \{(\alpha_{j*}, x_{j*})\}$.
If $|T_s| = n$, stop. The links in A_s form the SST.
If $|T_s| \neq n$, goto step 4.

Step 4: For all $x_j \notin T_s$ and $x_j \in \Gamma(x_{j*})$ update labels as follows:
If $\beta_j > c(x_{j*}, x_j)$, set $\beta_j = c(x_{j*}, x_j)$, $\alpha_j = x_{j*}$ and return to step 3.
If $\beta_j \leqslant c(x_{j*}, x_j)$ goto step 3.

3.3 Related problems

Up to now we have addressed ourselves to the problem of finding the SST of a graph, and in this respect described two algorithms that could be used. The applicability of these methods, however, is much wider than may be— at first sight—thought. Operation I on which these methods are based, has been proven without regard as to the sign of the link costs c_{ij} and hence the SST methods described are applicable to graphs with arbitrary positive negative or zero link costs. This immediately implies that the *longest* spanning tree of a graph could also be found by simply reversing the signs of the link costs and applying one of the above SST algorithms.

Moreover, in the proof of Operation I, no use of the fact that the total cost of a spanning tree is the sum of its link costs, has been made. All that has been assumed is that if a link of the tree with cost C is replaced by a link of cost $C' < C$ then the cost of the tree decreases. Thus, if the cost of a tree is represented by any monotonically increasing symmetric† function of its link costs, the spanning tree which minimizes this cost function would be the same as the SST (which minimizes the sum of the link costs). For example, if $C_1, C_2, \ldots, C_m$ are the costs of the m links of the graph G, then the spanning tree of G which minimizes $(C_{i_1}^3 + C_{i_2}^3 + \ldots + C_{i_{n-1}}^3)$ or $(C_{i_1} \times C_{i_2} \times \ldots \times C_{i_{n-1}})$ —where $C_{i_1}, C_{i_2}, \ldots, C_{i_{n-1}}$ are the costs of any $n-1$ links forming a spanning tree of G—is the same as the SST of G. In addition, since for the first of the above mentioned functions the power of 3 could be replaced by any other power $p > 0$, and since as $p \to \infty$:

$$\left[\sum_{l=1}^{n-1} C_{i_l}^p \right] \to \{ \max_{l=1, \ldots, n-1} [C_{i_l}] \}^p,$$

the spanning tree which minimizes the cost of the most costly link contained, is also the same as the SST of G.

3.4 Example [48]

In the graph G shown in Fig. 7.12 each vertex represents a person and a link (x_i, x_j) implies that x_i could communicate with x_j and vice versa. It is required

---

† A function is symmetric in variables $x_1, x_2, \ldots, x_m$ if the exchange of any two variables in the function leaves its value unchanged. Both the monotonicity and symmetry restrictions are forced by the condition that $C' < C$ should imply a decrease in the cost of the tree.

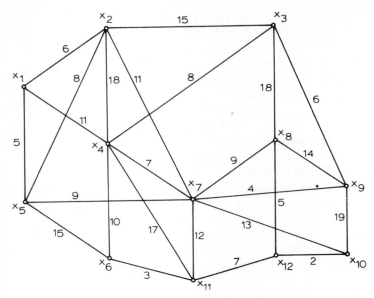

FIG. 7.12. Graph for example 3.4

that a confidential message be circulated amongst the 12 people in such a way as to minimize the probability of it becoming known to an outsider. For each transmission of the message from x_i to x_j there is a probability ρ_{ij} that the message may be intercepted by an outside person and these probabilities are given as percentages in Fig. 7.12. Obviously the paths of the message transmissions should form a spanning tree of G and what is then required is to find that spanning tree which minimizes: $1 - \Pi(1 - \rho_{ij})$, where the product is taken over those links forming the tree. Since this last function is both increasing and symmetric in ρ_{ij}, the required spanning tree is the same as the SST of G where ρ_{ij} are taken to be the link "costs" c_{ij}.

This problem will now be solved using the algorithm of Section 3.2.

Step 1. Take $x_s = x_1$, $T_1 = \{x_1\}$, $A_1 = \varnothing$.

Step 2. The labels of x_2, x_4 and x_5 are calculated as $[x_1, 6]$, $[x_1, 11]$ and $[x_1, 5]$ respectively, all other labels being $[0, \infty]$.

Step 3. The smallest β_j label is for x_5 and since $\alpha_5 = x_1$ link (x_1, x_5) is made. $T_1 = \{x_1, x_5\}$, $A_1 = \{(x_1, x_5)\}$.

Step 4. Update the labels of vertices x_2, x_6, x_7 to be:

for x_2: $\beta_2 = 6 < c(x_5, x_2)$ and no updating is necessary

for x_6: $\beta_6 = \infty > c(x_5, x_6) = 15$, hence the label of x_6 becomes $[x_5, 15]$

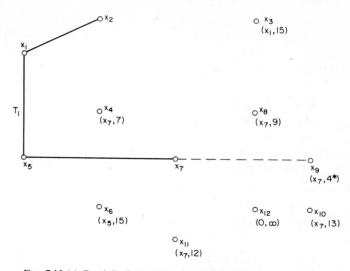

FIG. 7.13 (a). Partially formed tree T_1 with labels on vertices not in T_1
– – – – – – Next link to be added

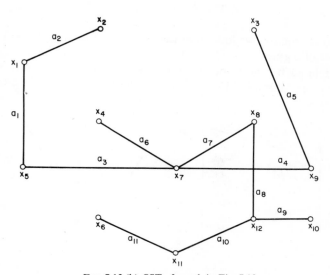

FIG. 7.13 (b). SST of graph in Fig. 7.12

for x_7: $\beta_7 = \infty > c(x_5, x_7) = 9$, and the label of x_7 becomes $[x_5, 9]$.

Since $x_4 \notin \Gamma(x_5)$ its label remains at $[x_1, 11]$ from the previous iteration.

Step 3. The labels are now: for x_2: $[x_1, 6]$, for x_4: $[x_1, 11]$, for x_6: $[x_5, 15]$, for x_7: $[x_5, 9]$. The smallest β_j is for x_2 and since $\alpha_2 = x_1$ link (x_1, x_2) is made.

$$T_1 = \{x_1, x_5, x_2\}, \qquad A_1 = \{(x_1, x_5), (x_1, x_2)\}$$

Step 4. Similarly, update the labels of vertices x_3, x_4, x_7 to be:
for x_3: $[x_2, 15]$, for x_4: $[x_1, 11]$ (no updating necessary)
for x_7: $[x_5, 9]$ (no updating necessary).
The label of $x_6 \notin \Gamma(x_2)$ remains at $[x_5, 15]$ from the previous iteration.

Step 3. The smallest β_j is for vertex x_7 and since $\alpha_7 = x_5$ link (x_5, x_7) is made.
$$T_1 = \{x_1, x_5, x_2, x_7\}, \qquad A_1 = \{(x_1, x_5), (x_1, x_2), (x_5, x_7)\}.$$

Step 4. The labels of the vertices are updated as before and are shown in Fig. 7.13(a) together with the links that have been added to the tree so far.

Continuing in this way the final SST is derived as shown in Fig. 7.13(b) and the links are numbered to show the order in which they were introduced into the tree.

The product $\Pi(1 - \rho_{ij})$ for the links of this tree is 0·5214 which gives a minimum probability of 47·86% for the message to be intercepted by an outside person.

4. The Steiner Problem

In the previous section we looked at the problem of finding the SST i.e. the shortest tree of a graph $G = (X, \Gamma)$ so that all the vertices of X are spanned. A closely related but very much more difficult problem, is known as the *"Steiner problem in graphs"* [27, 18]. In this problem, the shortest tree T is

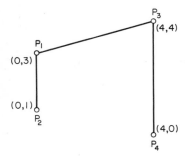

FIG. 7.14 (a). Shortest spanning tree
Length = 10·123

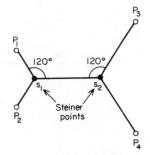

FIG. 7.14 (b). Shortest Steiner tree
Length = 9·196

required which spans a specified subset $P \subset X$ of the vertices of G. The other vertices in $X - P$ may be either spanned by the tree or not—as needed in order to minimize the length of T. Thus, the Steiner problem on a graph is equivalent to finding the shortest spanning tree of any subgraph $G' = (X', \Gamma)$ of G, with $P \subseteq X' \subseteq X$.

The *Euclidean Steiner problem* was, in fact, originally proposed as a problem in geometry [13, 14, 44, 24], where a set P of points on a Euclidean plane were to be connected by lines so that the total length of lines drawn was a minimum. If no two lines are allowed to meet anywhere but at the specified P points, then the problem becomes one of finding the SST of an equivalent graph of $|P|$ vertices, with a cost matrix calculated as the Euclidean distance matrix between the points in P. When, however, other "artificial" vertices (called Steiner points) can be introduced on the plane, the length of the SST of the resulting set $P' \supset P$ of points can be reduced even further. For example, considering the 4 points shown to scale in Fig. 7.14(a), the SST is as shown, whereas the introduction of two new points s_1 and s_2 in the middle, produces an SST spanning all 6 points, whose total length is less than that of the previous SST. (See Fig. 7.14(b).) Thus, in the original Steiner problem, as many Steiner points as necessary could be added anywhere in the plane, in order to produce the shortest tree spanning the specified set of P points. This resulting shortest tree is then called a *shortest Steiner tree*.

The Steiner problem on a Euclidean plane has been studied extensively [44, 24, 11, 12], and many properties of the shortest Steiner tree are known. Of these, the most important ones are [24], [11]:

(i) For a Steiner point s_i the degree $d(s_i) = 3$. It can be easily shown by geometrical considerations that the angle between links incident at any Steiner point must be 120°, and that exactly 3 links are incident at any Steiner point s_i. This point is, therefore, the "centre" (Steiner centre) of an imaginary triangle whose vertices are the other three points, to which s_i is linked in the shortest Steiner tree.

   Some of the points forming the vertices of this triangle may themselves be other Steiner points. For example in Fig. 7.14(b), Steiner point s_2 is the Steiner centre of the imaginary triangle with vertices p_3, p_4 and s_1.

(ii) For a vertex $p_i \in P$, $d(p_i) \leqslant 3$. If $d(p_i) = 3$, then the angle between any two of the three links incident at p_i must be 120°, and if $d(p_i) = 2$ the angle between the two links must be greater than or equal to 120°.

(iii) The number k of Steiner points in a shortest Steiner tree is $0 \leqslant k \leqslant n - 2$, where $n = |P|$.

Proof of the above properties can be found in [24].

Despite the attention that the Euclidean Steiner problem has received, only very small size problems indeed (i.e. ones with no more than about 10 points in *P*), can be solved optimally using the existing algorithms [44, 11], and hence the problem can be considered to be unsolved. For larger problems one must resort to a number of possible heuristics [7] [50].

A later version of the Steiner problem on a plane used *rectilinear*—rather than Euclidean—distances between points. The problem was first suggested by Hanan [28, 30] in connection with routing wires on printed circuit boards for electronic components. In this version of the problem, the distance between two points with coordinates (x_1, y_1), (x_2, y_2) is given by

$$d_{1,2} = |x_1 - x_2| + |y_1 - y_2|$$

Under these conditions it can be easily shown [44] that if, through each of the points in *P*, vertical and horizontal grid lines are drawn, then the solution to the Steiner problem can be found by considering only the intersections of these grid lines as possible Steiner point locations.

Thus, let a graph *G* be formed so that the set *X* of its vertices is the set of distinct grid line intersections and the links of *G* correspond to grid lines joining two intersection points. The Steiner problem on a plane with a rectilinear distance metric, then becomes the Steiner problem on the *finite* graph *G* as defined at the beginning of this section [54]. Figure 7.15(b) shows an example of a shortest Steiner tree of a 6-point rectilinear problem and, for comparison, Fig. 7.15(a) shows the SST of this problem.

The Steiner problem on a general nondirected graph has been studied by Hakimi [27] and by Dreyfus and Wagner [18] who gave exact algorithms

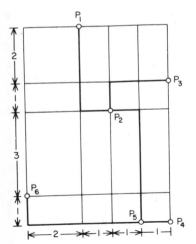

Fig. 7.15 (a). Shortest spanning tree
Length = 18

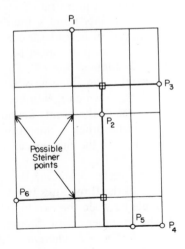

Fig. 7.15 (b). Shortest Steiner tree
Length = 15
⊞ Steiner points

for its solution. However, these algorithms are computationally inefficient procedures even though they are much better than the complete enumeration of SST's of all subgraphs G' of G. Nevertheless (just as is the case with the Euclidean-distance problems), the maximum size of a Steiner problem on a graph which can be solved in a reasonable computing time, is still not much above 10 vertices (in P). Thus, the Steiner problem on a graph can also be considered to be an unsolved problem and will not be discussed here further.

5. Problems P7

1. Prove that all three definitions of a spanning tree given in the Introduction are equivalent. Can you find other equivalent definitions?

2. Show that in a tree with more than one vertex there exists at least two vertices of degree 1.

3. Show that the determinant of any square submatrix of the incidence matrix of a graph has the values, $+1$, -1 or 0.

4. For a matrix of order $n \times m$, any square submatrix of order $\min(n, m) \times \min(n, m)$ is called a major submatrix. Show that if B_m is a major submatrix of the incidence matrix of a connected graph, then $|B_m|$ is nonzero ($+1$ or -1), if and only if the partial graph consisting of all links corresponding to the columns of B_m is a tree.

5. Find all the spanning trees of the graph shown in Fig. 7.16 and verify that their number is correct by using the result of theorem 1.

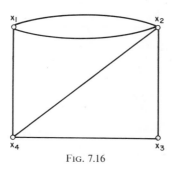

FIG. 7.16

6. Show that the matrix product $B_o \cdot B_o^t$ used in theorem 1 need not be obtained by multiplying B_o with B_o^t but may be derived directly from the graph as the $n \times n$ matrix $M = [m_{ij}]$ defined as follows: A diagonal entry m_{ii} is the degree of vertex x_i, and an entry m_{ij} is minus the number of parallel links between x_i and x_j. Only $(n-1)$ rows and columns of M need be constructed in this way to obtain $B_o \cdot B_o^t$.

F

7. Use theorem 1 to show that the number of spanning trees of a completely connected nondirected graph with n vertices is n^{n-2}.

8. For a directed graph let a matrix $M = [m_{ij}]$ to be defined as follows:

$$m_{ii} = d_t(x_i), \text{ the indegree of vertex } x_i,$$

$$m_{ij} = -k, \text{ where } k \text{ is the number of arcs}$$
$$\text{in parallel from } x_i \text{ to } x_j.$$

Show that a directed graph is a directed tree with root x_r if and only if: $m_{rr} = 0$,
$m_{ii} = 1$ for $i \neq r$, and the determinant of the minor submatrix resulting from the
erasure of the rth row and column of M has the value 1 (See Ref. [19]).

9. With M defined as in problem 8, use the above result to show that the number of
directed spanning trees with root x_r that a directed graph (without loops) has, is given
by the determinant of the minor submatrix resulting from the erasure of the rth row
and column of M (See Ref. [1], p. 163 and Ref. [2]).

10. Write an algorithm for enumerating all spanning trees of a graph without duplications using elementary tree transformations (See Ref. [42, 47, 8]).

11. Find the shortest spanning tree of the graph shown in Fig. 7.17 using both
Kruskal's and Prim's algorithms.

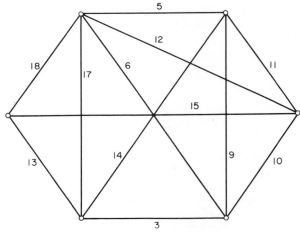

FIG. 7.17

12. For the complete graph on the vertex set $\{x_1, x_2, x_3, x_4\}$ as shown in Fig. 7.18,
and assuming the link costs to be the Euclidean distances, find:

(i) The SST,

(ii) The Steiner tree in the plane. (Use the properties of steiner points given in Section
4 and consider all possible topologies. (i.e. incidence matrices). Repeat the exercise
assuming the link costs to be the rectilinear distances between points.

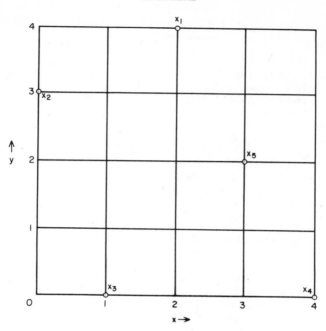

FIG. 7.18

13. Repeat exercise 12 when an extra vertex x_5 is added to the problem and note the increase in the computational effort involved.

6. References

1. Berge, C. (1962). "The Theory of Graphs", Methuen, London.
2. Bott, R. and Mayberry, J. P. (1954). Matrices and Trees, *In*: "Economic Activity Analysis", Wiley, New York.
3. Busacker, R. G. and Saaty, T. L. (1965). "Finite Graphs and Networks", McGraw-Hill, New York.
4. Cayley, A. (1874). On the mathematical theory of isomers, *Philosophical Magazine*, **67**, p. 444.
5. Cayley, A. (1897). Collected papers, *Quart. Jl. of Mathematics*, **13**, Cambridge, p. 26.
6. Chan, S. P. and Chan, S. G. (1968). Modifications of topological formulae, *IEEE Trans.*, **CT-15**, p. 84.
7. Change, S.-K. (1972). The generation of minimal trees with Steiner topology, *Jl. of ACM*, **19**, p. 699.
8. Chen, W.-K. (1971). "Applied Graph Theory", North-Holland, Amsterdam.
9. Chen, W.-K. (1966). On the directed trees and directed k-trees of a digraph and their generation, *Jl. of SIAM (Appl. Math.)*, **14**, p. 550.

10. Chen, W.K. and Li, H.-C. (1973). Computer generation of directed trees and complete trees, *Int. Jl. of Electronics*, **34**, p. 1.
11. Cockayne, E. J. (1970). On the efficiency of the algorithm for Steiner minimal trees, *Jl. of SIAM (Appl. Math.)*, **18**, p. 150.
12. Cockayne, E. J. and Melzak, Z. A. (1968). Steiner's problem for set terminals, *Quart. Applied Mathematics*, **26**, p. 213.
13. Courant, R. and Robbins, H. (1941). "What is mathematics", Oxford University Press, New York.
14. Coxeter, H. S. M. (1961). "Introduction to geometry", Wiley, New York.
15. Cummings, R. L. (1966). Hamilton circuits in tree graphs, *IEEE Trans.*, **CT-13**, p. 82.
16. De Bruijn, N. G. (1964). Polyas' theory of counting, *In*: "Applied Combinatorial Mathematics", Beckenbach, Ed., Wiley, New York.
17. Dijkstra, E. W. (1959). A note on two problems in connection with graphs, *Numerische Mathematik*, **1**, p. 269.
18. Dreyfus, S. E. and Wagner, R. A. (1972). The Steiner problem in graphs, *Networks*, **1**, 195.
19. Even, S. (1973). "Algorithmic combinatorics", Macmillan, New York.
20. Fidler, J. K. and Horrocks, D. H. (1973). On the generation of k-trees, *Int. Jl. of Electronics*, **34**, p. 185.
21. Floyd, R. W. (1962). TREESORT-Algorithm 113, ACM Collected Algorithms.
22. Floyd, R. W. (1964). TREESORT 3-Algorithm 245, ACM Collected Algorithms.
23. Fu. Y. (1967). Application of linear graph theory to printed circuits, Proc. Asilomar Conf. on Systems and Circuits
24. Gilbert, E. N. and Pollack, H. O. (1968). Steiner minimal trees, *Jl. of SIAM (Appl. Math.)*, **16**, p. 1.
25. Glover, F. and Klingman, D. (1970). Locating stepping-stone paths in distribution problems via the predecessor index method, *Transp. Sci.*, **4**, p. 220.
26. Gower, J. C. and Ross, G. J. S. (1969). Minimum spanning trees and single linkage cluster analysis, *Applied Statistics*, **18**, p. 54.
27. Hakimi, S. L. (1971). Steiner's problem in graphs and its implications, *Networks*, **1**, p. 113.
28. Hanan, M. (1966). On Steiner's problem with rectilinear distance, *Jl. of SIAM (Appl. Math.)*, **14**, p. 255.
29. Hanan, M. (1972). A counterexample to a theorem of Fu on Steiner's problem, *IEEE Trans.*, **CT-19**, p. 74.
30. Hanan, M. and Kurzberg, J. M. (1972). Placement techniques, *In*: "Design automation of digital systems," Breuer, Ed., Prentice Gall, New Jersey.
31. Holzmann, C. A. and Harary, F. (1972). On the tree graph of a matroid, *Jl. of SIAM (Appl. Math.)*, **22**, p. 187.
32. Johnson, E. (1962). Networks and basic solutions, *Ops. Res.*, **14**, p. 89.
33. Kamae, T. (1967). The existence of Hamiltonian circuits in tree graphs, *IEEE Trans.*, **CT-14**, p. 279.
34. Kershenbaum, A. and Van Slyke, R. (1972). Computing minimum spanning trees efficiently, Proc. of the Ann. Conf. of ACM, Boston, p. 518.
35. Kevin, V. and Whitney, M. (1972). Algorithm 422—Minimal spanning tree, *Comm. of ACM*, **15**, p. 273.
36. Kirchhoff, G. (1847). *In*: "Annalen der Physik and Chemie," **72**, p. 497.
37. Kishi, G. and Kajitani, Y. (1968). On Hamilton circuits in tree graphs, *IEEE Trans*, **CT-15**, p. 42.

38. Kishi, G. and Kajitani, Y. (1968). On the realization of tree graphs, *IEEE Trans.*, **CT-15**, p. 271.

39. Knuth, D. E. (1968). "The art of computer programming, Vol. 1/Fundamental algorithms", Addison Wesley, Reading, Massachusetts.

40. Kruskal, J. B. Jr. (1956). On the shortest spanning subtree of a graph and the traveling salesman problem, *Proc. American Mathematical Soc.*, **7**, p. 48.

41. Loberman, H. and Weinberger, A. (1957). Formal procedures for connecting terminals with a minimum total wire length, *Jl. of ACM*, **4**, p. 428.

42. Mayeda, W. (1972). "Graph Theory", Wiley-Interscience, New York.

43. Mayeda, W., Hakimi, S. L., Chen, W.-K. and Deo, N. (1968). Generation of complete trees, *IEEE Trans.*, **CT-15**, p. 101.

44. Melzak, Z. A. (1961). On the problem of Steiner, *Canadian Mathematical Bulletin*, **4**, p. 335.

45. Moon, J. W. (1967). Various proofs of Cayley's formula for counting trees, *In*: "A seminar on graph theory", Harary, Ed., Holt, Rinehart and Winston, New York.

46. Obruca, A. (1964). Algorithm 1—MINTREE, *Computer Bulletin*, p. 67.

47. Paul, A. J. Jr. (1967). Generation of directed trees and 2-trees without duplication, *IEEE Trans,* **CT-14**, p. 354.

48. Prim, R. C. (1957). Shortest connection networks and some generalizations, *Bell Syst. Tech. Jl.,* **36**, p. 1389.

49. Riordan, J. (1958). "An Introduction to Combinatorial Analysis", Wiley, New York.

50. Scott, A. (1971). "Combinatorial programming, spatial analysis and planning", Methuen, London.

51. Shank, H. (1968). Note on Hamilton circuits in tree graphs, *IEEE Trans.*, **CT-15**, p. 86.

52. Srinivasan, V. and Thompson, G. L. (1972). Accelerated algorithms for labelling and relabelling of trees with applications to distribution problems, *Jl. of ACM*, **19**, p. 712.

53. Trent, H. M. (1954). A note on the enumeration and listing of all possible trees in a connected linear graph, *Proc. Nat. Acad. Sci. U.S.A.*, **40**, p. 1004.

54. Yang, Y. Y. and Wing, O. (1972). Suboptimal algorithm for a wire routing problem, *IEEE Trans.*, **CT-19**, p. 508.

Chapter 8

Shortest Paths

1. Introduction

For a given arc-weighted graph $G = (X, \Gamma)$ with arc costs given by the matrix $C = [c_{ij}]$, the *shortest path problem* is the problem of finding the shortest path from a specific starting vertex $s \in X$ to a specific ending vertex $t \in X$, provided that such a path exists i.e. provided $t \in R(s)$, where $R(s)$ is the reachable set of the vertex s as defined in Chapter 2. The elements c_{ij} of the cost matrix C can be positive, negative or zero provided that no circuit of G exists whose total cost is negative. If such a circuit Φ does exist and x_i is a vertex on this circuit, then by proceeding from s to x_i traversing the circuit Φ an arbitrarily large number of times and finally proceeding to t, will result in a path with an arbitrarily small ($\rightarrow -\infty$) cost so that a best path is not uniquely defined.

If on the other hand such circuits exist but are excluded from consideration, then finding the shortest (elementary) path between s and t becomes equivalent to the problem of finding the shortest Hamiltonian path of the graph with s and t as its ends. This can be seen from the fact that, if an arbitrarily large number L is subtracted from each entry c_{ij} of the cost matrix C to produce a new cost matrix $C' = [c'_{ij}]$ with all c'_{ij} negative, then the shortest path from s to t—excluding negative circuits—must, by necessity, be Hamiltonian i.e. pass through all other vertices. Since the cost of any Hamiltonian path under the cost matrix C' is equal to its cost under C minus a constant term $(n - 1) . L$, it follows that the shortest (elementary) path from s to t under C' is the shortest Hamiltonian path from s to t under the initial matrix C. The problem of finding the shortest Hamiltonian path is of very much greater complexity than the shortest path problem and is discussed separately in Chapter 10. We will therefore assume here that all the circuits of G have non-negative total costs, which also implies that if G contains nondirected arcs (links) these cannot have negative costs.

The following problems are immediate generalizations of the above mentioned shortest path problem.

(i) For a specific starting vertex s find the shortest path between s and *all* other vertices $x_i \in X$.

and

(ii) Find the shortest path between *all pairs* of vertices.

It will be noted in the following sections, that almost all methods which solve the s-to-t shortest path problem also derive (during the process) the shortest paths from s to x_i ($\forall x_i \in X$) i.e. they also solve problem (i) above with very little extra computational cost. Problem (ii) on the other hand, can be solved either by applying n times an algorithm which solves problem (i), taking at each iteration a different vertex as the starting vertex s, or can be solved by a special single pass algorithm.

In this Chapter we will give general algorithms for the solution of the above mentioned problems, and particular algorithms for the special case where all c_{ij} are non-negative. These special cases occur often enough in practice, (for example in cases where the c_{ij} represent physical distances), to warrant the description of these special algorithms. We will assume that the matrix C does not satisfy the triangularity conditions i.e. c_{ij} is not less than $c_{ik} + c_{kj}$ for all i, j, and k, otherwise the shortest path between x_i and x_j is always the single arc (x_i, x_j) and the problem becomes nonexistent. In particular, if an arc (x_i, x_j) does not exist in G, then its cost will be assumed to have been set to ∞.

What is often required in practice is not simply the shortest but also the second, third etc. shortest paths in a graph. With this information one could then decide on the best path to choose using also criteria which are either difficult to incorporate directly into the algorithms or which are subjective in nature. Moreover, the second, third, etc. shortest paths can be used in a sensitivity analysis of the shortest path problem. In this chapter we give a recent algorithm for the computation of the K shortest elementary paths between two specified vertices in a general graph.

The chapter also discusses the problems of finding the maximum reliability and maximum capacity paths in graphs. These problems are related to the shortest path problem although the path characteristic (say cost) is not the sum of the characteristics (costs) of the arcs forming the path but other functions of the arc characteristics. These problems can be either reformulated as shortest path problems, or the methods described for shortest path problems recast in order to solve the maximum capacity or maximum reliability path problems directly.

A case is discussed in which arc capacities and reliabilities are combined to form a maximum expected capacity path problem, and although this particular problem cannot be solved by the shortest path techniques, an iterative algorithm using these techniques as basic steps is shown to offer an effective means of obtaining the optimal answer.

Shortest path problems in which the paths are restricted or constrained in some way [4], [12], [23] are not dealt with in this chapter since the labelling methods which form the basis of all the unconstrained shortest path algorithms described here are not directly applicable. These constrained problems are often of such difficulty that algorithms can only solve optimally problems which are several orders of magnitude smaller (as far as the number of vertices are concerned), than the equivalent unconstrained path problems. The important constrained problems such as the problem of the shortest Hamiltonian path mentioned earlier are considered in separate chapters.

2. The Shortest Path Between Two Specified Vertices s and t

In the first instance we will describe a very simple and efficient algorithm to solve this problem for the case $c_{ij} \geqslant 0$ ($\forall i, j$) and later extend the method to the general case of $c_{ij} \geqslant 0$ with the proviso that no negative cost circuits exist.

2.1 Case of non-negative cost matrix

The most efficient algorithm for the solution of the s–t shortest path problem was given initially by Dijkstra [10]. In general, the method is based on assigning temporary labels to vertices, the label on a vertex being an upper bound on the path length from s to that vertex. These labels are then continuously reduced by an iterative procedure and at each iteration exactly one of the temporary labels becomes permanent indicating that it is no longer an upper bound but the exact length of the shortest path from s to the vertex in question. The details of the method are as follows:

2.1.1 DIJKSTRA'S ALGORITHM ($c_{ij} \geqslant 0$).

Let $l(x_i)$ be the label on vertex x_i.

Initialization
Step 1. Set $l(s) = 0$ and mark the label as permanent. Set $l(x_i) = \infty$ for all $x_i \neq s$ and mark these labels temporary. Set $p = s$.

Updating of labels
Step 2. For all $x_i \in \Gamma(p)$ and which have temporary labels, update the labels according to:

$$l(x_i) = \min\left[l(x_i), l(p) + c(p, x_i)\right] \tag{8.1}$$

Fixing a label as permanent
Step 3. Of all temporarily labelled vertices find x_i^* for which $l(x_i^*) = \min\left[l(x_i)\right]$.
Step 4. Mark the label of x_i^* permanent and set $p = x_i^*$.
Step 5. (i) (*If only the path from s to t is desired*)

If $p = t$, $l(p)$ is the required shortest path length. Stop.
If $p \neq t$, go to step 2.

(ii) (*If the path from s to every other vertex is required*)
If all the vertices are permanently labelled, then the labels are the lengths of the shortest paths. Stop.
If some labels are temporary go to step 2.

The proof that the above algorithm indeed produces the shortest paths is quite simple and an outline of this proof is given below.

Let us suppose that the permanent labels at some stage are shortest path lengths. Let S_1 be the set of vertices with these labels, whereas S_2 is the set of the vertices which have temporary labels. At the end of step 2 of every iteration, the temporary label $l(x_i)$ is the shortest path from s to x_i which passes entirely through vertices in the set S_1. (Because only one vertex enters S_1 at each iteration, the updating of $l(x_i)$ requires only the one comparison given by step 2.)

Let the shortest path from s to x_i^* not pass entirely through S_1 but contain at least one vertex from S_2 and let $x_j \in S_2$ be the first such vertex on this path. Since c_{ij} were assumed non-negative, the part of the path from x_j to x_i^* must have a non-negative cost Δ (say) so that $l(x_j) < l(x_i^*) - \Delta < l(x_i^*)$. This, however, contradicts the assertion that $l(x_i^*)$ is the smallest temporary label, and hence the shortest path to x_i^* passes entirely through vertices in S_1 and $l(x_i^*)$ is therefore its length.

Since S_1 is initially set to $\{s\}$ and at every iteration x_i^* is added to S_1, the assumption that $l(x_i)$ are shortest path lengths, $\forall x_i \in S_1$, is valid at every iteration and hence, by induction, the answer produced by the algorithm is optimal.

For the case of a n-vertex completely connected graph where the shortest paths between s and all other vertices are required, the algorithm involves $n(n-1)/2$ additions and comparisons at step 2 and another $n(n-1)/2$ comparisons at step 3. Additionally, at steps 2 and 3 it is necessary to determine which vertices are temporarily labelled which requires an extra $n(n-1)/2$ comparisons. These figures are also upper bounds on the number of operations necessary to find the shortest path from s to a specified t, and can in fact be realized if t happens to be the last vertex to be permanently labelled. (In Reference [22] Johnson describes a sorting method whereby the number of operations at step 3 is reduced further.)

Once the shortest path lengths from s are obtained as the final values of the vertex labels, the paths themselves can be obtained by a recursive application of eqn (8.2) below. Thus if x_i' is the vertex just before x_i in the shortest path from s to x_i, then for any given vertex x_i, x_i' can be found as that one of the remaining vertices for which:

$$l(x_i') + c(x_i', x_i) = l(x_i) \tag{8.2}$$

If the shortest path from s to any x_i is unique, then the arcs (x_i', x_i) on the shortest paths form a directed tree (see the previous chapter), with s as its root. If there is more than one "shortest" path from s to any other vertex, then eqn (8.2) will be satisfied by more than one vertex x_i' for some x_i. In that case either an arbitrary choice can be made (if only one shortest path between s and x_i is required), or all of the arcs (x_i', x_i) appearing in any of the shortest paths can be considered in which case the totality of such arcs do not form a directed tree but a general graph called the *base relative to s* or *s-base*† for short.

2.1.2 EXAMPLE. Consider the graph shown in Fig. 8.1 where nondirected links are to be considered as two arcs of equal cost in opposite directions. Let the cost matrix C_1 be as shown below. It is required to find all the shortest paths from vertex x_1 to any other vertex. We will use the Dijkstra algorithm and attach a $(+)$ sign to a label whenever the label is permanent, otherwise the label is considered temporary.

	x_1	x_2	x_3	x_4	x_5	x_6	x_7	x_8	x_9
x_1		10					3	6	12
x_2	10		18				2		13
x_3		18		25		20			7
x_4			25		5	16	4		
$C_1 = x_5$				5		10			
x_6			20		10		14	15	9
x_7		2		4		14			24
x_8	6				23	15			5
x_9	12	13				9	24	5	

The algorithm proceeds as follows:

Step 1: $l(x_1) = 0^+$, $l(x_i) = \infty \forall x_i \neq x_1$, $p = x_1$

† Note that there is no relationship between the *s-base*, and the *basis* of a graph defined earlier in Chapter 2.

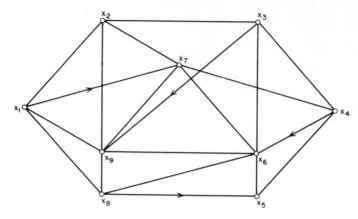

FIG. 8.1. Graph for example 2.1.2

First iteration
Step 2: $\Gamma(p) = \Gamma(x_1) = \{x_2, x_7, x_8, x_9\}$—all with temporary labels. Take x_2 first, eqn (8.1) gives:

$$l(x_2) = \min[\infty, 0^+ + 10] = 10$$

similarly $l(x_7) = 3$, $l(x_8) = 6$, $l(x_9) = 12$
Step 3: $\min[\underbrace{10}_{\widetilde{x}_2}, \underbrace{3}_{\widetilde{x}_7}, \underbrace{6}_{\widetilde{x}_8}, \underbrace{12}_{\widetilde{x}_9}, \underbrace{\infty}_{x_3, x_4, x_5, x_6}] = 3$ corresponding to x_7

Step 4: x_7 is now permanently labelled, $\boxed{l(x_7) = 3^+}$, $p = x_7$
Step 5: Not all vertices are permanently labelled hence goto step 2. At the beginning of the next iteration the labels are shown in Fig. 8.2(a).

Second iteration
Step 2: $\Gamma(p) = \Gamma(x_7) = \{x_2, x_4, x_6, x_9\}$, all with temporary labels. From eqn (8.1):

$$l(x_2) = \min[10, 3^+ + 2] = 5$$

similarly $l(x_4) = 7$, $l(x_6) = 17$, $l(x_9) = 12$.
The labels are now as shown in Fig. 8.2(b)
Step 3: $\min[\underbrace{5}_{\widetilde{x}_2}, \underbrace{7}_{\widetilde{x}_4}, \underbrace{17}_{\widetilde{x}_6}, \underbrace{6}_{\widetilde{x}_8}, \underbrace{12}_{\widetilde{x}_9}, \underbrace{\infty}_{\widetilde{x}_3, \widetilde{x}_5}] = 5$ corresponding to x_2

Step 4: x_2 is now permanently labelled, $\boxed{l(x_2) = 5^+}$, $p = x_2$
Step 5: Goto step 2

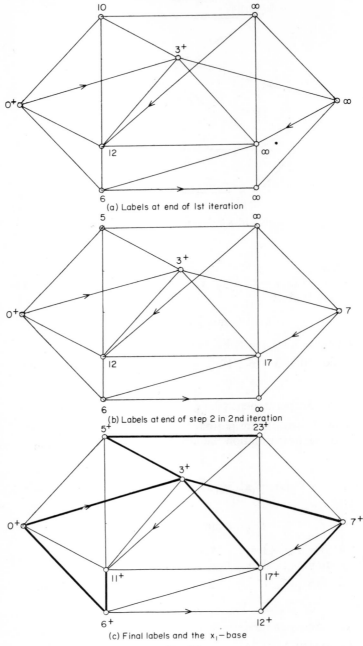

(a) Labels at end of 1st iteration

(b) Labels at end of step 2 in 2nd iteration

(c) Final labels and the x_1–base

FIG. 8.2

Third iteration

Step 2: $\Gamma(p) = \Gamma(x_2) = \{x_1, x_3, x_7, x_9\}$—only $\{x_3, x_9\}$ with temporary labels. Hence from equation (8.1):

$$l(x_3) = \min[\infty, 5^+ + 18] = 23$$

and similarly

$$l(x_9) = 12$$

Step 3: $\min[\,23\,,\,\,7\,\,,\,\,17\,,\,\,6\,\,,\,\,12\,,\,\,\infty\,] = 6$ corresponding to x_8

$$\widetilde{x}_3 \quad \widetilde{x}_4 \quad \widetilde{x}_6 \quad \widetilde{x}_8 \quad \widetilde{x}_9 \quad \widetilde{x}_5$$

Step 4: x_8 is now permanently labelled, $\boxed{l(x_8) = 6^+}$ $p = x_8$

Step 5: Goto step 2.

Continuing in this way, we obtain the final labels shown in Fig. 8.2(c). To find the shortest path between a vertex (say x_2) and the initial vertex x_1 we make use of eqn (8.2) iteratively. Thus taking $x_i = x_2$ the vertex x_2'— which is the vertex just before x_2 on the shortest path from x_1 to x_2—is that which satisfies the equation:

$$l(x_2') + c(x_2', x_2) = l(x_2) = 5$$

The only vertex x_2' which satisfies the above equation is x_7. Proceeding now with the second application of eqn (8.2) starting with $x_i = x_7$ we find the vertex x_7' just before vertex x_7 on the shortest path from x_1 to x_2 to be that which satisfies:

$$l(x_7') + c(x_7', x_7) = l(x_7) = 3$$

The only vertex x_7' which satisfies the above equation is x_1 itself so that the shortest path from x_1 to x_2 is (x_1, x_7, x_2). The resulting x_1-base given all shortest paths from x_1, is a tree and is shown with heavy lines in Fig. 8.2(c).

2.2 Case of general cost matrix

The previously given algorithm of Dijkstra only applies when $c_{ij} \geqslant 0$ for all i and j. In cases, however, where the matrix C represents costs then profitable arcs would have negative "costs" associated with them. In this case the procedure given below could be used to calculate the shortest paths between vertex s and all other vertices. The method is again iterative and based on vertex labelling, where at the end of the kth iteration the labels represent the values of those shortest paths (from s to all other vertices) which contain $k + 1$ or fewer arcs. Unlike the Dijkstra algorithm no label is considered final in this process until all are. The method was originally proposed in the mid-1950's by Ford [14]. Moore [26] and Bellman [2] and is essentially as follows:

2.2.1 Algorithm for General Cost Matrix.
Let $l^k(x_i)$ be the label on vertex x_i at the end of the $(k + 1)$st iteration.

Initialization
Step 1. Set $S = \Gamma(s)$, $k = 1$, $l^1(s) = 0$, $l^1(x_i) = c(s, x_i)$ for all $x_i \in \Gamma(s)$, and $l^1(x_i) = \infty$ for all other x_i.

Updating of labels
Step 2. For every vertex $x_i \in \Gamma(s)$, $(x_i \neq s)$, update its label according to the expression:

$$l^{k+1}(x_i) = \min \left[l^k(x_i), \min_{x_j \in T_i} \{l^k(x_j) + c(x_j, x_i)\} \right] \tag{8.3}$$

where $T_i = \Gamma^{-1}(x_i) \cap S$. (The set S now contains all vertices whose currently shortest paths from s are of cardinality k).

The set T_i contains those vertices for which the currently shortest paths from s are of cardinality k, (i.e. those vertices in S), and for which an arc to vertex x_i exists. Note that if $x_i \notin \Gamma(S)$, the shortest path from s to x_i cannot possibly be of cardinality $k + 1$ and no change to the label of x_i is necessary.

For those vertices $x_i \notin \Gamma(S)$ set $l^{k+1}(x_i) = l^k(x_i)$.

Termination test
Step 3. (a) If $k \leqslant n - 1$ and $l^{k+1}(x_i) = l^k(x_i)$ for all x_i, then the optimal answer has been obtained and the labels are the lengths of the shortest paths. Stop.

(b) If $k < n - 1$ but $l^{k+1}(x_i) \neq l^k(x_i)$ for some x_i, then goto step 4.

(c) If $k = n - 1$ and $l^{k+1}(x_i) \neq l^k(x_i)$ for some x_i, then a negative cost circuit exists in the graph and the problem has no solution. Stop.

Preliminaries for next iteration
Step 4. Update the set S as:

$$S = \{x_i \mid l^{k+1}(x_i) \neq l^k(x_i)\} \tag{8.4}$$

(The set S now contains all vertices whose currently shortest paths from s are of cardinality $k + 1$.)

Step 5. Set $k = k + 1$ and goto step 2.

Once the shortest path lengths from s to every other vertex are obtained, it is again an easy matter to find the paths themselves by repeated application of eqn (8.2). Alternatively, the paths could be obtained immediately if, in addition to the $l^k(x_i)$ labels, another label $\theta^k(x_i)$ is stored for each vertex during the computations, where $\theta^k(x_i)$ is the vertex just before vertex x_i on the shortest path from s to x_i during the kth iteration. One could start with $\theta^1(x_i) = s$ $\forall x_i \in \Gamma(s)$ and $\theta^1(x_i) =$ arbitrary (say 0) for all other x_i. The

$\theta^k(x_i)$ labels could then be updated after eqn (8.3) so that $\theta^{k+1}(x_i) = \theta^k(x_i)$ if the first term in the square brackets of eqn (8.3) is smallest; or $\theta^{k+1}(x_i) = x_j$ if the second term in the brackets is smallest. If $\theta(x_i)$ is the vector of the θ labels at the end of the algorithm, the shortest path from s to x_i is obtained in reverse order as $s, \ldots, \theta^3(x_i), \theta^2(x_i), \theta(x_i), x_i$: where $\theta^2(x_i)$ is written for $\theta(\theta(x_i))$ etc.

The proof that the answer produced by the above algorithm is indeed optimal is quite simple. This proof is not given here, but is based on the optimality principle of dynamic programming and the fact that if no optimal path of k arcs exists, then no optimal path of $k + 1$ arcs can exist either [4]. This algorithm can be used for the case of a non-negative cost matrix although in general it is much inferior to the Dijkstra algorithm. For the case of a completely connected graph of n vertices, the algorithm requires of the order of n^3 operations (additions and comparisons), as compared with order n^2 operations required by the Dijkstra method. Various improvements to the above algorithm which were proposed by Yen, reduce the computing effort by a factor of four although the dependence on the number of vertices in the graph is still cubic.

2.2.2 EXAMPLE. Consider the graph of Fig. 8.3 where once more nondirected links are to be considered as two equal cost arcs in opposite directions. The arc costs are shown next to the arcs and consist of both positive and negative

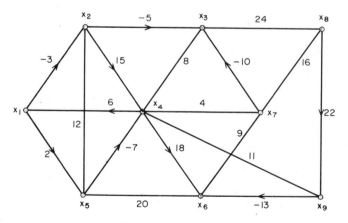

FIG. 8.3. Graph for example 2.2.2

numbers. It is required to find the shortest paths from x_1 to all other vertices, (provided that the graph does not contain any negative cost circuits) or indicate that negative cost circuits exist if they do.

The algorithm proceeds as follows:

Initialization

Step 1: $s = x_1$, $S = \{x_2, x_5\}$, $l^1(x_1) = 0$, $l^1(x_2) = -3$, $l^1(x_5) = 2$, $l^1(x_i) = \infty$ for all other x_i. Set $k = 1$.

First iteration

Step 2: $\Gamma(S) = \{x_2, x_3, x_4, x_5, x_6\}$. Therefore:

for x_2: $T_2 = \{x_1, x_5\} \cap \{x_2, x_5\} = \{x_5\}$ and from eqn (8.3):

$$l^2(x_2) = \min\left[-3, \underbrace{\{l^1(x_5) + c(x_5, x_2)\}}_{x_j = x_5}\right]$$

$$= \min[-3, (2 + 12)]$$

$$= -3$$

for x_3: $T_3 = \{x_2, x_4, x_7, x_8\} \cap \{x_2, x_5\} = \{x_2\}$,

$$l^2(x_3) = \min[\infty, \underbrace{(-3 - 5)}_{x_j = x_2}] = -8$$

for x_4: $T_4 = \{x_2, x_3, x_5, x_7, x_9\} \cap \{x_2, x_5\} = \{x_2, x_5\}$

$$l^2(x_4) = \min[\infty, \min\{\underbrace{(-3 + 15)}_{x_j = x_2}, \underbrace{(2 - 7)}_{x_j = x_5}\}] = -5$$

for x_5: $T_5 = \{x_1, x_2, x_6\} \cap \{x_2, x_5\} = \{x_2\}$

$$l^2(x_5) = \min[2, \underbrace{(-3 + 12)}_{x_j = x_2}] = 2$$

for x_6: $T_6 = \{x_4, x_5, x_7, x_9\} \cap \{x_2, x_5\} = \{x_5\}$

$$l^2(x_6) = \min[\infty, \underbrace{(2 + 20)}_{x_j = x_5}] = 22$$

The labels $l^2(x_i)$ are now: $[0, -3, -8, -5, 2, 22, \infty, \infty, \infty]$ for $x_i = x_1$, $x_2, \ldots, x_9$ respectively.

Step 3(b): Goto step 4.

Step 4: $S = \{x_3, x_4, x_6\}$

Step 5: $k = 2$, goto step 2.

Second iteration

Step 2: $\qquad \Gamma(S) = \{x_1, x_3, x_4, x_5, x_6, x_7, x_8, x_9\};$

for x_3: $\qquad T_3 = \{x_2, x_4, x_7, x_8\} \cap \{x_3, x_4, x_6\} = \{x_4\}$

$$l^3(x_3) = \min [-8, \underbrace{(-5 + 8)}_{x_j = x_4}] = -8$$

for x_4: $\qquad T_4 = \{x_2, x_3, x_5, x_7, x_9\} \cap \{x_3, x_4, x_6\} = \{x_3\}$

$$l^3(x_4) = \min [-5, \underbrace{(-8 + 8)}_{x_j = x_3}] = -5$$

for x_5: $\qquad T_5 = \{x_1, x_2, x_6\} \cap \{x_3, x_4, x_6\} = \{x_6\}$

$$l^3(x_5) = \min [2, \underbrace{(22 + 20)}_{x_j = x_6}] = 2$$

for x_6: $\qquad T_6 = \{x_4, x_5, x_7, x_9\} \cap \{x_3, x_4, x_6\} = \{x_4\}$

$$l^3(x_6) = \min [22, \underbrace{(-5 + 18)}_{x_j = x_4}] = 13$$

for x_7: $\qquad T_7 = \{x_4, x_6, x_8\} \cap \{x_3, x_4, x_6\} = \{x_4, x_6\}$

$$l^3(x_7) = \min [\infty, \min \{\underbrace{(-5 + 4)}_{x_j = x_4}, \underbrace{(22 + 9)}_{x_j = x_6}\}] = -1$$

for x_8: $\qquad T_8 = \{x_3, x_7\} \cap \{x_3, x_4, x_6\} = \{x_3\}$

$$l^3(x_8) = \min [\infty, \underbrace{(-8 + 24)}_{x_j = x_3}] = 16$$

for x_9: $\qquad T_9 = \{x_4, x_8\} \cap \{x_3, x_4, x_6\} = \{x_4\}$

$$l^6(x_9) = \min [\infty, \underbrace{(-5 + 11)}_{x_j = x_4}] = 6$$

The labels $l^3(x_i)$ are now: $[0, -3, -8, -5, 2, 13, -1, 16, 6]$ for $x_i = x_1$, $x_2, \ldots, x_9$ respectively.

Step 3(b): Goto step 4.

Step 4: $S = \{x_6, x_7, x_8, x_9\}$

Step 5: $k = 3$, go to step 2.

ETC.

Continuing in this way we obtain the results shown in summary below.

Third iteration

Step 2:
$$\Gamma(S) = \{x_3, x_4, x_5, x_6, x_7, x_8, x_9\}$$
$$T_3 = \{x_7, x_8\}, \quad l^4(x_3) = -11$$
$$T_4 = \{x_7, x_9\}, \quad l^4(x_4) = -5$$
$$T_5 = \{x_6\}, \quad l^4(x_5) = 2$$
$$T_6 = \{x_7, x_9\}, \quad l^4(x_6) = -7$$
$$T_7 = \{x_6, x_8\}, \quad l^4(x_7) = -1$$
$$T_8 = \{x_7\}, \quad l^4(x_8) = 15$$
$$T_9 = \{x_8\}, \quad l^4(x_9) = 6$$

The vector of the labels $l^4(x_i)$ is therefore: $[0, -3, -11, -5, 2, -7, -1, 15, 6.]$

Step 4: $S = \{x_3, x_6, x_8\}$

Fourth iteration

Step 2:
$$\Gamma(S) = \{x_3, x_4, x_5, x_7, x_8, x_9\}$$
$$T_3 = \{x_8\}, \quad l^5(x_3) = -11$$
$$T_4 = \{x_3\}, \quad l^5(x_4) = -5$$
$$T_5 = \{x_6\}, \quad l^5(x_5) = 2$$
$$T_7 = \{x_6, x_8\}, \quad l^5(x_7) = -1$$
$$T_8 = \{x_3\}, \quad l^5(x_8) = 13$$
$$T_9 = \{x_8\}, l^5(x_9) = 6$$

The vector of labels $l^5(x_i)$ is therefore: $[0, -3, -11, -5, 2, -7, -1, 13, 6]$.

Step 4: $S = \{x_8\}$

Fifth iteration

Step 2:
$$\Gamma(S) = \{x_3, x_7, x_9\}$$
$$T_3 = \{x_8\}, \quad l^6(x_3) = -11$$
$$T_7 = \{x_8\}, \quad l^6(x_7) = -1$$
$$T_9 = \{x_9\}, \quad l^6(x_9) = 6$$

Step 3(a): Stop.

The vector of labels $l^6(x_i)$ is the same as $l^5(x_i)$ and hence these labels are the shortest path lengths. The actual paths themselves are obtained from an iterative equation similar to (8.2) and the resulting x_1-base is shown by the heavy lines in Fig. 8.4.

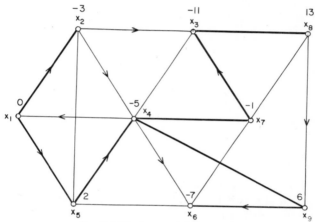

FIG. 8.4. Final vertex labels and the x_1-base

3. The Shortest Paths Between all Pairs of Vertices

When the shortest paths between all pairs of vertices of a graph are required, an obvious way for obtaining the answers is to apply the algorithms of the last section n times, each time with a different vertex as the starting vertex s. For the case of a complete graph with a non-negative cost matrix C, the resulting calculation time would be proportional to n^3 whereas for a general cost matrix it would be proportional to n^4. This last relationship, would rule out the possibility of solving large scale shortest path problems of this type by repeated application of the algorithm in Section 2.2.

In this section we describe a completely different approach to the problem of finding the shortest paths between all pairs of vertices. The method given applies to both non-negative and quite general cost matrices and requires a computation time proportional to n^3. The method, when applied to graphs with non-negative cost matrices, is in general about 50% faster, [11], than the application of the Dijkstra algorithm n times. This procedure was described for the first time by Floyd [13] and elaborated upon by Murchland [27]. It is based on a sequence of n transformations (iterations) of the initial cost matrix C, so that at the kth iteration, the matrix represents shortest path distances between every pair of vertices with the restriction that the path between x_i and x_j (for any x_i and x_j) contains only vertices from the restricted set $\{x_1, x_2, \ldots, x_k\}$ as intermediates.

3.1 Floyd's algorithm (for general cost matrices)

We will assume that the cost matrix has been initialized so that $c_{ii} = 0$ for all $i = 1, 2, \ldots, n$, and $c_{ij} = \infty$ whenever arc (x_i, x_j) is not in the graph G.

Initialization
Step 1. Set $k = 0$

An iteration
Step 2. $k = k + 1$

Step 3. For all $i \neq k$ such that $c_{ik} \neq \infty$ and all $j \neq k$ such that $c_{kj} \neq \infty$, perform the operation:

$$c_{ij} = \min\left[c_{ij}, (c_{ik} + c_{kj})\right] \tag{8.5}$$

Termination test
Step 4. (a) If any $c_{ii} < 0$, then a negative cost circuit containing vertex x_i exists in G, and no solution is possible. Stop.

(b) If all $c_{ii} \geqslant 0$, and $k = n$, the solution has been reached, and $[c_{ij}]$ gives the lengths of all the shortest paths, Stop.

(c) If all $c_{ii} \geqslant 0$ but $k < n$, return to step 2 and continue.

The proof of optimality of the answers obtained by this algorithm is quite simple [21], [27] and will be left for the reader. The basic operation of eqn (8.5) in the algorithm described above is called the *triple operation* and has wider application to problems of a nature similar to that of the shortest path problem. Such problems are discussed in later sections.

The shortest paths themselves can once more be obtained from the shortest path lengths using a recursive relation similar to that of eqn (8.2). Alternatively, a bookkeeping mechanism suggested by Hu [21] could be used to record (concurrently with the shortest path lengths) information about the paths themselves. This last method is similar to that used in Section 2.2.1 and is especially useful in cases where what is required is to find negative cost circuits in the graphs, if any such circuits exist. The technique involves the storage and updating of a second $n \times n$ matrix $\Theta = [\theta_{ij}]$ in addition to the cost matrix C. The entry θ_{ij} implies that θ_{ij} is the vertex just before vertex x_j on the shortest path from x_i to x_j. The matrix Θ is initialized so that $\theta_{ij} = x_i$ for all x_i and x_j.

Following eqn (8.5) in step 3 of the algorithm one would then introduce the updating of matrix Θ as follows:

$$\theta_{ij} = \begin{cases} \theta_{kj}, & \text{if } (c_{ik} + c_{kj}) < c_{ij} \text{ in the square brackets of eqn (8.5)} \\ \text{unchanged}, & \text{if } c_{ij} \leqslant (c_{ik} + c_{kj}). \end{cases}$$

At the end of the algorithm the shortest paths can be obtained immediately from the final Θ matrix. Thus, if the shortest path between any two vertices x_i and x_j is required this path is given by the vertex sequence:

$$x_i, x_v, \ldots, x_\gamma, x_\beta, x_\alpha, x_j$$

where $x_\alpha = \theta_{ij}$, $x_\beta = \theta_{i\alpha}$, $x_\gamma = \theta_{i\beta}$ etc. until finally $x_i = \theta_{iv}$.

It should perhaps the pointed out here that had all c_{ii} been initialized to ∞, (instead of at 0), at the start of the algorithm, then the final values of c_{ii} would be the cost of the shortest circuit through vertex x_i. It is also easy to see that from the state of the matrix Θ at the iteration during which an entry c_{ii} becomes negative, one can identify the negative cost circuit corresponding to that c_{ii}. This facility is used in Chapter 11, to perform a basic step in an algorithm for finding minimum cost flows in graphs, and is useful in many other applications as mentioned in the next section.

4. The Detection of Negative Cost Circuits

The problem of detecting negative cost circuits in a general graph is important both as a problem in its own right and as a basic step in larger algorithms (see Section 4.1 below and also Chapter 11). This section discusses the problem in somewhat greater detail than that given in the last section.

It was explained in Section 3.1 that Floyd's algorithm for finding shortest paths between every pair of vertices could be used to detect negative cost circuits in a graph. Additionally, in cases where the graph contains a vertex s from which all other vertices of the graph are reachable, then the algorithm of Section 2.2.1 (for finding all shortest paths from a given vertex s) could also be used to detect negative cost circuits as indicated by step 3(c) of that section. If not all vertices of the graph G can be reached from s (for example if G is a nondirected graph composed of two or more connected components) then the algorithm of Section 2.2.1 would terminate (as of course it should) with finite labels only on the vertices of the component containing s, and ∞ for the labels on the vertices in the other components. In this case any negative cost circuits which may exist in these other components would go undetected. However, in many important applications of the negative cost circuit algorithm (see for example Chapter 11, Section 5), a vertex s reaching all other vertices of G is readily available and the algorithm of Section 2.2.1 is—in such cases—computationally superior to Floyd's algorithm in detecting negative cost circuits.

The termination rules of the algorithm of Section 2.2.1 are described so as to minimize the computations involved in finding shortest paths from s when no negative cost circuits exist in the graph. These rules could be modified as follows to detect negative cost circuits sooner.

After every relabelling of a vertex x_i from a particular vertex x_{j*} according to eqn (8.3) one could check if x_i is on the currently shortest path (which could be derived from the current θ labels), from s to x_{j*}. If so, it implies that x_{j*} has been labelled via x_i and the fact that now $l^k(x_{j*}) + c(x_{j*}, x_i) < l^k(x_i)$ means that the part of the current shortest path from x_i to x_{j*} plus the arc (x_{j*}, x_i) must form a negative cost circuit and the algorithm can terminate. If, on the other hand, the label on vertex x_i is either unchanged by eqn (8.3), or receives a new label from a vertex x_{j*} but x_i is not on the current shortest path from s to x_{j*}, then the algorithm can continue to step 3(a) as before. One should now note that the above modification renders step 3(c) of the algorithm in Section 2.2.1 redundant, since a negative cost circuit is now identified as soon as it is generated rather than at the end of the procedure.

4.1 Optimal circuits in doubly-weighted graphs [24, 9]

A problem which arises in a variety of contexts, involves a graph G whose arcs (x_i, x_j) are weighted by two numbers; a cost c_{ij} and another number b_{ij} (say). The problem is then to find a circuit Φ for which the objective function

$$z(\Phi) = \frac{\sum\limits_{(x_i, x_j) \in \Phi} c_{ij}}{\sum\limits_{(x_i, x_j) \in \Phi} b_{ij}}$$

is a minimum (or maximum).

For example, consider the problem of operating a ship or airplane on a route network and suppose that c_{ij} is the "profit" and b_{ij} is the "time" necessary to traverse an arc (x_i, x_j). The problem of finding which route-cycle to operate in order to maximize the rate of profit is then a problem of the form just described. A more realistic problem of this type involving vehicle capacities etc. is given in [9].

Other problems which can be formulated as problems of finding optimal circuits in doubly-weighted graphs are: The scheduling of parallel computations [31] and interference in industrial processes [6].

The problem of finding a circuit Φ in a doubly weighted graph for which the ratio $z^*(\Phi)$ is a minimum, can be solved using an algorithm for detecting negative cost circuits in graphs as follows. Let us assume that the weights c_{ij} and b_{ij} are arbitrary real numbers (positive, negative or zero) but are subject to the restriction that $\sum\limits_{(x_i, x_j) \in \Phi} b_{ij} > 0$, for all circuits Φ of G. (In most practical situations, e.g. those mentioned above, $c_{ij} \gtrless 0$ but $b_{ij} \geqslant 0$ for all i and j).

Let us choose a trial value z^k for the objective function $z(\Phi)$ and consider the graph G with modified costs:

$$c_{ij}^k = c_{ij} - z^k b_{ij}$$

An attempt to determine negative cost circuits in G under the cost matrix $[c_{ij}^k]$ can lead to one of three possible results.

A. There exists a negative cost circuit Φ^- for which

$$\sum_{(x_i,\, x_j) \in \Phi^-} c_{ij}^k < 0.$$

B. There exists no negative cost circuit and

$$\sum_{(x_i,\, x_j) \in \Phi} c_{ij}^k > 0 \text{ for all circuits } \Phi.$$

C. Thrre exists a zero cost circuit, (but no negative cost one) i.e.

$$\sum_{(x_i,\, x_j) \in \Phi^o} c_{ij}^k = 0 \text{ for some circuit(s) } \Phi^o.$$

In case A we can say that z^* (the minimum value of z) is less than z^k since:

$$\sum_{(x_i.\, x_j) \in \Phi^-} c_{ij}^k \equiv \sum_{(x_i,\, x_j) \in \Phi^-} c_{ij} - z^k \sum_{(x_i,\, x_j) \in \Phi^-} b_{ij} < 0$$

can only be true if:

$$\frac{\sum\limits_{(x_i,\, x_j) \in \Phi^-} c_{ij}}{\sum\limits_{(x_i,\, x_j) \in \Phi^-} b_{ij}} < z_k$$

which obviously implies that z^* must also be less than z^k.

Similarly, in case B we can say that $z^* > z^k$ and in case C that $z^* = z^k$.

Thus, a binary search procedure immediately suggests itself as follows: Start with an initial try z^1; if it is too large (i.e. case A applies) try $z^2 < z^1$; if it is too small (i.e. case B applies) try $z^2 > z^1$. As soon as upper and lower bounds (z_u and z_l respectively) on the value z^* are established, proceed by trying $z^k = (z_u + z_l)/2$ and replacing z_u by z^k if case A results, or replacing z_l by z^k if case B results from this kth trial. Since the number of trials is proportional to the number of significant digits of accuracy required—i.e. proportional to $\log 1/\eta$, where η is the fraction of uncertainty—and since each trial (negative circuit determination or calculation of complete distance matrix) requires order n^3 operations, the solution to the above double weighted problem requires $O[n^3 \log 1/\eta]$ operations.

5. The K Shortest Paths Between Two Specified Vertices

In Section 2 methods were given for finding the shortest path from s to t in a general graph G. In many practical applications however, what may be required is that shortest path having some particular specified attribute. This problem may, of course, be treated either as a shortest path problem with the attribute specified in constraints [4], or as a multi-objective problem in which the objective is some combination of path length and the particular path attribute. However. this complication will, in general, greatly increase the computational effort involved, and a much simpler practical alternative is simply to list the K shortest paths from s to t and then choose from this list that path with the required particular attribute. Although this method is not equivalent to considering the attribute directly (an example of this is given in Section 7 later in this chapter), it is applicable even in cases where the attribute is only loosely defined or even subjective in nature. The method presupposes that the K shortest paths between s and t, of a graph G can be calculated reasonably efficiently, and this is the subject matter of the present section.

We will assume here that only elementary paths are to be considered. Thus, although the shortest path must, by necessity (assuming that the graph contains no negative cost circuits), be elementary; the second, third, etc. shortest paths need not be so-even in cases where all the c_{ij} are positive. The problem of finding the K shortest paths without requiring them to be elementary is much simpler and iterative methods of solution similar to the ones given earlier for the shortest paths have been described by Hoffman and Pavley [20], Sakarovitch [32], Bellman and Kalaba [3] and others. However, modification of these methods to produce elementary paths is not at all easy, and since almost all practical applications of a K shortest path algorithm require paths to be elementary, we will restrict ourselves to describing a method due to Yen [35] which produces the K shortest *elementary* paths.

Let $P^k = s,\ x_2^k,\ x_3^k,\ \ldots,\ x_{q_k}^k,\ t$; be the kth shortest path from s to t where $x_2^k, x_3^k, \ldots, x_{q_k}^k$ are respectively the 2nd, 3rd, $\ldots, q_k$th vertex on the kth shortest path. Also let P_i^k be a "deviation from path P^{k-1} at point i". By this is meant that P_i^k is the shortest of the paths that coincide with P^{k-1} from s up to the ith vertex and then deviate to a vertex that is different from any of the $(i+1)$st vertices of those (previously generated), jth shortest paths P^j ($j = 1, 2, \ldots, k-1$) that have the same initial subpaths from s to the ith vertex as does P^{k-1}. P_i^k finally reaches t by a shortest subpath not passing through any of the vertices $s, x_2^{k-1}, x_3^{k-1}, \ldots, x_i^{k-1}$ which formed the first part of P_i^k. It should thus be noted here that P_i^k is, by necessity, elementary.

The first subpath $s, x_2^k, x_3^k, \ldots, x_i^k$ (which is the same as $s, x_2^{k-1}, x_3^{k-1}, \ldots, x_i^{k-1}$) of P_i^k is called its root R_i^k and the second subpath $x_i^k, \ldots, t$ of P_i^k is called its spur S_i^k.

The algorithm starts by finding P^1 using a suitable s to t shortest path algorithm as described in Section 2. This path is placed in a list L_o (which is to contain the k-shortest paths). In general in order to find P^k the shortest paths $P^1, P^2, \ldots, P^{k-1}$ must have already been determined. A description of the algorithm is given below.

5.1 Description of the algorithm

Initialization

Step 1. Find P^1. Set $k = 2$. If there is only one shortest path P^1 enter it into list L_o and goto step 2. If there is more than one but less than K enter one into L_o and the rest into another list L_1. Goto step 2. If there are K or more shortest paths P^1 the problem is finished Stop.

Find all deviations

Step 2. Find all deviations P_i^k of the $(k - 1)$-shortest path P^{k-1} for all $i = 1$, $2, \ldots, q_{k-1}$ by executing steps 3 to 6 for each i.

Step 3. Check if the subpath consisting of the first i vertices of P^{k-1} coincides with the subpath consisting of the first i vertices of any $P^j (j = 1, 2, \ldots, k - 1)$. If so set $c(x_i^{k-1}, x_{i+1}^j) = \infty$, otherwise make no changes. (During the algorithm x_1 will refer to vertex s.) Goto step 4.

Step 4. Use a shortest path algorithm to find the shortest path, S_i^k, from x_i^{k-1} to t—excluding from consideration vertices $s, x_2^{k-1}, x_3^{k-1}, \ldots, x_i^{k-1}$. If there are more than one shortest paths take any one and denote it S_i^k.

Step 5. Form P_i^k by joining $R_i^k (\equiv s, x_2^{k-1}, x_3^{k-1}, \ldots, x_i^{k-1})$ with S_i^k and place P_i^k into list L_1.

Step 6. Replace the elements of the cost matrix changed at step 3 with their initial values and return to step 3.

Choose shortest deviation

Step 7. Find the shortest path in list L_1. Denote this path P^k and move it from L_1 to L_0. If $k = K$ the algorithm is finished and L_0 is the required list of K shortest paths. If $k < K$ set $k = k + 1$ and return to step 2.

If more than one (say h) shortest paths exist in L_1 choose an arbitrary one to transfer to L_0 and continue as above, unless h plus the number of paths already in L_0 is K or more in which case the algorithm is finished.

The justification of the above algorithm follows immediately from the obvious fact [11], [20] that P^k must be a deviation at the ith step (for some $i \geqslant 1$), from one of the shorter paths $P^1, P^2, \ldots, P^{k-1}$. Therefore, all that is necessary is to generate all shortest deviations from each P^j and scan the list in order to find the shortest one, this being P^k. It should be noted, that at the kth iteration all shortest deviations for the $P^j, j = 1, 2, \ldots, k - 2$, already exist in L_1 so that only the deviations from P^{k-1} are needed in order to complete the list.

The reason for setting the costs $c(x_i^{k-1}, x_{i+1}^j) = \infty$ at step 3, for those P^j whose first subpath of i vertices coincides with the first i-vertex subpath of P^{k-1}, is to avoid regenerating P^j as a deviation (at point i) of P^{k-1} which would otherwise be the case since the cost of path $P^j \leqslant$ the cost of P^{k-1} if $j < k - 1$.

Although in step 5 of the algorithm every generated P_i^k is placed in list L_1, it is quite apparent that this list need not contain more than the $K - k + 1$ shortest P_i^k's at the kth iteration. The computationally most expensive step of the algorithm is step 4 requiring $O(n^2)$ or $O(n^3)$ operations per execution depending on whether all $c_{ij} \geqslant 0$ or whether $c_{ij} \gtrless 0$. Since this step must be executed q_k times at iteration k and since $q_k \propto n$ and the number of iterations is K, the algorithm requires of the order of Kn^3 or Kn^4 operations to find the K shortest paths in graphs with non-negative and general cost matrices respectively.†

6. The Shortest Path Between Two Specified Vertices in the Special Case of a Directed Acyclic Graph

The methods given in the previous sections of this chapter could be applied to quite general graphs. However, a class of graphs which often arises in practical situations that require the solution of shortest-path type problems, is the class of directed acyclic graphs. These graphs appear in PERT (Project Evaluation Research Task), and CPM (Critical Path Method), diagrams as follows.

Suppose that a large project is to be embarked upon and this project is made up of a large number of activities. We can represent each activity by a vertex of a graph and draw an arc from vertex x_i to vertex x_j to indicate that activity i must precede activity j. With each arc we associate a cost c_{ij} representing the minimum delay in time that is necessary between the beginning of activity i and the beginning of activity j. Thus, let us say that the whole project is the construction of a building; activity i may, for example, be the building of the walls, activity j the placement of the window frames, activity k the placement of the wiring ducts on the wall etc. Obviously in this example there would be arcs from x_i to x_j and also from x_i to x_k. The minimum time delay c_{ij} between commencing the building of the walls and the placement of the window frames may, however, well be different from the delay c_{ik} between the building of the walls and the ducting. If, for example, the window frames are wooden and the walls need to be dry before these are placed whereas the same does not hold true for the placement of the wiring ducts, then $c_{ij} > c_{ik}$.

† It is easy to see that all shortest paths from s to t remain the same relative to each other when all arc costs c_{ij} are replaced by $c'_{ij} = c_{ij} + h_i - h_j$, where the h's are arbitrary numbers associated with the vertices of the graph. With this observation D. Knuth (in a private communication) has pointed out that by taking $h_i = d(s, x_i)$ one always has $c'_{ij} \geqslant 0$. Hence, since the distances $d(s, x_i)$ can be obtained in $0(n^3)$ operations in the general case, it is possible—using this initial cost transformation—to calculate the K shortest paths in $0(Kn^3)$ operations for the general cost case-

Such a graph is, quite obviously, both directed and acyclic since the existence of a circuit containing any vertex x_i would lead to the illogical result that there is a positive time delay (equal to the cycle length) between the beginning of activity i and itself.

What is required is to find the minimum time necessary to complete the project, i.e. to find the *longest path* in the graph (constructed as mentioned above) between a vertex s representing the beginning and another vertex t representing the end. This longest path is called a *critical path* since the activities lying on this path are the ones which determine the overall completion time for the project, and any delay in starting any activity on the critical path will be reflected by a delay in the whole project.

The similarity of this problem to the shortest path problem between s and t is quite obvious, and one could, for example, solve the present problem by using the algorithm of Section 2.1.1 (since all $c_{ij} \geqslant 0$), and replacing all its *min* operations with *max*. However, because of the special structure of the graph this would be a very wasteful method and the following algorithm can be used instead in order to find the longest path. The algorithm is described in terms of the longest path since this is the form that is most often required in practice, although the shortest path path problem in a directed acyclic graph can also be solved by the same algorithm when all *max* operations are replaced by *min*.

6.1 Algorithm for finding the longest (critical) path in a directed acyclic graph

It is assumed here that the vertices are numbered in such a way so that an arc (x_i, x_j) is always directed from a vertex numbered x_i to a higher numbered vertex x_j. For an acyclic graph, this numbering is, quite obviously, always possible and very easy to achieve. The initial vertex is then numbered 1 and the final vertex numbered n.

To label vertex x_j with $l(x_j)$—the longest path from 1 to x_j—perform the operation:

$$l(x_j) = \max_{x_i \in \Gamma^{-1}(x_j)} [l(x_i) + c_{ij}] \tag{8.7}$$

then continue to label vertex $(x_j + 1)$ using eqn (8.7), and so on until the final vertex n is labelled by $l(n)$. In eqn (8.7) $l(1)$ is initially set to zero. When labelling vertex x_j the labels $l(x_i)$ are all known for the vertices $x_i \in \Gamma^{-1}(x_j)$ since, according to the initial vertex numbering, this implies that $x_i < x_j$ and hence the x_i vertices have already been labelled by the algorithm.

The label $l(n)$ is the length of the longest path from 1 to n. The arcs forming the path itself may be found in the usual way by tracing backwards, so that arc (x_i, x_j) is on this path if and only if: $l(x_j) = l(x_i) + c_{ij}$. Starting with x_j equal to n, x_j is set at each step equal to the value of x_i (say x_i^*) satisfying this last equality until $x_i^* = 1$, i.e. the initial vertex has been reached.

The label $l(x_j)$ on any vertex x_j quite obviously represents the longest path from 1 to x_j, i.e. the earliest possible starting time of the activity represented by x_j. Thus, if (x_i, x_j) is an arc in the graph, then $l(x_j) - l(x_i) - c_{ij}$ (a nonnegative quantity as can be seen from eqn (8.7)) is the longest possible time by which activity i may be delayed without affecting the starting time of activity j. From what was said above, one should note that if activities i and j are two consecutive activities on the longest path, then $l(x_j) - l(x_i) - c_{ij} = 0$.

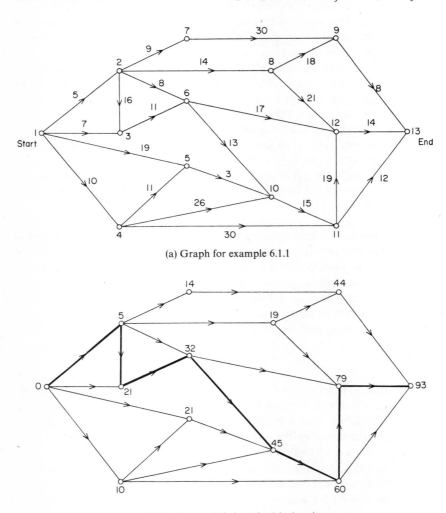

(a) Graph for example 6.1.1

(b) Final vertex labels and critical path

Fig. 8.5

6.1.1 EXAMPLE. Consider the PERT diagram shown in Fig. 8.5(a), where the numbers next to the arcs are the time durations c_{ij} in days. What is the shortest possible period for completion of this project? (Assume that vertex 13 is a dummy end vertex corresponding to an activity with zero time duration.) The vertices of the graph are already numbered so that an arc (x_i, x_j) exists only if $x_j > x_i$. Immediate application of eqn (8.7) to the graph of Fig. 8.5(a) results in the vertex labels shown in Fig. 8.5(b), and hence the critical path is that shown in heavy lines. The length of this path, and hence the shortest possible completion period for the project is 93 days.

If any one of the activities 1, 2, 3, 6, 10, 11, 12, 13 on the critical path is delayed by D days, the whole project will be delayed by D days. On the other hand an activity such as 8 (not on the critical path) can be delayed by up to $79 - 19 - 21 = 39$ days without any effect on the completion time of the entire project, since, only when a delay of $D > 39$ days occurs will the label of vertex 12 (which is on the critical path) be affected.

PERT–CPM has been described in this section in terms of the longest path problem in a graph where vertices represent activities and arcs the precedence relations with the costs c_{ij} on arc (x_i, x_j) as the time delay between the beginning of activity i and that of activity j. In practice the representation of PERT diagrams is somewhat different, with arcs being used to represent activities and vertices being abstract "events" signifying the beginning or end of an activity. Although such a representation is in itself incomplete (because precedence relationships can not necessarily be fully expressed in this way), this shortcoming is removed by the addition of dummy activities to express the correct precedence relations [34]. On the other hand, this last representation has one important advantage in the sense that, usually, a much simpler diagram results in practice. This is because the time delay c_{ij} between the beginnings of activities i and j is generally constant for a given i and quite independent of the following activity, with only a few exceptions which can always be treated by the addition of some dummy activities. In such a situation, the representation used in practice produces a simpler graph (even with the added dummies), whereas the representation used earlier in this section remains unchanged. Another difference in the practical use of PERT is that the time durations of the activities are considered to be stochastic variables rather than deterministic ones as was done here.

The project scheduling problem as treated by PERT–CPM does not include constraints as to the availability of resources and does not consider the resource requirements of the activities. Instead, the "resource levelling"— i.e. the smoothing of the resource requirements during the project—is done after a critical path is determined, by delaying non-critical activities by suitable amounts, within the limits explained earlier. If, on the other hand, resource constraints are explicitly included in the scheduling phase, the problem changes from the very easy one described above to a quite difficult problem [1, 30, 37], and which can only be solved (optimally) for sizes of 3 or 4 orders of magnitude lower than the unconstrained problem.

7. Problems Related to the Shortest Path

In Floyd's algorithm (see Section 3 of the present chapter), for calculating shortest paths between all pairs of vertices, use was made of the triple operation given by eqn (8.5). The application of this operation n times on the cost matrix $[c_{ij}]$, ensured that the final values in the matrix were the correct shortest path lengths. This triple operation, is a particular case of the more general triple operation:

$$z_{ij} = \text{OPT} \left[z_{ij}, z_{ik} \otimes z_{kj} \right] \tag{8.8}$$

where z is the path objective to be optimized (minimized or maximized), and "$\otimes$" is a general operation. The only restriction on the applicability of the triple operation to a variety of path finding problems in graphs, is imposed by the operator $\otimes$. Thus if a path from x_i to x_j passes through any intermediate vertex x_k, and z is the path characteristic on which the optimization is based, then $\otimes$ must satisfy the condition:

$$z_{ij} = z_{ik} \otimes z_{kj} \tag{8.9}$$

Some examples on the use of the triple operation are given below.

7.1 The most reliable path

In the shortest path problem the "length" of a path was taken as the sum of the costs of the arcs forming the path. Consider now the case where the arc "costs" represent arc reliabilities. The reliability of a path from s to t composed of the arcs in the set P is then given by:

$$\rho(P) = \prod_{(x_i, x_j) \in P} \rho_{ij} \tag{8.10}$$

where ρ_{ij} is the reliability of arc (x_i, x_j), i.e. the probability of it being existent; (or—in the case of a physical system— the probability of the corresponding entity being in good working order).

The problem of finding the most reliable path from s to t can be transformed into the s to t shortest path problem by taking the "cost" c_{ij} of an arc (x_i, x_j) to be $c_{ij} = -\log \rho_{ij}$. Since, taking logarithms of both sides of eqn (8.8), we have:

$$\log \rho(P) = \sum_{(x_i, x_j) \in P} \log \rho_{ij} = - \sum_{(x_i, x_j) \in P} c_{ij},$$

the shortest path from s to t under the cost matrix $[c_{ij}]$ will also be the most reliable path under the reliability matrix $[\rho_{ij}]$ and the reliability of this path is the antilogarithm of its length.

An alternative formulation of the problem of the most reliable path is by the direct application of the triple-operation where:

$$\rho_{ij} = \max \left[\rho_{ij}, \rho_{ik} \times \rho_{kj} \right]$$

is used as the operation. The initial starting matrix $[\rho_{ij}]$ is taken as the arc reliability matrix, where zero is entered when no arc exists. This formulation will obviously yield the most reliable paths between all pairs of vertices.

7.2 The largest-capacity path

In this problem each arc (x_i, x_j) of the graph has associated with it a capacity q_{ij}, and what is required is that path from s to t which has the largest capacity. The capacity of a path P is, of course, determined by that arc in P which has the smallest capacity, i.e.

$$Q(P) = \min_{(x_i, x_j) \in P} [q_{ij}] \tag{8.11}$$

THEOREM 1.† *The capacity of the largest-capacity path from s to t is equal to:*

$$\min_{K} \left\{ \max_{(x_i, x_j) \in K} [q_{ij}] \right\}$$

where K is any s-to-t cut set of arcs.

Proof. Let $\hat{Q}$ be the capacity of the largest-capacity path. Since every path from s to t must contain at least one arc from every s-to-t cut, every cut K satisfies the relation:

$$\max_{(x_i, x_j) \in K} [q_{ij}] \geqslant \hat{Q} \tag{8.12}$$

Moreover, since there must be at least one arc of capacity less than or equal to $\hat{Q}$ in every s-to-t path, and since taking one such arc from each s-to-t path forms an s-to-t cut (by definition), there exists at least one cut, K' say, for which:

$$\max_{(x_i, x_j) \in K'} [q_{ij}] \leqslant \hat{Q} \tag{8.13}$$

Thus, the cut $\hat{K}$ which produces the minimum of the expression

$$\min_{K} \left\{ \max_{(x_i, x_j) \in K} [q_{ij}] \right\} \tag{8.14}$$

---

† One should note the similarity of theorem 1 with the maximum flow/minimum cut theorem of Chapter 11; it could in fact be considered as a *minimax* version of this latter theorem.

must satisfy inequalities (8.12) and (8.13) simultaneously, i.e. equality must apply; hence the theorem.

An obvious way of finding the largest capacity path based on theorem 1 is as follows:

Step 1. Start with the s-to-t cut $\overline{K} \equiv (\{s\}, X - \{s\})$, and find the largest capacity $\overline{Q}$ of any arc in $\overline{K}$.

Step 2. Form the partial graph $G' = (X, A')$ where $A' = \{(x_i, x_j) | q_{ij} \geq \overline{Q}\}$.

Step 3. Find the reachable set $R'(s)$ of vertices reachable from s via arcs in A'. (See Chapter 2 for the computation of $R'(s)$.)

Step 4. If $t \in R'(s)$, $\hat{Q} = \overline{Q}$ and any s-to-t path in the partial graph G' will have the largest capacity possible, i.e. $\hat{Q}$. If $t \notin R'(s)$, goto step 5.

Step 5. Redefine the cut $\overline{K}$ to be $(R'(s), X - R'(s))$, and find the largest capacity $\overline{Q}$ of an arc in this new cut. (The current value of $\overline{Q}$ is less than the previous one by definition of the set $R'(s)$ at step 3 and the set of arcs A' at step 2.) Return to step 2.

In the event of the graph G being nondirected, an even simpler procedure for calculating the largest capacity path was given by Frank and Frisch [16], and is as follows.

Let K_1 be an s-to-t cut set of links, and let $\overline{Q}$ be the largest capacity of any link in K_1. If now any link (x_i, x_j) with capacity $q_{ij} \geq \overline{Q}$ is "shorted", i.e. vertices x_i and x_j are replaced by a single vertex x with

$$\Gamma(x) = \Gamma(x_i) \cup \Gamma(x_j), \quad \Gamma^{-1}(x) = \Gamma^{-1}(x_i) \cup \Gamma^{-1}(x_j)$$

and link (x_i, x_j) removed, then the resulting graph G_1 will have the same largest capacity path as the original graph G. The validity of this statement follows immediately from the fact that $\overline{Q}$ is an upper bound on $\hat{Q}$ (according to eqn 8.12) and hence links with $q_{ij} \geq \overline{Q}$ cannot possibly affect the optimal solution i.e. cannot be in the cut set $\hat{K}$. Thus, since by "shorting" a link (x_i, x_j) only cuts containing (x_i, x_j) are affected (eliminated), the cut set $\hat{K}$ of graph G (or all cuts corresponding to the minimum of eqn (8.14), if more than one such cuts exist), is also a cut to the transformed graph G_1.

What was done above for the original graph G can now be repeated for graph G_1 by choosing another s-to-t cut K_2 shorting all links of G_1 with capacity greater than or equal to the largest capacity of any link in K_2, thus producing graph G_2 etc. The process is finished when s and t are shorted. Every s-to-t path in the graph $\hat{G}$ formed by the vertices of G and those links which have been shorted during the procedure, now has the maximum possible capacity $\hat{Q}$. The restriction of the applicability of the algorithm to only nondirected graphs is introduced implicitly by the shorting process since it is assumed that a path of capacity greater than or equal to $\overline{Q}$ exists between any two or more vertices which have been replaced by a single vertex and this is not necessarily so if the shorted arcs are directed.

7.2.1 EXAMPLE. The problem is to find the largest-capacity s-to-t path (s) for the nondirected graph shown in Fig. 8.6(a), where the numbers next to the links refer to the link capacities.

Choosing an arbitrary s-to-t cut K_1 as shown dotted in Fig. 8.6(a), the maximum-capacity link in K_1 is (x_3, x_6) with capacity $q_{3.6} = 16$. Shorting all links with capacity $\geqslant 16$ produces the graph G_1 shown in Fig. 8.6(b). Let

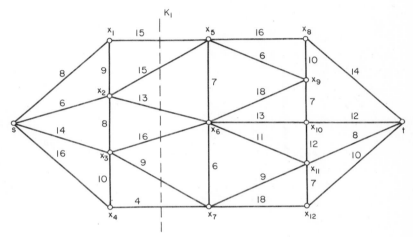

FIG. 8.6 (a). Graph for example 7.2.1

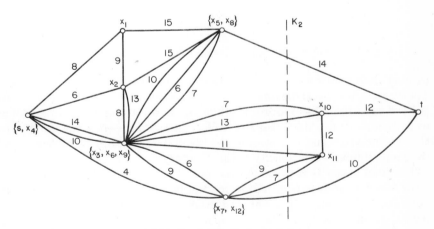

FIG. 8.6 (b). Graph after first contraction

G

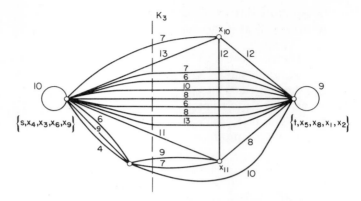

FIG. 8.6 (c). Graph after second contraction

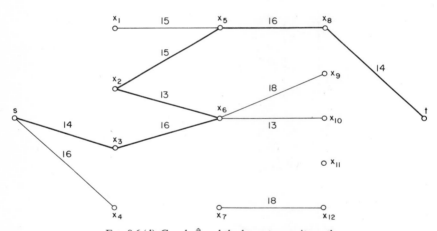

FIG. 8.6 (d). Graph $\hat{G}$ and the largest capacity path

us now choose an s to t cut K_2 of graph G_1 as shown dotted in Fig. 8.6(b). The largest capacity of the links in K_2 is 14 and hence we proceed to short all links of capacity ≥ 14 to form the graph G_2 shown in Fig. 8.6(c). Proceeding to pick K_3—shown dotted in Fig. 8.6(c)—as the next s-to-t cut, we obtain $\hat{Q} = 13$ and the s and t vertices in the resulting graph (formed from G_2 by shorting all links with capacity ≤ 13) are shorted.

The final graph $\hat{G}$ is therefore that partial graph of G in which only the links which have been shorted during the procedure are present; (i.e. only links (x_i, x_j) with $q_{ij} \geq 13$). This last graph is shown in Fig. 8.6(d) and it is seen that only one s to t largest-capacity path exists as indicated by the heavy lines; the critical link in this path being (x_2, x_6) of capacity 13.

If the largest-capacity paths are required between every pair of vertices in the graph, then the triple operation of eqn (8.8) can be used directly in the form:

$$q_{ij} = \max \left[q_{ij}, \min \{ q_{ik}, q_{kj} \} \right] \tag{8.15}$$

since the operator of eqn (8.15) satisfies the applicability conditions imposed by eqn (8.9). The starting matrix $[q_{ij}]$ is the initial arc capacity matrix (with zero entries where arcs do not exist), and the final $[q_{ij}]$ matrix will then give the capacity of paths between all pairs of vertices.

7.3 The path with the largest expected capacity

Consider a graph G in which every arc (x_i, x_j) has two numbers ρ_{ij} and q_{ij} associated with it representing the reliability and capacity of the arc respectively. The problem of finding the path from s to t with the greatest expected capacity, is then a combination of the last two path problems discussed under 7.1 and 7.2 above. If the expected capacity of a path P is written as $e(P)$, the problem is then to find that P which minimizes the expression:

$$e(P) = \prod_{(x_i,\, x_j) \in P} \rho_{ij} \cdot \left\{ \min_{(x_i,\, x_j) \in P} [q_{ij}] \right\} \tag{8.16}$$

In this case the triple operation cannot be used, since no operator $\otimes$ can be found which (for a path from x_a to x_b via vertex x_c) satisfies $e_{ab} = e_{ac} \otimes e_{cb}$. The non-existence of such an operator can be illustrated as follows:

$$e_{ab} = \prod_{(x_i,\, x_j) \in a \to b} \rho_{ij} \cdot \min_{(x_i,\, x_j) \in a \to b} [q_{ij}]$$

$$= \prod_{(x_i,\, x_j) \in a \to c} \rho_{ij} \cdot \prod_{(x_i,\, x_j) \in c \to b} \rho_{ij} \cdot \min \left\{ \min_{(x_i,\, x_j) \in a \to c} [q_{ij}], \min_{(x_i,\, x_j) \in c \to b} [q_{ij}] \right\}$$

where $a \to b$ etc. is written for the path from x_a to x_b and also for the set of arcs in this path.
Writing:

$$\bar{\rho}_{ac} = \prod_{(x_i,\, x_j) \in a \to c} \rho_{ij}, \qquad \bar{\rho}_{cb} = \prod_{(x_i,\, x_j) \in c \to b} \rho_{ij}$$

$$\bar{q}_{ac} = \min_{(x_i,\, x_j) \in a \to c} [q_{ij}], \qquad \bar{q}_{cb} = \min_{(x_i,\, x_j) \in c \to b} [q_{ij}]$$

and

$$e_{ac} = \bar{\rho}_{ac} \cdot \bar{q}_{ac}, \qquad e_{cb} = \bar{\rho}_{cb} \cdot \bar{q}_{cb}$$

we get:

$$e_{ab} = \min \left[\bar{\rho}_{ac} e_{cb}, \bar{\rho}_{cb} e_{ac} \right] \tag{8.17}$$

It is apparent from the form of this expression that since the path reliabilities $\bar{\rho}_{ac}$ and $\bar{\rho}_{cb}$ cannot both be written in terms of the path objectives e_{ac} and e_{cb} and the elementary arc properties ρ_{ij} and q_{ij}, it is not possible to find an operator to satisfy the general condition of eqn (8.9) and hence it is impossible to use the triple operation directly.

It can, in fact, be demonstrated that the optimal path from x_a to x_b via x_c is independent of:

(i)   the optimal paths from x_a to x_c and x_c to x_b
(ii) the largest-capacity paths from x_a to x_c and x_c to x_b
(iii) the greatest-reliability paths from x_a to x_c and x_c to x_b or any combination of these.

This can be seen from Fig. 8.7 which shows the above three types of paths from x_a to x_c and from x_c to x_b, the optimum value of e which can be derived from any combination of these paths being 4.20. The figure also shows a path from x_a to x_c and another from x_c to x_b neither of which are—in any of the above three ways—optimal between their respective terminal vertices, but whose totality is the optimal path from x_a to x_b and has a value of 4.48.

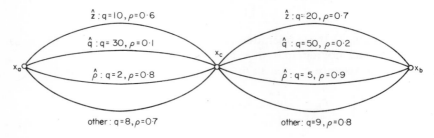

FIG. 8.7

7.3.1 A METHOD OF FINDING THE LARGEST EXPECTED CAPACITY PATH. An iterative method of finding the largest expected capacity path from a specified vertex s to another vertex t of a graph $G = (X, A)$ will now be given. The method is based on progressively eliminating arcs from the graph when these cannot possibly form part of the optimal path, until the graph becomes disconnected.

Begin by finding the path $P_{\hat{\rho}}$ of greatest reliability between s and t. Let $Q_{\hat{\rho}}$ be the capacity of this path, i.e.

$$Q_{\hat{\rho}} = \min_{(x_i, x_j) \in P_{\hat{\rho}}} [q_{ij}]$$

where $P_{\hat{\rho}}$ will also be used for the set of arcs forming the path. If the set of arcs $A_0 \equiv \{(x_i, x_j) | (x_i, x_j) \in A, \ q_{ij} \leqslant Q_{\hat{\rho}}\}$ are removed from A to form the set

$A' = A - A_0$, then the graph $G' = (X, A')$ which is a partial graph of G, either contains the optimal path of G, or $P_{\hat{\rho}}$ is the optimal path. This can be seen as follows:

The optimal path of G must have a reliability less than or equal to that of $P_{\hat{\rho}}$ by definition. Hence it its value (expected capacity) is to be greater than the value of $P_{\hat{\rho}}$, the path must have a capacity greater than $Q_{\hat{\rho}}$ and therefore cannot use any of the arcs in A_0.

One can now find the maximum reliability path, $P'_{\hat{\rho}}$, of the graph G'. The capacity of this path, $Q'_{\hat{\rho}}$, is greater than $Q_{\hat{\rho}}$ but its reliability is less than or equal to the reliability of $P_{\hat{\rho}}$. If the value of $P'_{\hat{\rho}}$ is greater than that of $P_{\hat{\rho}}$, $P'_{\hat{\rho}}$ is kept as the best path so far otherwise it is rejected. Once more the set of arcs

$$A_1 = \{(x_i, x_j) | (x_i, x_j) \in A', q_{ij} \leqslant Q'\}$$

can be removed from A' to yield $A'' = A' - A_1$ and the partial graph $G'' = (X, A'')$. By the same argument as above either G'' contains the optimal path or the best path so far is the optimal one. This procedure is continued until the partial graph becomes G^l which:

either: becomes disconnected (no path exists) between s and t

or: the reliability of $P^l_{\hat{\rho}}$ times the capacity of the largest capacity path in G (which can be calculated just once at the beginning), is less than the value of the best path so far in which case, quite obviously, no better solution can be obtained by any partial graph of G^l.

At the end of the procedure the best path so far is the optimal path.

Two factors affect the efficiency of the above method.

First, is the rate with which arcs are removed from the graph. In the worst case, when the most reliable path uses the smallest capacity arc at every stage, $m-k$ stages (of computing the best reliability path) would be needed, where m is the number of arcs in G and k is the number of arcs in the optimal path. The best case occurs when the first partial graph G' is disconnected and only one calculation is necessary.

The second factor is the method used to recalculate the most reliable path in the partial graphs at each stage. In this respect it should be noted that although the removal of arcs may, in general, eliminate several arcs from the s-base of the previous graph (giving the most reliable paths from s to all the other vertices), that part of the s-base containing s together with its vertex labels remain unchanged, and therefore need not be recalculated.

7.3.2 EXAMPLE. The problem is to find the s to t path of largest expected capacity in the graph shown in Fig. 8.8, where each arc is labelled (a, b), a being its capacity and b its reliability.

The greatest reliability path $P_{\hat{\rho}}$ in this graph is shown by the heavy lines

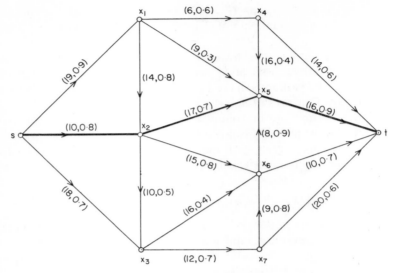

FIG. 8.8. Graph for example 7.3.2

in Fig. 8.8 and $\hat{\rho} = 0.504$, $Q_{\hat{\rho}} = 10$ and hence the expected capacity of this path is $e = 5.04$. Removing from G all arcs with capacity $\leqslant 10$, we obtain the graph G' as shown in Fig. 8.9(a). The greatest reliability path $P'_{\hat{\rho}}$ in this graph is shown in heavy lines in the above figure, and for this path $\hat{\rho} = 0.454$,

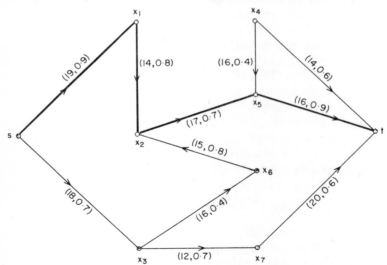

FIG. 8.9 (a). Graph G' for example 7.3.2

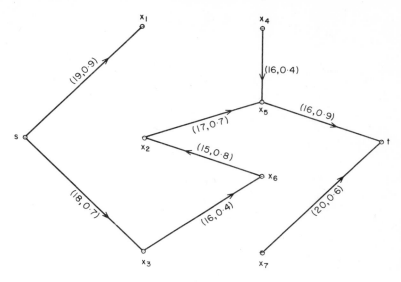

FIG. 8.9 (b). Graph G'' for example 7.3.2

$Q'_\rho = 14$ and hence the expected capacity of this path is $e = 14 \times 0{\cdot}454 = 6{\cdot}36$ which is better than the previous best (5·04) and hence replaces it. Removing from G' all arcs with capacity $\leqslant 14$, we obtain G'' as shown in Fig. 8.9(b). There is only one path in this graph with $\hat{\rho} = 0{\cdot}141$, $Q''_\rho = 15$ and hence an expected capacity of 2·12 which is worse than the previous best. Removing all arcs with capacity $\leqslant 15$ from G'' disconnects s from t so that the best answer so far,—i.e. path $(s, 1, 2, 5, t)$ with expected capacity 6·36—is the optimal answer.

8. Problems P8

1. Find the shortest paths from vertex 1 to all other vertices of the graph in Fig. 8.10. Hence draw the 1-base.

2. Use the method of Section 2.2 to find the shortest paths from vertex 1 to all other vertices of the graph shown in Fig. 8.11.

3. Repeat problem 2 with the cost of link (4, 5) changed from 8 to -8. Identify any negative circuits that may exist as soon as they are detected.

4. The graph of Fig. 8.12 shows a permissible route network for a ship to operate. Each arc is marked (a, b), "a" being the profit resulting from operating this "leg" of route and "b" being the time required to do so. Find the most profitable (in terms of the rate of return of capital), route for the ship to operate.

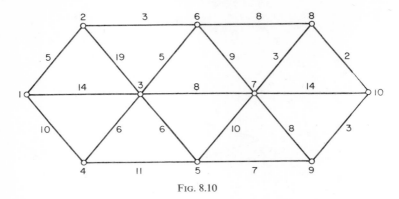

FIG. 8.10

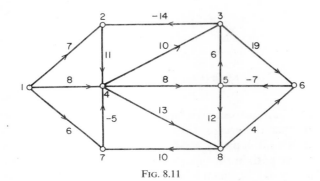

FIG. 8.11

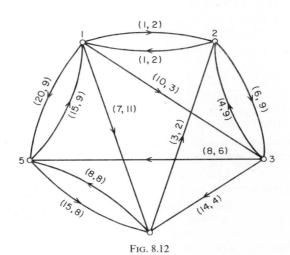

FIG. 8.12

5. For the graph shown in Fig. 8.10 find the four shortest elementary paths from vertex 1 to vertex 10. Hence find the shortest elementary path of cardinality 5 from 1 to 10.

6. A PERT diagram is shown in Fig. 8.13, where the duration of activity i is given by the number attached to all the arcs emanating from the corresponding vertex i. Find the critical path and the maximum possible delay in the commencement of activity 5 so that the whole project is not delayed.

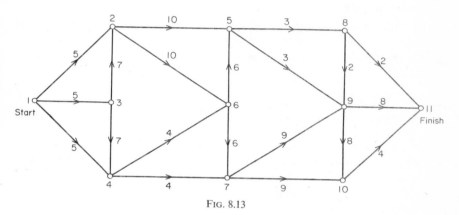

FIG. 8.13

7. In the graph shown in Fig. 8.14, each link is marked (a, b) where "a" is the capacity and "b" the reliability of the link. Find:
   (a) The maximum-capacity path from vertex 1 to 10.
   (b) The maximum-reliability path from vertex 1 to 10.
   (c) The maximum expected capacity path from vertex 1 to 10.

8. For the case where all arc costs are non-negative, modify the Dijkstra algorithm so that labelling proceeds both from vertex s and from vertex t simultaneously. Pay particular attention to the termination rule (See Ref. [29, 28]).

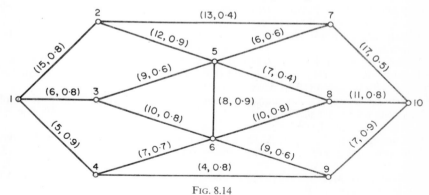

FIG. 8.14

9. Prove that the answer obtained by the algorithm of Floyd described in Section 3.1, is optimal and show that its growth is $O(n^3)$. (See Ref. [7]).

10. Let T be a directed spanning tree with root s of a graph $G = (X, A)$, and let $l(x_i)$ denote the distance from vertex s to vertex x_i measured along the tree path. Show that T is an s-base if and only if every arc $(x_i, x_j) \in A$ satisfies: $l(x_j) \leqslant l(x_i) + c_{ij}$.

11. The theorem in problem 10 above provides the basis for an iterative procedure for finding the shortest paths from s to all other vertices of a graph as follows: (Here we assume that all $x_i \in X$ are reachable from s).

Step 1: Start with T as any directed spanning tree of G, and calculate the vertex labels $l(x_i)$ as the distances of the vertices x_i from the root s along the tree paths.

Step 2: If every arc $(x_i, x_j) \in A$ satisfies:

$$l(x_j) \leqslant l(x_i) + c_{ij},$$

stop. The current T is the s-base of G. Otherwise go to step 3.

Step 3: If for some arc (x_i', x_j') we have:

$$l(x_j') > l(x_i') + c_{i'j'}$$

add arc (x_i', x_j') to the tree and remove from the tree that arc (x, x_j') whose final vertex is x_j' to produce a new tree T.

Step 4: Update the labels $l(x_k)$ of all vertices x_k reachable from x_j' along paths of the tree T. Return to step 2.

Show that the above algorithm is finite and show that its rate of growth for a complete graph of n vertices is $O(n^3)$. Are there any restrictions on the costs c_{ij} imposed by the algorithm? Use this algorithm to verify the answer obtained for problem 2. (See Ref. [14]).

9. References

1. Balas, E. (1969). Project scheduling with resources constraints, IBM New York Scientific Center, Report No. 320–2960.
2. Bellman, R. (1958). On a routing problem, *Quart. of Applied Mathematics*, **16**, p. 87.
3. Bellman, R. and Kalaba, R. (1960). On k^{th}-best policies, *Jl. of SIAM*, **8**, p. 582.
4. Berry, R. C. (1971). A constrained shortest path algorithm, Paper presented at the 39th National ORSA Meeting, Dallas, Texas.
5. Brucker, P. (1972). All shortest distances in networks with a large number of strong components, Presented at the 41st National Meeting Research Soc. of America, New Orleans.
6. Cunninghame-Green, R. A. (1962). Describing industrial processes with interference and approximating their steady-state behaviour, *Opl. Res. Quart.*, **13**, p. 95.
7. Dantzig, G. B. (1966). All shortest routes in a graph, *In:* "Int. Symp. on Theory of Graphs", Dunod, Paris, p. 91.

8. Dantzig, G. B., Blattner, W. O. and Rao, M. R. (1966). All shortest routes from a fixed origin in a graph. *In:* "Int. Symp. on Theory of Graphs", Dunod, Paris, p. 85.
9. Dantzig, G. B., Blattner, W. O. and Rao, M. R. (1966). Finding a cycle in a graph with minimum cost to time ratio with applications to a ship routing problem, *In:* "Int. Symp. on Theory of Graphs", Dunod, Paris, p. 77.
10. Dijkstra, E. W. (1959). A note on two problems in connection with graphs, *Numerische Mathematik*, 1, p. 269.
11. Dreyfus, S. E. (1969). An appraisal of some shortest path algorithms, *Ops. Res.*, 17, p. 395.
12. Ermolev, Yu. M. (1966). Shortest admissible paths I, *Kibernetika*, 2, p. 88.
13. Floyd, R. W. (1962). Algorithm 97—Shortest path, *Comm. of ACM*, 5, p. 345.
14. Ford, L. R. Jr. (1946). Network flow theory, Rand Corporation Report P-923.
15. Fox, B. (1969). Finding a minimal cost to time ratio circuit, *Ops. Res.*, 17, p. 546.
16. Frank, H. and Frisch, I. T. (1971). "Communication, Transmission and Transportation Networks", Addison-Wesley, Reading, Massachusetts.
17. Gill, A. and Traiger, T. (1968). Computation of optimal paths in finite graphs, Proc. 2nd Int. Conf. on Comp. Methods in Optimization Problems, San Remo, Italy.
18. Hitchener, L. E. (1968). A comparative investigation of the computational efficiency of shortest path algorithms, Operations Research Centre, University of California, Berkeley, Report ORC 68–17.
19. Hoffman, A. J. and Winograd, S. (1971). On finding all shortest distance in a directed network, IBM, Thomas J. Watson Research Center, Report RC 3613.
20. Hoffman, W. and Pavley, R. (1959). A method for the solution of the Nth best path problem, *Jl. of ACM*, 6, p. 506.
21. Hu, T. C. (1969). "Integer programming and network flows", Addison-Wesley, Reading, Massachusetts.
22. Johnson, E. L. (1972). On shortest paths and sorting, Proc. of Annual Conf. of ACM, Boston, p. 510.
23. Joksch, H. C. (1966). The shortest route problem with constraints, *Jl. of Math. Anal. and Appl.*, 14, p. 191.
24. Lawler, E. L. (1966). Optimal cycles in doubly weighted directed linear graphs, *In:* "Int. Symp. on Theory of Graphs", Dunod, Paris, p. 209.
25. Minieka, E. (1971). All k-shortest paths in a graph, Internal Report, Department of Statistics, Trinity College, Dublin.
26. Moore, E. F. (1957). The shortest path through a maze, Proc. Int. Symp. on the Theory of Switching, Part II, p. 285.
27. Murchland J. D. (1965). A new method for finding all elementary paths in a complete directed graph, London School of Economics, Report LSE-TNT-22.
28. Murchland, J. D. (1967). The once-through method of finding all shortest distances in a graph from a single origin, London Graduate School of Business Studies, Report LBS-TNT-56.
29. Nicholson, T. A. J. (1966). Finding the shortest route between two points in a network, *The Computer Jl.*, 9, p. 275.
30. Raimond, J. F. (1969). Minimaximal paths in disjunctive graphs by direct search, *IBM Jl. of Res. and Dev.*, 13, p. 391.
31. Reiter, R. (1968). Scheduling parallel computation, *Jl. of ACM*, 15, p. 590.
32. Sakarovitch, M. (1966). The k-shortest routes and k-shortest chains in a graph,

Operations Research Centre, University of California, Berkeley, Report ORC 66-32.

33. Shapiro, J. F. (1968). Shortest route method for finite state space dynamic programming problems, *Jl. of SIAM (Appl. Maths.)*, **16**, p. 1232.

34. Taha, H. A. (1971). "Operations Research", Macmillan, New-York.

35. Yen, J. Y. (1971). Finding the k-shortest, loopless paths in a network, *Man. Sci.*, **17**, p. 712.

36. Yen, J. Y. (1971). On the efficiencies of algorithms for detecting negative loops in networks, Santa Clara Business Review, p. 52.

37. Zaloom, V. (1971). On the resource-constrained project sheduling problem, *AIIE Trans.*, **3**, p. 302.

Chapter 9

Circuits, Cut-sets and Euler's Problem

1. Introduction

The purpose of this Chapter is to study circuits in graphs and to explore some of their properties and relationships to other concepts (such as that of a tree) which were introduced earlier. Two particular types of circuits in graphs, the Eulerian and Hamiltonian circuits, are of special interest due to their frequent occurrence in problems found in practice. Of these, the former is considered in this Chapter and the latter considered in the next Chapter.

In the present Chapter we will be concerned with circuits in both directed and nondirected graphs or a slight generalization of graphs called multi-graphs. These are graphs in which more than one arc (x_i, x_j) may exist between two given vertices x_i and x_j. If the largest number of arcs in "parallel" is s, the graph is called an s-graph. Thus, Fig. 9.1 shows a representation of a 3-graph. Obviously an ordinary graph can also be considered as a 1-graph.

s-Graphs occur very frequently in practice whenever a graph is meant to represent a physical system in natural science, management science, engineering and so on. The 3-graph of Fig. 9.1, for example, represents the molecular structure of the organic chemical compound acrylonitrile. In electrical engineering the graphs representing electrical networks are almost without exception s-graphs since many electrical components can be in parallel. In problems of reliability of equipment or communication networks, vital components or communication lines are often in duplicate, triplicate and so on and the resulting redundancies improve the system reliability. The graphs of such systems are also s-graphs.

2. The Cyclomatic Number and Fundamental Circuits

Consider G to be a nondirected s-graph having n vertices, m links and p connected components.

Define the number $\rho(G)$ as:

$$\rho(G) = n - p \qquad (9.1)$$

$\rho(G)$ is then the total number of links in the spanning trees of each of the p connected components of G.

The number $v(G)$ defined by:

$$v(G) = m - \rho(G) = m - n + p \qquad (9.2)$$

is called the *cyclomatic number* (sometimes also referred to as the *nullity* or the *first Betti number*), and the number $\rho(G)$ is called the *cocyclomatic number*.

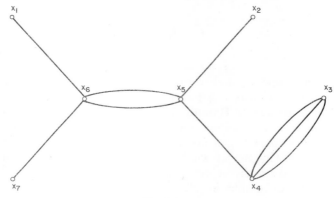

FIG. 9.1

In electric network theory (and indeed any representation of a lumped-parameter system) the numbers $\rho(G)$ and $v(G)$ have direct physical significance. Thus, the cyclomatic number is the largest number of independent circuits in the graph of the electric network, i.e. it is the largest number of independent circular currents that can flow in the network; and the cocyclomatic number is the largest number of independent potential differences between the nodes of the electric network.

Consider, for example, the electric network shown in Fig. 9.2(a). The non-directed graph G of this network consists of a single connected component ($p = 1$) and is as shown in Fig. 9.2(b) where a spanning tree T is shown in heavy lines (see Chapter 6).

Now the addition of any link (x_i, x_j) of G not in T to the links of T will close precisely one (elementary) circuit consisting of the links of T on the (unique) path from x_j to x_i and of the link just added. For example the addition of link $a_3 = (x_1, x_2)$ will close the circuit $(x_1, x_7, x_6, x_3, x_2, x_1)$.

Since there are m links in the graph G and $(n - 1)$ of these are in T, the number of such circuits that can be formed is $m - n + 1$ which is the cyclomatic number of G. In the examples of Fig. 9.2 the number of such circuits is $15 - 7 + 1 = 9$ and are shown in the figure as $\Phi_1, \Phi_2, \ldots, \Phi_9$ by dotted lines. All these circuits are obviously independent of each other since each one has at least one link not found in any other circuit. In general, the $\nu(G)$ circuits formed by the addition of any link of G not in T to the links of T are called *fundamental circuits*. If the family of these circuits is denoted by Φ, where $\Phi = \{\Phi_1, \Phi_2, \ldots, \Phi_9\}$ in the example, then any other circuit in the graph not in Φ can be expressed as a linear combination of the circuits in Φ if the following convention is adopted [21].

Let every fundamental circuit Φ_i, $i = 1, \ldots, \nu(G)$, be represented by an m-dimensional vector where the jth element is 1 (or 0) depending on whether the jth link is (or is not) part of this circuit. Then, if $\oplus$ is used for modulo 2 addition, any circuit Φ_k can be expressed as the mod 2 sum of fundamental circuits. Thus in Fig. 9.2(b) the circuit $\Phi_{10} = (a_3, a_{11}, a_{14}, a_6, a_8, a_1)$ can be

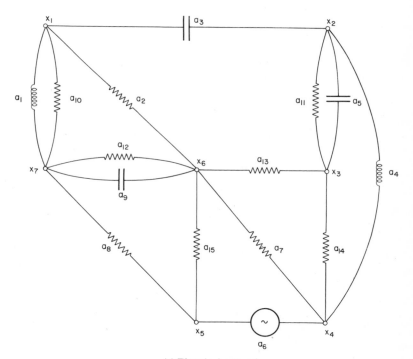

(a) Electrical network

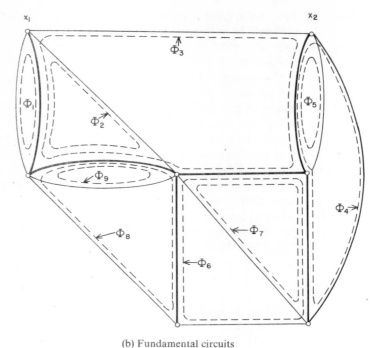

(b) Fundamental circuits

FIG. 9.2

expressed as:

$$\Phi_{10} = \Phi_1 \oplus \Phi_3 \oplus \Phi_6 \oplus \Phi_8$$

$$= (100000000100000) \oplus (001000000111100)$$

$$\oplus (000001000000111) \oplus (000000010001001)$$

$$= (101001010010010)$$

However, the reverse of the above statement is not true, i.e. any mod 2 sum of fundamental circuits is not necessarily a single circuit but may represent two or more circuits. For example $\Phi_2 \oplus \Phi_3 \oplus \Phi_6 \oplus \Phi_7$ produces the two circuits $(a_1, a_2, a_{11}, a_{13})$ and (a_6, a_7, a_{15}). Thus, if we wish to generate all circuits of a graph G and take all $2^{\nu(G)} - 1$ combinations of fundamental circuits to add mod 2, then some of the results will not be valid circuits. Moreover, if a given combination, say $\Phi_i \oplus \Phi_j \oplus \Phi_k \oplus \ldots$, is invalid one cannot discard other combinations containing it, since the mod 2 addition of $\Phi_i \oplus \Phi_j \oplus \Phi_k \oplus \ldots$ with another combination say $\Phi_\alpha \oplus \Phi_\beta \oplus \Phi_\gamma \oplus \ldots$

(which could by itself also be invalid) may well lead to a circuit. A method for overcoming this difficulty so that invalid combinations of fundamental circuits can be discarded as soon as they are generated, was described by Welch [26], and depends on ordering the fundamental circuits according to a specific set of rules.

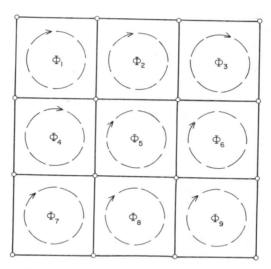

FIG. 9.3. Independent but not fundamental circuits

It should also be noted that although the number of fundamental circuits is given by $v(G)$ the circuits themselves are not unique and they depend on the spanning tree T originally chosen. In fact, sets of $v(G)$ circuits of a graph G can be found which cannot be derived by adding links to a tree as mentioned above, but which are nevertheless sets of independent circuits. Such sets of circuits are not, however, referred to as "fundamental". Figure 9.3 shows a set of $v(G) = 9$ independent circuits of a graph G which cannot be derived by adding links to any spanning tree of G and which is, therefore, not a fundamental set.

3. Cut-sets

The concept of a cut-set is very closely related to that of a circuit and is defined as follows:

Definition. If the vertices of a nondirected graph $G = (X, A)$ are partitioned into two sets X_o and $\tilde{X}_o$ (where $X_o \subset X$ and $\tilde{X}_o$ is the complement of X_o in X), then the set of links of G whose terminal vertices lie one in X_o and the other in $\tilde{X}_o$, is called a *cut-set* of G.

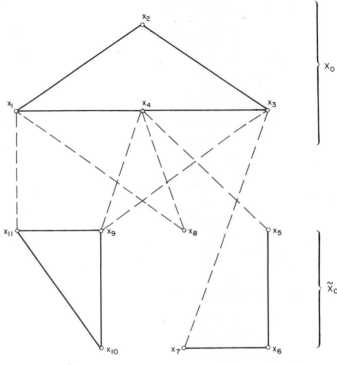

FIG. 9.4. ---- Cut-set $(X_0, \tilde{X}_0)$

The set of links in the cut-set can be represented by the vertex set doublet $(X_o, \tilde{X}_o)$. Thus the partial graph $G_p = (X, A - (X_o, \tilde{X}_o))$, formed from G by the removal of the links in the cut-set, is a disconnected graph composed of at least two connected components. In the graph of Fig. 9.4 the set of links shown dotted forms a cut-set given by $X_o = \{x_i | i = 1, \ldots, 4\}$ and $\tilde{X}_o = \{x_i | i = 5, \ldots, 11\}$. The resulting partial graph is then composed of four connected components i.e.:

$$\langle x_1, x_2, x_3, x_4 \rangle, \langle x_5, x_6, x_7 \rangle, \quad \langle x_8 \rangle \quad \text{and} \quad \langle x_9, x_{10}, x_{11} \rangle.$$

A set S of links, whose removal from the graph G causes the partial graph $G_p = (X, A - S)$ to be disconnected, and there exists no subset $S' \subset S$ for which $G'_p \equiv (X, A - S')$ is also disconnected, is called a *proper cut-set*. This set can also be represented alternatively by a doublet $(X_o, \tilde{X}_o)$ although now the partial graph formed by the removal of a proper cut-set from G will disconnect G into *exactly* two connected components one based on the vertex set X_o and the other on the set $\tilde{X}_o$. In general, every cut-set is the union of a number of proper cut-sets. In the graph of Fig. 9.4, for example, the cut-set shown by the dotted links is the union of the three proper cut-sets: $\{(x_1, x_{11}), (x_4, x_9), (x_3, x_9)\}$, $\{(x_1, x_8), (x_4, x_8)\}$ and $\{(x_3, x_7), (x_4, x_5)\}$. In the remainder of this Chapter, and in the Chapters that follow we will use the word cut-set to imply proper cut-set, unless otherwise stated.

The dual relationship between the spanning tree of a graph G and a cut-set becomes apparent when it is recalled that a tree is a minimal set of links which connects all the vertices of G, whereas a cut-set is a minimal set of links which disconnects some vertices from others. From this observation it is quite obvious that any spanning tree of G must have at least one link in common with every proper cut-set.

The cut-set was defined above for a nondirected graph. When the graph $G = (X, A)$ is directed, then the cut-set of G is defined as the set of arcs corresponding to the links forming the cut-set of the non-directed counterpart $\bar{G}$ of G. Some of the arcs in the cut-set of G will be directed from vertices in X_o to vertices in $\tilde{X}_o$ and the set of these arcs will be written as $(X_o \rightarrow \tilde{X}_o)$, whereas the set of arcs directed from vertices in $\tilde{X}_o$ to vertices in X_o will be written as $(\tilde{X}_o \rightarrow X_o)$. The arc cut-set $(X_o, \tilde{X}_o)$ of a directed graph is therefore given by $(X_o, \tilde{X}_o) = (X_o \rightarrow \tilde{X}_o) \cup (\tilde{X}_o \rightarrow X_o)$.

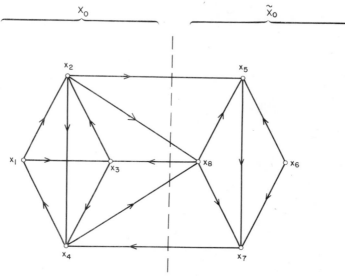

FIG. 9.5. Directed cut-sets

For the graph of Fig. 9.5, for example, the set of arcs $\{(x_2, x_5), (x_2, x_8),$ $(x_8, x_3), (x_4, x_8), (x_7, x_4)\}$ is a cut-set separating the set of vertices $X_o = \{x_1, x_2, x_3, x_4\}$ from the set of vertices $\tilde{X}_o = \{x_5, x_6, x_7, x_8\}$. This cut-set is composed of the arc subsets $(X_o \to \tilde{X}_o) = \{(x_2, x_5), (x_2, x_8), (x_4, x_8)\}$ and $(\tilde{X}_o \to X_o) = \{(x_8, x_3), (x_7, x_4)\}$.

The set of fundamental circuits of a nondirected graph G was defined in the previous section as the $v(G)$ circuits formed by the addition of a link not in a spanning tree T to the links of the tree. In a similar manner, and in view of the relationship between cut-sets and trees mentioned earlier, the *fundamental cut-sets* with respect to a spanning tree T are defined as the $n - 1$ cut-sets each one of which contains one and only one link which is in the tree T.

The following theorem gives the relationship between fundamental cut-sets and fundamental circuits, and suggests a way by which the fundamental cut-sets can be derived.

THEOREM 1. *If T is a spanning tree of a nondirected graph G, the fundamental cut-set determined by a link a_i of T is composed of a_i and those links of G not in T, which when added to T lead to fundamental circuits containing a_i.*

Proof. If link a_i is removed from the spanning tree T, T is decomposed into two subtrees T_1 and T_2 (see Fig. 9.6). Any link whose terminal vertices are one in T_1 and the other in T_2 must be in the fundamental cut-set since the addition of any such link to the links of T_1 and T_2 will form another spanning

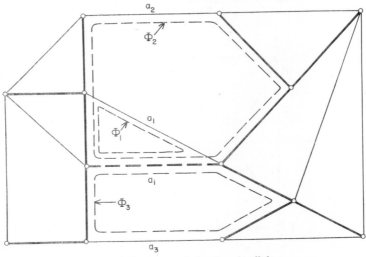

FIG. 9.6. Fundamental circuits using link a_i

tree of G, and hence any set not including such links will not be a cut-set. The set of such links plus the link a_i is a cut-set, since their removal separates the graph into two subgraphs one based on the vertices of T_1 and the other based on the vertices of T_2. This cut-set is, therefore, a fundamental cut-set. Moreover, since link a_i is only contained on those paths of T starting from a vertex in T_1 and terminating at a vertex in T_2, only links whose terminal vertices are one in T_1 and the other in T_2 (e.g. links a_1, a_2 and a_3 in the graph of Fig. 9.6) will close a fundamental circuit containing a_i. Hence the theorem.

4. Circuit and Cut-set Matrices

Let T be a spanning tree of a nondirected graph G. The *fundamental circuit matrix* of G is then a $v(G)$-row, m-column matrix $\Phi = [\phi_{ij}]$, where ϕ_{ij} is 1 if link a_j is part of circuit Φ_i, and zero otherwise. If the links not in the tree T are numbered consecutively from 1 to $v(G)$ and the links of T numbered from $v(G) + 1$ to m, then the circuit matrix Φ takes the form:

$$\Phi = (\mathbf{I} \,|\, \Phi_{12})$$

where $\mathbf{I}$ is the unity matrix. This is because each circuit Φ_i contains one and only one link not in T, and the circuits can always be numbered with the number of the non-tree link that they contain, so that all the ones in the first $v(G) \times v(G)$ submatrix of Φ are on the diagonal.

The *fundamental cut-set matrix* is similarly defined as the $(n-1)$-row, m-column matrix $\mathbf{K} = [k_{ij}]$ where k_{ij} is 1 if link a_j is part of cut-set K_i and zero otherwise. Thus, with the same link numbering as before, the matrix $\mathbf{K}$

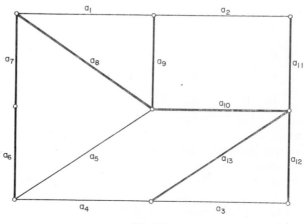

FIG. 9.7

takes the form:

$$\mathbf{K} = (\mathbf{K}_{11}|\mathbf{I}),$$

since now each fundamental cut-set contains one and only one link from the links of the tree T.

For the graph shown in Fig. 9.7, for example, and with the spanning tree T chosen as shown by the heavy lines, the fundamental circuit matrix is:

	a_1	a_2	a_3	a_4	a_5	a_6	a_7	a_8	a_9	a_{10}	a_{11}	a_{12}	a_{13}
Φ_1	1	0	0	0	0	0	0	1	1	0	0	0	0
Φ_2	0	1	0	0	0	0	0	0	1	1	1	0	0
$\Phi = \Phi_3$	0	0	1	0	0	0	0	0	0	0	0	1	1
Φ_4	0	0	0	1	0	1	1	1	0	1	0	0	1
Φ_5	0	0	0	0	1	1	1	1	0	0	0	0	0

and the fundamental cut-set matrix is:

	a_1	a_2	a_3	a_4	a_5	a_6	a_7	a_8	a_9	a_{10}	a_{11}	a_{12}	a_{13}
K_1	0	0	0	1	1	1	0	0	0	0	0	0	0
K_2	0	0	0	1	1	0	1	0	0	0	0	0	0
K_3	1	0	0	1	1	0	0	1	0	0	0	0	0
$\mathbf{K} = K_4$	1	1	0	0	0	0	0	0	1	0	0	0	0
K_5	0	1	0	1	0	0	0	0	0	1	0	0	0
K_6	0	1	0	0	0	0	0	0	0	0	1	0	0
K_7	0	0	1	0	0	0	0	0	0	0	0	1	0
K_8	0	0	1	1	0	0	0	0	0	0	0	0	1

where cut-set K_i corresponds to the tree link numbered $v(G) + i = 5 + i$.

There are some interesting relationships between the fundamental circuit, cut-set and incidence matrices when all arithmetic operations are considered modulo 2. We will consider here nondirected graphs with no loops.

THEOREM 2. *The incidence matrix* $\mathbf{B}$ *and the transpose* Φ^t *of the fundamental circuit matrix* Φ *are orthogonal, i.e.* $\mathbf{B} \cdot \Phi^t = 0$.

THEOREM 3. *The fundamental circuit matrix* Φ *and the transpose* K^t *of the fundamental cut-set matrix* K *are orthogonal i.e.* $\Phi . K^t = 0$.

The above two theorems are immediate consequences of the obvious facts that:

    (i) Each vertex in a circuit is incident with an even (two in the case of an elementary circuit) number of links in the circuit.

and (ii) Each circuit cut by a cut-set has an even number of links in common with the cut-set.

Theorem 2 follows from (i) and theorem 3 from (ii) when it is remembered that all operations are done mod 2 so that, for example, $2 \equiv 0 \pmod 2$. Moreover, the proofs [21, 17, 6] make no use of the fact that the circuits or cut-sets are fundamental, and the theorems are valid for any circuit and cut-set matrices defined in a similar fashion.

From theorem 3 we can write:

$$\Phi . K^t = (I \,|\, \Phi_{12}) . \left(\frac{K^t_{11}}{I} \right)$$

$$= K^t_{11} + \Phi_{12} = 0$$

Hence:

$$K^t_{11} = -\Phi_{12} = \Phi_{12} \qquad (9.3)$$

since $-1 = 1 \pmod 2$. The fundamental cut-set matrix can, therefore, be derived immediately once the fundamental circuit matrix is known and *vice versa*. The validity of eqn (9.3) can be checked with the Φ and K matrices given above for the graph shown in Fig. 9.7.

5. Eulerian Circuits and the Chinese Postman's Problem

Definition. Given a nondirected s-graph G, an *Eulerian circuit* (path) is a circuit (path) which traverses every link of G once and only once.

Obviously not all graphs have Eulerian circuits (see for example the graph in Fig. 9.8), but if an Eulerian circuit (or path) exists it means that the graph can be drawn on paper by following this circuit (or path) and without lifting the pencil from the paper. The 3-graph of Fig. 9.9 has an Eulerian circuit given by the sequence: (starting from vertex x_1)

$$a_1 a_2 a_3 a_4 a_{15} a_{14} a_{13} a_{12} a_{11} a_{16} a_{17} a_{10} a_9 a_8 a_5 a_7 a_6$$

The direction of traversal of each link is shown by the arrows in the figure.

Euler—in his celebrated *Konigsberg bridge problem*—was the first person to consider the existence, or otherwise, of such circuits in graphs. Konigsberg

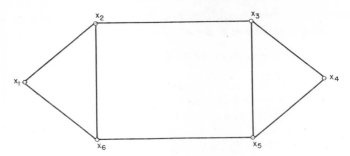

FIG. 9.8. Graph without Eulerian circuit

(now Kaliningrad) is a town built on both banks of the river Pregel and on two islands in the river. The river banks and the two islands are connected by seven bridges as shown by the map in Fig. 9.10(a). The question—posed in 1736—was whether it was possible starting from any point to go for a walk and cross every bridge once and once only before returning to that point. If each river bank and island is represented by a vertex and each bridge by a link, the map of Fig. 9.10(a) can be represented by the 2-graph of Fig. 9.10(b), and the problem now becomes one of deciding whether an Eulerian circuit exists in this graph. Euler concluded that the graph contained no Eulerian circuit and his study is considered to have marked the beginning of graph theory.

The basic theorem on the existence of an Eulerian circuit is as follows:

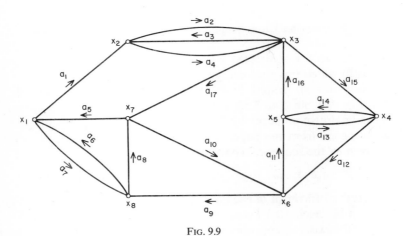

FIG. 9.9

THEOEREM 4(a). *A connected, nondirected s-graph G contains an Eulerian circuit (path) if and only if the number of vertices of odd degree is 0 (0 or 2).*

Proof. We will prove the theorem for the case of a circuit, the case of a path being completely analogous.

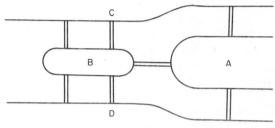

(a) Map of Koningsberg

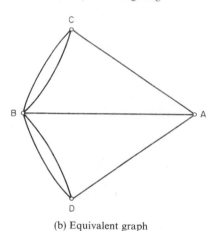

(b) Equivalent graph

FIG. 9.10

Necessity. Any Eulerian circuit must use one link to arrive and a different link to leave a vertex since any link must be traversed exactly once. Hence, if G contains an Eulerian circuit all vertex degrees must be even.

Sufficiency. Let G be a connected nondirected s-graph with even vertex degrees. Start a path at some arbitrary vertex x_0 and proceed along a previously unused link to the next vertex until the path returns to x_0 and a circuit is completed. If all links have been used the desired Eulerian circuit has been established. If some links have not been used, then let Φ be the circuit just completed. Since G is connected, Φ must have passed through some vertex, say x_i, which is the terminal vertex of some hitherto unused link. If all links used by Φ are removed, then the resulting graph would still have all vertex degrees even since Φ must use an even number of links (0 is an even number), incident at every vertex.

Starting again from x_i we could traverse a circuit Φ' starting and finishing at x_i. Once more if all remaining links are used by Φ' we are finished. The part of Φ from x_o to x_i followed by circuit Φ', followed by the part of Φ from x_i to x_o, would be the required Eulerian circuit. If some links are still unused, let the union of Φ and Φ' described above be the new circuit Φ. We could again find a vertex x_j met by Φ which is the terminal of some unused link. We could then proceed to form a new circuit Φ' starting at x_j and so on until finally all links are used in this way and an Eulerian circuit Φ has been obtained. Hence the theorem.

Obviously if G is not connected (excepting isolated vertices), there is no Eulerian circuit since there is no path from one of its components to another. In the case of an Eulerian path it is also obvious that if its two end vertices are p and q then these are the two vertices with the odd degrees.

If G is a directed s-graph a theorem completely analogous to theorem 4(a) above holds, i.e.

THEOREM 4(b). *A connected directed s-graph G contains an Eulerian circuit (path) if and only if the indegrees $d_t(x_i)$ and the outdegrees $d_o(x_i)$ of the vertices satisfy the condition:*

for the case of a circuit: $d_t(x_i) = d_o(x_i), \forall\, x_i \in X$

for the case of a path: $\quad d_t(x_i) = d_o(x_i) \,\forall\, x_i \neq p$ or q.

$$d_t(q) \;=\; d_o(q) + 1$$

$$and\; d_t(p) \;=\; d_o(p) - 1$$

where p is the initial and q the final vertex of the Eulerian path.

A very simple rule for tracing an Eulerian circuit in a nondirected graph (if such a circuit exists) was given by Fleury [16]. The rule can be easily extended to directed graphs and is as follows: (See also problems 7 and 8).
Rule for finding an Eulerian circuit. Start from any vertex p and each time a link has been traversed erase it. Never traverse a link if at that particular moment the removal of this link will divide the graph into two connected components (excluding isolated vertices).

5.1 Some related problems
A number of questions related to that of whether a nondirected graph G has an Eulerian circuit or not come immediately to mind. For example:

(i) What is the smallest number of paths or circuits necessary so that every link of G is contained in exactly one path or circuit. Obviously if G has an Eulerian path or circuit the answer to these questions is one.†

(ii) If the links of G are weighted with positive costs, find a circuit which will traverse every link of G *at least* once and for which the total cost of traversal (being the sum of $n_j c(a_j)$ where n_j is the number of times a link a_j is traversed and $c(a_j)$ is its cost), is a minimum. Obviously, if G contains an Eulerian circuit any such circuit is optimal since each link is traversed once and only once, and its cost is then $\sum_{j=1}^{m} c(a_j)$.

Problem (ii), above, is called the *Chinese postman's* problem [18] and its solution has a number of potential applications, e.g.:

(a) REFUSE COLLECTION. Let us consider the problem of collecting household refuse, and assume that a specified city area is to be served by a single vehicle. We will take links in the graph to represent roads and take vertices as road junctions so that if $c(a_j)$—the cost of link a_j—is assumed to be the corresponding road distance, then the problem of refuse collection from all the streets in the area is one of finding a circuit in G which traverses every link of G at least once. What is then required is to find that circuit involving the least mileage travelled. In reality the vehicle capacity and length of the working day may place restrictions on the number of streets that a vehicle can serve in any one day, and what may in fact be required is not a single circuit but a number Q of circuits, to be operated one a day, say, for a Q-day period; the circuits being chosen so as not to violate the above-mentioned constraints [1, 2, 4, 11, 22].

(b) DELIVERY OF MILK OR POST. The problems of delivery of milk or post are two more cases where a route is required which should traverse each and every street at least once and the problem is to find that route which minimizes the route mileage (time, cost, etc.).

(c) THE INSPECTION OF ELECTRIC POWER, TELEPHONE OR RAILWAY LINES. The problems of inspection of distributed systems (of which the above are only a small sample) have the inherent requirement that all "components" have to be inspected, and they are therefore problems of type (ii) above or slight variations thereof.

Other applications include the spraying of salt-grit on main roads in winter in order to prevent ice formation, the best method of operation of an auto-

† This question is recognized as one of finding the least number of link-disjoint subgraphs (in this case paths or circuits) in order to "cover" the links of the graph G. These types of problems have also been discussed as variations of the *set covering problem in Chapter 3.*

matic draughting machine, the cleaning of offices and corridors in large office blocks, even the best way to tour a museum!

5.2 An algorithm for the Chinese postman's problem

The Chinese postman's problem with all link costs assumed unity was considered by Bellman and Cook using dynamic programming [3]. The more general problem with arbitrary link costs was formulated and solved as a matching problem (with a shortest path subproblem) by Edmonds [9], Edmonds and Johnson [10], Busacker and Saaty [7] and Christofides [8].

Consider the non-directed graph $G = (X, A)$. Of the vertices of X some vertices (say in the set X^+) will have even degrees and some (in the set $X^- \equiv X - X^+$) will have odd degrees. Now the sum of the degrees d_i of all the vertices $x_i \in X$ is equal to twice the number of links in A (since each link adds unity to the degrees of its two end vertices), and is therefore an even number 2m. Hence:

$$\sum_{x_i \in X} d_i = \sum_{x_i \in X^+} d_i + \sum_{x_i \in X^-} d_i = 2m$$

and since $\sum_{x_i \in X^+} d_i$ is even $\sum_{x_i \in X^-} d_i$ is also even, which means that since all d_i in this last expression are odd the number $|X^-|$ of vertices of odd degree is even.

Let M be a set of paths of G (μ_{ij} say) between end vertices x_i and x_j, ($x_i, x_j \in X^-$), so that no two paths have any end vertex the same, i.e. the paths are between disjoint pairs of vertices of X^- and constitute a pairwise vertex matching. The number of paths μ_{ij} in M is $\frac{1}{2}|X^-|$, and since $|X^-|$ was shown to be always even, this number is always integer as, of course, it should be. Suppose now that all the links forming a path μ_{ij} are added into G as *artificial* links in parallel with the links of G already there. (In the first instance this means that all links of G forming μ_{ij} are now doubled.) This is done for every path $\mu_{ij} \in M$ and the resulting s-graph is called $G^-(M)$. Since some links of G may appear in more than one path μ_{ij}, some links of $G^-(M)$ may (after all the paths μ_{ij} have been added in), be in triplicate, quadruplicate etc.

We can now state the following theorem:

THEOREM 5. *For any circuit traversing G, there is some choice of M for which* $G^-(M)$ *possesses an Eulerian circuit corresponding to the circuit of G. The correspondence is such that if a circuit traverses a link* (x_i, x_j) *of G l times, there are l links (one real and* $(l-1)$ *artificial) between* x_i *and* x_j *in* $G^-(M)$, *each of which links is traversed exactly once by the Eulerian circuit of* $G^-(M)$; *and conversely.*

Proof. If a circuit traverses G, then at least one link incident at every odd-degree vertex x_i must (by theorem 4(a)) be traversed twice. (A link traversed twice can be considered as two parallel links one real and one artificial both of which are traversed once.) Let this link be chosen as (x_i, x_k). If the degree d_k of x_k in G is odd then the addition of the aritificial link earlier will make d_k even and only this link need be traversed twice as far as x_i and x_k are concerned. If, however, d_k is even then the addition of the artificial link will now make d_k odd and a second link leading from x_k must also be traversed twice (i.e. another artificial link added). The argument continues from x_k until a vertex of odd degree is reached as mentioned above. Thus, in order to satisfy the condition of traversibility at x_i a whole path from x_i to some other odd-degree vertex x_r of X^- must be traversed twice. This automatically satisfies the traversibility condition of vertex x_r. Similarly for all other vertices x_i of X^-, which means that a whole set M of paths of G, as defined earlier, must be traversed twice, and since this means that every link of $G^-(M)$ must be traversed once the theorem follows.

The algorithm for the solution of the Chinese postman's problem follows immediately from the above theorem, since all that is now necessary is to find that set of paths M^* (matching the vertices of odd degree) which produces the least additional cost. The least cost circuit traversing G would then have a cost equal to the sum of the costs of the links of G plus the sum of the costs of the links in the paths of M^*. This is the same as the sum of the costs of all the links—real and artificial—of the graph $G^-(M^*)$. A description of the algorithm now follows:

5.2.1 DESCRIPTION OF THE ALGORITHM. *Step 1.* Let $[c_{ij}]$ be the link cost matrix of graph G. Using a shortest path algorithm (see Chapter 8) form the $|X^-|$ by $|X^-|$ matrix $\mathbf{D} = [d_{ij}]$ where d_{ij} is the cost of the least-cost path from a vertex $x_i \in X^-$ to another vertex $x_j \in X^-$.

Step 2. Find that pairwise matching M^* of the vertices in X^- which produces the least cost according to the cost matrix $\mathbf{D}$. (This can be done efficiently by a minimum matching algorithm described in Chapter 12.)

Step 3. If vertex x_α is matched to another vertex x_β identify the least cost path $\mu_{\alpha\beta}$ (from x_α to x_β) corresponding to the cost $d_{\alpha\beta}$ of step 1. Insert artificial links in G corresponding to links in $\mu_{\alpha\beta}$ and repeat for all other paths in the matching M^* to obtain the s-graph $G^-(M^*)$.

Step 4. The sum of the costs from matrix $[c_{ij}]$ of all links in $G^-(M^*)$—taking the cost of an artificial link to be the same as the cost of the real link in parallel with it—is the minimum cost of a circuit traversing G. The number of times that this circuit traverses a link (x_i, x_j) being the total number of links in parallel between x_i and x_j in $G^-(M^*)$.

It should be noted here that since at step 2 we are using a minimum matching, no two shortest paths μ_{ij} and μ_{pq} in such a matching (of, say, x_i to x_j and x_p to x_q) can now have any link in common; for if they have link (x_a, x_b) in common as shown in Fig. 9.11, then a matching of x_i to x_q (using subpaths x_i to x_b and x_b to x_q), and a matching of x_p to x_j (using subpaths x_p to x_a and x_a to x_j),

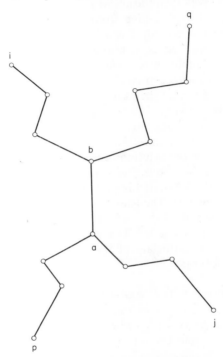

FIG. 9.11. Paths μ_{ij} and μ_{pq} having common link (a, b)

produces an overall matching of cost $2c_{ab}$ less than the original matching, which is contrary to the assumption that this matching is minimal. This means that the graph $G^-(M^*)$ does not have more than two links in parallel between any two vertices, i.e. the optimal circuit never traverses any link of G more than twice.

5.2.2 EXAMPLE. Consider the graph G of Fig. 9.12 having 12 vertices and 22 links where the link cost is given next to each link. The problem is to find a circuit which traverses all the links of G at least once and has minimum cost.

The set X^- of odd-degree vertices for this graph is $\{1, 3, 4, 6, 8, 9\}$.

Using Dijkstra's shortest path algorithm (see Chapter 8) we find matrix

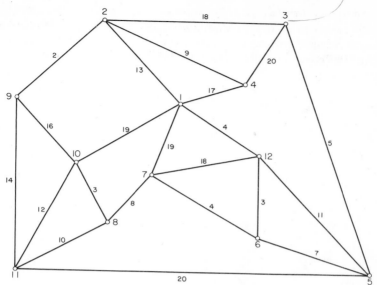

FIG. 9.12. Graph for example 5.2.2

D of step 1 as:

	1	3	4	6	8	9
1	—	19	17	7	19	15
3	19	—	20	12	24	20
4	17	20	—	24	30	11
6	7	12	24	—	12	22
8	19	24	30	12	—	19
9	15	20	11	22	19	—

$\mathbf{D} =$

At step 2 the minimal matching algorithm (see Chapter 12) matches the following vertices: (two other matchings are also minimal)

> 1 with 6, path: 1–12–6,   cost 7
>
> 3 with 8, path: 3–5–6–7–8, cost 24
>
> 9 with 4, part: 9–2–4,   cost 11

The graph $G^-(M^*)$ at the end of step 3 is, therefore, as shown in Fig. 9.13, where the artificial links are shown dotted. From this figure it is seen that the optimum circuit traversing G traverses links (9, 2) (2, 4) (3, 5) (5, 6) (1, 12) (12, 6) (7, 6) and (8, 7) twice and all other links once. The cost of this circuit is therefore 294 units.

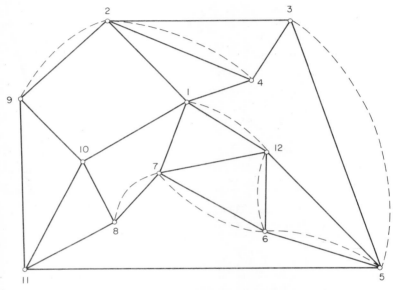

FIG. 9.13. Graph $G^-(M^*)$ for example 5.2.2
————— Real links
————— Artificial links

Once the graph $G^-(M^*)$ is found, an Eulerian circuit of this graph and hence the corresponding optimum circuit traversing the original graph G, can be constructed immediately by applying Fleury's rule mentioned earlier. For the graph of Fig. 9.13, for example, one possible Eulerian circuit obtained from the above rule is given by the vertex sequence: (starting from vertex 1)

1　4, 3, 5, 3, 2, 4, 2, 1, 12, 5, 6, 7, 8, 10, 1, 7, 6, 12, 6, 5, 11, 9, 2, 9, 10,

11, 8, 7, 12, 1.

This circuit is also the corresponding optimum traversal circuit of the original graph G of Fig. 9.12. Other optimal circuits which traverse the links of G in a different order to the one shown above—but which traverse the same set of 8 links twice—are also possible. Thus, once the links of G that must be traversed twice by an optimal circuit are determined, there may, in general, be more than one such circuit traversing G.

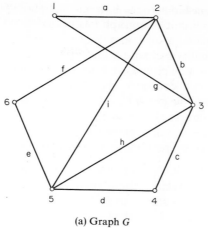

(a) Graph G

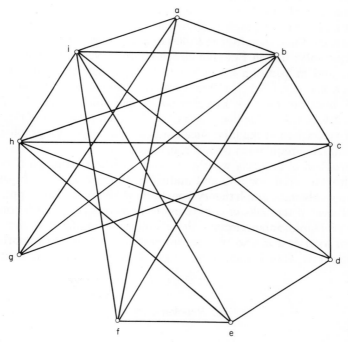

(b) Line graph G_l

Fig. 9.14

H

5.3 Relationship between Eulerian and Hamiltonian circuits

The Hamiltonian circuit of a graph has been defined in Chapter 1, as an elementary circuit passing once and only once through every vertex of G. With this kind of duality (substitute link for vertex and *vice versa*) between Eulerian and Hamiltonian circuits, it is not surprising to find a close relationship between these two concepts when applied to a nondirected graph G and its corresponding *line graph* G_l defined as follows:

Definition. The line graph G_l has as many vertices as there are links in G. A link between two vertices of G_l exists if and only if the links of G corresponding to these two vertices are adjacent (i.e. incident on the same vertex of G).

If G is the graph of Fig. 9.14(a), for example, then G_l is as shown in Fig. 9.14(b).

The following two relationships between Eulerian and Hamiltonian circuits can then be easily shown to be true [15].

(i) If G has an Eulerian circuit then G_l has both Eulerian and Hamiltonian circuits.

(ii) If G has a Hamiltonian circuit then G_l also has a Hamiltonian circuit.

The converse of the above two statements are not true as can be readily demonstrated. For the graph G shown in Fig. 9.14(a) the degrees of all its vertices are even and therefore an Eulerian circuit exists. Thus, G_l—shown in Fig. 9.14(b)—also has an Eulerian circuit (since its vertex degrees are again all even), and in addition has a Hamiltonian circuit given by the vertex sequence a, g, c, d, e, f, b, h, i, a. If now link b is removed from G, the resulting graph G' does not have an Eulerian circuit but still has the Hamiltonian circuit 1, 2, 6, 5, 4, 3, 1. The line graph G'_l of G' is then the graph of Fig. 9.14(b) but with vertex b (and the links incident at b) removed. This graph still has a Hamiltonian circuit given by a, f, e, d, c, g, h, i, a.

The problem of deciding whether a given nondirected graph G contains a Hamiltonian circuit or not, and finding such a circuit if one exists, is an important problem in graph theory from both the theoretical and applications points of view. If the links of G have associated costs $[c_{ij}]$ and G contains a number of Hamiltonian circuits, the problem of finding that circuit with minimum total cost is also of great interest. Because of the importance of these questions, Hamiltonian circuits are discussed separately in the next Chapter.

6. Problems P9

1. For the nondirected graph shown in Fig. 9.15 find the cyclomatic and cocyclomatic numbers, and the fundamental circuits relative to some spanning tree.

2. For the graph of Fig. 9.15 write the fundamental circuit and fundamental cut-set matrices Φ and K respectively, and verify theorems 2 and 3 for this example.

3. Using Theorem 2 derive an expression for the fundamental circuit matrix $\boldsymbol{\Phi}$ in terms of the incidence matrix $\mathbf{B}$.

4. Prove that a set K of links of a nondirected graph G such that K has an even number of links in common with every circuit, is a cut-set (not necessarily proper).

5. Prove that a nondirected graph G has an Eulerian circuit if and only if G is a union of link-disjoint circuits.

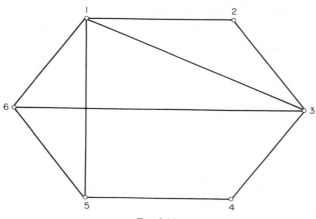

FIG. 9.15

6. If the columns of matrices $\mathbf{B}$, $\boldsymbol{\Phi}$ and $\mathbf{K}$ are arranged in the same order of links (relative to a spanning tree T from which the fundamental circuits and cut-sets are formed), and then partitioned as : $\mathbf{B} = [\mathbf{B}_{11}, \mathbf{B}_{12}]$, $\boldsymbol{\Phi} = [\mathbf{I}, \boldsymbol{\Phi}_{12}]$ and $\mathbf{K} = [\mathbf{K}_{11}, \mathbf{I}]$, show that:

$$\mathbf{K}_{11} = \boldsymbol{\Phi}^t_{12} = \mathbf{B}_{12}^{-1} \cdot \mathbf{B}_{11}$$

and

$$\boldsymbol{\Phi} = [\boldsymbol{\Phi}^t_{12}, \mathbf{I}] = \mathbf{B}_{12}^{-1} \cdot \mathbf{B}$$

and hence that one can start with any one of the three matrices $\mathbf{B}$, $\boldsymbol{\Phi}$ and $\mathbf{K}$ and find the others.

7. Show that Fleury's Rule always traces an Eulerian circuit of a graph if such a circuit exists.

8. Write an algorithm based on Fleury's Rule which traces *all* Eulerian circuits of a graph.

9. Show that for a directed graph $G = (X, A)$ with degrees $d_0(x_i) = d_t(x_i)$, $\forall x_i \in X$, the number of distinct Eulerian circuits is given by:

$$\Delta_r \cdot \prod_{x_i \in X} (d_0(x_i) - 1)!$$

where Δ_r is the number of directed spanning trees of G with root x_r and is independent of the choice of x_r. (See Ref. [24].)

10. Use the concept of Eulerian circuits to describe an algorithm for traversing a labyrinth from a starting point s to a finishing point t without traversing any corridor twice in the same direction. Assume that the traverser has zero memory and is only allowed to use two types of markings (e.g. 0 or 1) to mark the entrances and exits of corridors. (See Ref. [23], [13] and [14].)

11. Solve the Chinese postman's problem for the weighted nondirected graph shown in Fig. 9.16. (Use an algorithm from Chapter 9 to calculate shortest paths and solve the matching problem by complete enumeration.)

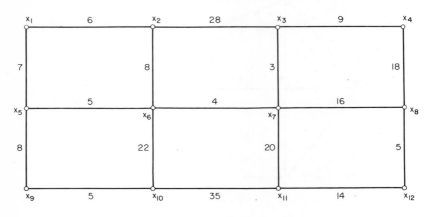

FIG. 9.16

7. References

1. Altman, S., Beltrami, E., Rappaport, S. and Schoepfle, G. (1971). A nonlinear programming model for household refuse collection, *IEEE Trans.*, **SMC-1**, p. 289.
2. Altman, S., Bhagat, N. and Bodin, L. (1971). Extensions of the Clarke and Wright algorithm for routing garbage trucks, Presented at the 8th International TIMS Meeting, Washington D.C.
3. Bellman, R. and Cooke, K. L. (1969). The Konigsberg bridges problem generalized, *Jl. of Math. Anal. and Appl.*, **25**, p. 1.
4. Beltrami, E. J. and Bodin, L. D. (1972). Networks and vehicle routing for municipal waste collection, Proc. 10th Annual Allterton Conference on Circuit and Systems Theory, University of Illinois, p. 143.
5. Berge, C. and Ghouila-Houri, A. (1965). "Programming, game and transportation networks", Methuen, London.
6. Berge, C. (1962). "Theory of graphs", Methuen, London.
7. Busacker, R. G. and Saaty, T. L. (1965). "Finite graphs and networks", McGraw-Hill, New York.
8. Christofides, N. (1973). The optimum traversal of a graph, *Omega—The Int. Jl. of Management Science*, **1**, p. 1.

9. Edmonds, J. (1965). The Chinese postman's problem, Bulletin of the Operations Research Soc. of America, **13**, Supplement 1, p. B-73.

10. Edmonds, J. and Johnson, E. (in press). Matching, Euler tours and the Chinese postman's problem.

11. Eilon, S., Watson-Gandy and Christofides, N. (1971). "Distribution Management: Mathematical modelling and practical analysis", Griffin, London.

12. Even, S. (1973). "Algorithmic Combinatorics", Macmillan, New York.

13. Fraenkel, A. S. (1970). Economic traversal of labyrinths, *Mathematical Magazine*, **43**, p. 125.

14. Fraenkel, A. S. (1971). Economic traversal of labyrinths (Correction), *Mathematical Magazine*, **44**, p. 1.

15. Harary, F. and Nash-Williams, C. St. J. A. (1965). On Eulerian and Hamiltonian graphs and line graphs, *Canadian Mathematical Bulletin*, **8**, p. 701.

16. Kaufmann, A. (1967). "Graphs, dynamic programming and finite games", Academic Press, New York.

17. Kim, W. H. and Chien, R. T-W. (1962). "Topological analysis and synthesis of communication networks", Columbia University Press, New York.

18. Kwan, M-K. (1962). Graphic programming using odd or even points, Chinese Mathematics, **1**, 273.

19. Lovasz, L. (1968). On covering of graphs, *In*: "Theorie des Graphs", Dunod, Paris, p. 231.

20. Marshall, C. W. (1971). "Applied graph theory", Wiley, New York.

21. Seshu, S. and Reed, M. B. (1961). "Linear graphs and electrical networks", Addison-Wesley, Reading, Massachusetts.

22. Stricker, R. (1970). Public sector vehicle routing—the Chinese postman problem, Ph.D. Thesis, Electrical Eng. Dept., M.I.T.

23. Tarry, G. (1895). Le problème des labyrinths, *Nouvelles Ann. de Math.*, **14**, p. 187.

24. Van Aardenne-Ehrenfest, T. and de-Bruijn, N. G. (1951). Circuits and trees in oriented linear graphs, *Simon Stevin*, **28**, p. 203.

25. Voss, H. J. (1968). Some properties of graphs containing k independent circuits, *In*: Theorie des Graphes, Dunod, Paris, p. 321.

26. Welch, J. T. (1966). A mechanical analysis of the cyclic structure of undirected linear graphs, *Jl. of ACM*, **13**, p. 205.

CHAPTER 10

Hamiltonian Circuits, Paths and The Travelling Salesman Problem

1. Introduction

In a number of industries, especially the chemical and pharmaceutical industries, the following basic sheduling problem arises: A number (say n) of items is to be manufactured using a single processing facility or reaction vessel. The facility (vessel), may or may not have to be reset (cleaned), after item p_i has been manufactured, (but before production of p_j is started); depending on the item combination (p_i, p_j). The cost of resetting the facility is constant regardless of the item p_i that has just been produced or the item p_j that is to follow; and, of course, there is no cost incurred if no resetting of the facility is required.† Suppose that these n items are to be manufactured in a continuous cyclic manner, so that after the production of the last of the n items the manufacture of the first item in the fixed cycle is started again.

The problem then arises as to whether a cyclic production sequence for p_i $(i = 1, 2, ..., n)$ can be found which requires no resetting of the facility. The answer to this question depends on whether a graph G whose vertices represent the items and where the existence of an arc (x_i, x_j) implies that p_j can follow the production of item p_i on the facility without resetting, possesses a Hamiltonian circuit or not. (A Hamiltonian circuit is a circuit passing once and only once through each vertex of the graph. See Chapter 1.)

If a cyclic production sequence requiring no resetting of the facility cannot be found, what is the manufacturing sequence which incurs the least additional resetting cost, i.e. which requires the smallest number of necessary resettings? The answer to this question can also be derived by an iterative application of an algorithm which determines a Hamiltonian circuit in a graph, and the way that this can be done is discussed later on in the Chapter.

Thus, a method for finding whether a graph contains a Hamiltonian circuit

---

† The above problem can arise in two ways; either the resetting costs are in reality independent of the items, or, alternatively, detailed cost data is not available and an average constant value is taken as an approximation.

or not, has direct applications in problems of sequencing or scheduling of operations. Equally important, however, is the use of such a method as a basic step in algorithms for the solution of other, seemingly unrelated, graph theory problems.

In this Chapter we will be dealing with two questions:

Problem (i). Given a general graph G, find a Hamiltonian circuit of G, (or alternatively all the circuits), if one or more such circuits exist.

Problem (ii). Given a *complete* graph G whose arcs have arbitrary costs $\mathbf{C} = [c_{ij}]$ associated with them, find that Hamiltonian circuit, (or path), with the least total cost. The problem of finding the least cost Hamiltonian circuit is widely known in the literature as the *travelling salesman problem* [1, 2, 15]. It should be noted that if G is not complete, it can be considered as a complete graph with infinities inserted as the costs of the non-existent arcs.

Solutions to the travelling salesman problem and its variants have a large number of practical applications in many diverse fields. For example, consider the problem where a vehicle leaves a central depot to deliver goods to a given number of customers and return back to the depot. The cost of the trip is proportional to the total distance travelled by the vehicle so that, given the distance matrix between customers, the least cost trip is the solution to the corresponding travelling salesman problem. Similar types of problem occur in collecting mail from mail boxes, scheduling school buses through a number of stops, etc. The problem generalizes quite readily to one where more than one vehicle performs the deliveries, (or collections), although this problem can also be reformulated into a larger size single travelling salesman problem [9]. Other applications include the scheduling of operations on machines [5], the design of electricity supply networks [3], the operation of sequential machines [17] etc.

It is quite obvious that problem (i) above is a special case of problem (ii). Thus, for any given graph G, random finite costs can be allocated to the arcs of G to form a travelling salesman problem. If the solution to this problem— i.e. the shortest Hamiltonian circuit—has a finite value, then this solution is indeed a Hamiltonian circuit of G (i.e. the answer to problem (i)). If on the other hand the solution has infinite value, then G does not possess a Hamiltonian circuit. However, an alternative interpretation can be placed on problem (i) as follows. Consider once more a *complete* graph G_1 with a general arc cost matrix $[c_{ij}]$, and consider the problem of finding that hamiltonian circuit of G_1 whose *longest* arc is a minimum. This problem could be called the *minimax travelling salesman problem* because of the minimax nature of its objective, as compared with the classical travelling salesman problem which could in the present terms also be called the *minisum* problem. We will now show that problem (i) above is in fact equivalent to the minimax travelling salesman problem.

In the complete graph G_1 mentioned above, we can certainly find a Hamiltonian circuit. Let this circuit be Φ_1 and the cost of the longest arc in Φ_1 be $\hat{c}_1$. Remove from G_1 any arc whose cost is greater than or equal to $\hat{c}_1$ to obtain the graph G_2. Find a Hamiltonian circuit say Φ_2 in G_2, and let the cost of the longest arc in Φ_2 be $\hat{c}_2$. Again remove from G_2 any arc with cost greater or equal to $\hat{c}_2$ to form G_3 and continue in the same way until a graph G_{m+1} is found which contains no Hamiltonian circuit. The Hamiltonian circuit Φ_m in G_m (with cost $\hat{c}_m$, say) is then, by definition, the solution to the minimax travelling salesman problem, since the lack of a Hamiltonian circuit in G_{m+1} implies that no Hamiltonian circuit exists in G_1 which does not use at least one arc with cost greater than or equal to $\hat{c}_m$. Thus, an algorithm for finding a Hamiltonian circuit in a graph also solves the minimax travelling salesman problem. Conversely, if we possess an algorithm for solving this latter problem, a Hamiltonian circuit in an arbitrary graph G can be found by constructing a complete graph G_1 on the same set of vertices as G, placing unit costs on the arcs that correspond to arcs in G and infinite costs on all other arcs. If the solution to the minimax travelling salesman problem for G_1 has a finite cost (in fact unity cost), then a corresponding Hamiltonian circuit in G has been found; if the solution has infinite cost then no Hamiltonian circuit in G exists. Hence, since it has been demonstrated that the ability to find a Hamiltonian circuit in a graph implies the ability to solve the minimax travelling salesman problem and vice versa, the two problems can be considered equivalent.

In view of the fact that both problems (i) and (ii) above occur frequently in practical situations, and (as we shall see later) it is much easier to solve problem (i) as a separate problem rather than as a subproblem of problem (ii); we will discuss the two problems separately in Parts I and II of the present Chapter.

PART I

2. Hamiltonian Circuits in a Graph

Given a graph G, there exists no easy criterion or algebraic method for deciding whether G contains a Hamiltonian circuit or not. Existing criteria such as those given by Pósa [29], Nash-Williams [25] and Ore [26] are theoretically interesting but are too loose to be of value for arbitrary graphs encountered in practice (See problem P10-1). The algebraic methods which have appeared in the literature for determining Hamiltonian circuits, are unable to deal with problems of more than a few tens of vertices since they require large amounts of both computer memory and time. More successful is the enumerative technique of Roberts and Flores [30, 31] which does not have large computer storage requirements but whose time requirement still increases exponentially with the number of vertices in the graph. However, another implicit enumeration method [35, 6] has a very low order monomial dependence of computation time with the number of vertices for most types of graphs, and can therefore be used to find Hamiltonian circuits in very large graphs. In the present section we will describe an algebraic method and two enumerative techniques.

2.1 An algebraic method

This method is based on the work of Yau [37], Danielson [11], and Dhawan [14] and involves the generation of all elementary paths by successive matrix multiplications.

The "internal vertex product" of a path $x_1, x_2, x_3, \ldots, x_{k-1}, x_k$ is defined as the sequence of vertices $x_2, x_3, \ldots, x_{k-1}$ excluding the two terminal vertices x_1 and x_k. The "modified variable adjacency matrix" $\mathbf{B} = [\beta(i, j)]$ is an $n \times n$ matrix where $\beta(i, j) = x_j$ if there is an arc from x_i to x_j and zero otherwise. Now suppose we have a matrix $\mathbf{P}_l = [p_l(i, j)]$, where $p_l(i, j)$ is the sum of the internal vertex products of all the elementary paths of cardinality l between vertices x_i and x_j, for $x_i \neq x_j$. Assume $p_1(i, i) = 0$ for all i. The ordinary algebraic matrix product $\mathbf{B} \cdot \mathbf{P}_l \equiv \mathbf{P}'_{l+1} = [p'_{l+1}(s, t)]$ is now given by:

$$p'_{l+1}(s, t) = \sum_k \beta(s, k) \cdot p_l(k, t) \qquad (10.1)$$

i.e. $p'_{l+1}(s, t)$ is the sum of the inner products of all paths from x_s to x_t of cardinality $l + 1$. Since all paths from x_k to x_t represented by the inner vertex products of $p_l(k, t)$ are elementary, the only non-elementary paths that can result from expression (10.1) are those whose inner vertex products in $p_l(k, t)$ contain vertex x_s. Thus, if all terms containing x_s are eliminated from $p'_{l+1}(s, t)$ (and this can be done with a very simple check), $p_{l+1}(s, t)$ will result. The matrix $\mathbf{P}_{l+1} = [p_{l+1}(s, t)]$ with all diagonal terms set to 0, is then the matrix of all elementary paths of cardinality $l + 1$.

Continuing in this way $\mathbf{B} \cdot \mathbf{P}_{l+1}$ will produce $\mathbf{P}_{l+2}$ etc. until the path matrix $\mathbf{P}_{n-1}$ is generated giving all the Hamiltonian paths (which must have a cardinality of $n - 1$), between all pairs of vertices. The Hamiltonian circuits are then immediately given by the paths of $\mathbf{P}_{n-1}$ and those arcs of G which join the end with the initial vertex of each path. Alternatively, the Hamiltonian circuits are given by the inner vertex product terms in any diagonal cell of the matrix $\mathbf{BP}_{n-1}$. (All diagonal entries of this matrix are the same.)

The initial value of the matrix $\mathbf{P}$ (i.e. $\mathbf{P}_1$), can quite obviously be taken as the adjacency matrix $\mathbf{A}$ of the graph with all diagonal elements set to zero.

2.1.1 EXAMPLE. Consider the graph shown in Fig. 10.1 whose adjacency matrix is given by:

		a	b	c	d	e
	a	0	1	0	1	0
	b	0	0	0	1	1
$\mathbf{A} =$	c	0	1	0	0	1
	d	0	0	1	0	0
	e	1	0	1	0	0

and the modified variable adjacency matrix given by:

		a	b	c	d	e
	a	0	b	0	d	0
	b	0	0	0	d	e
$\mathbf{B} =$	c	0	b	0	0	e
	d	0	0	c	0	0
	e	a	0	c	0	0

Set $\mathbf{P}_1 \equiv \mathbf{A}$.

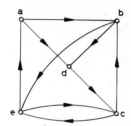

FIG. 10.1. Graph for example 2.1.1

abdcea

adcbea

$\mathbf{P}'_2 = \mathbf{B} \cdot \mathbf{P}_1$ now becomes:

$$
\mathbf{P}'_2 = \quad
\begin{array}{c|ccccc}
 & a & b & c & d & e \\
\hline
a & 0 & 0 & d & b & b \\
b & e & 0 & d+e & 0 & 0 \\
c & e & 0 & \underline{e} & b & b \\
d & 0 & c & 0 & 0 & c \\
e & 0 & a+c & 0 & a & \underline{c} \\
\end{array}
$$

$\mathbf{P}_2$ is the same as $\mathbf{P}'_2$ with the underlined terms set to 0.

$\mathbf{P}'_3 = \mathbf{B} \cdot \mathbf{P}_2$ becomes:

$$
\mathbf{P}'_3 = \quad
\begin{array}{c|ccccc}
 & a & b & c & d & e \\
\hline
a & \underline{be} & dc & bd+be & 0 & dc \\
b & 0 & \underline{dc+ea+ec} & 0 & ea & dc \\
c & be & ea+\underline{ec} & \underline{bd+be} & ea & 0 \\
d & ce & 0 & 0 & \underline{cb} & cb \\
e & \underline{ce} & 0 & ad & ab+cb & \underline{ab+cb} \\
\end{array}
$$

P_3 is the same as P'_3 with the underlined terms set to 0. Similarly:

	a	b	c	d	e
a	$\underline{dce}$	0	0	$\underline{bea}$	$bdc + bdc$
b	dce	0	ead	$\underline{eab} + \underline{ecb}$	$\underline{dcb}$
$P'_4 = c$	0	0	$\underline{ead}$	$bea + eab$ $+ ecb$	$\underline{bdc}$
d	cbe	cea	0	$\underline{cea}$	0
e	$\underline{cbe}$	$adc + \underline{cea}$	$abd + abe$	$\underline{cea}$	$\underline{adc}$

The matrix P_4 of Hamiltonian paths is therefore:

	a	b	c	d	e
a	0	0	0	0	$bdc + dcb$
b	dce	0	ead	0	0
$P_4 = c$	0	0	0	$bea + eab$	0
d	cbe	cea	0	0	0
e	0	adc	abd	0	0

The Hamiltonian paths *abdce* and *adcbe* corresponding to element (1,4) of the matrix above lead to the Hamiltonian circuits:

abdcea

and   *adcbea*

when the return arc (e, a) is added. All other Hamiltonian paths in P_4 lead to the same two Hamiltonian circuits and these are therefore the only two such circuits in G.

The disadvantages of this method are quite obvious. Thus, as the matrix multiplication process continues (i.e. as l increases), each element of the matrix P_l consists of more and more terms up to some critical value of l beyond which the number of terms starts to decrease again. This is because for small values of l and for graphs usually encountered in practice, the number of paths of cardinality $l + 1$ is likely to be greater than the number of paths of cardinality l, whereas for large values of l the reverse is likely to be the case. Moreover, since the length of each inner vertex product term itself is increased by one whenever l is increased by one, the storage requirements of the matrix P_l increase very rapidly indeed up to a maximum for some critical value of l beyond which the requirements start to decrease again.

A slight modification to the above method reduces both its storage and time requirements by large factors. Since we are interested only in the Hamiltonian circuits, and as mentioned earlier these can be obtained from the inner product terms in any diagonal cell of the matrix $\mathbf{B} \cdot \mathbf{P}_{n-1}$, then only the cell $p_{n-1}(1, 1)$ is needed. This can be obtained by calculating and keeping at every stage not the entire matrix $\mathbf{P}_l$ but only the first column of $\mathbf{P}_l$. Thus the first column of $\mathbf{P}_{l+1}$ is derived, (after the deletions of the non-elementary paths mentioned earlier), by multiplying $\mathbf{B}$ with the first column of $\mathbf{P}_l$. This modification reduces both the time and storage requirement of the method by a factor of n. However, even with these modifications a computer program [14] written in PL1 to take advantage of both the string manipulation and variable storage allocation facilities offered by this language, was unable to find all the Hamiltonian circuits for nondirected graphs of more than about 20 vertices and an average vertex degree of more than 4, on an IBM 360/65 computer with a memory of 120,000 bytes. Morover, even for the above mentioned size of graph, the method used virtually all memory locations and the running time was 1·8 minutes, not an insignificant time for such a small graph.

2.2 The enumeration method of Roberts and Flores

Contrary to the algebraic methods which attempt to find all Hamiltonian circuits at once and hence have to store all paths which might conceivably form part of such circuits, the enumeration method considers one path at a time which is continuously extended until such time as: either a Hamiltonian circuit is obtained, or it becomes apparent that the path will not lead to a Hamiltonian circuit. The path is then modified in a systematic way (which will ensure that in the end all possibilities are exhausted), and the search for a Hamiltonian circuit continues. In this way only a very small amount of storage is required for the search and the Hamiltonian circuits are found one at a time.

The following enumerative scheme which uses the usual backtracking technique was initially suggested by Roberts and Flores [30, 31]. The method starts by forming a $k \times n$ matrix $\mathbf{M} = [m_{ij}]$ where element m_{ij} is the ith vertex (x_q say) for which an arc (x_j, x_q) exists in the graph $G = (X, \Gamma)$. The vertices x_q in the set $\Gamma(x_j)$ can be arbitrarily arranged to form the entries of the jth column of the $\mathbf{M}$ matrix. The number of rows k of the matrix $\mathbf{M}$ is then the largest outdegree of a vertex.

The method now proceeds as follows. An initial vertex (say x_1) is chosen as the starting vertex and forms the first entry of the set S which will store the search path at any one time. The first vertex, (say vertex a) in column x_1 is added to S. Then the first feasible vertex (say vertex b) in column a is added to S, then the first feasible vertex (say vertex c) in column b is added to S and so on, where by "feasible" we mean a vertex which is not already in S. Two possibilities now exist which will prevent any vertex being entered into $S = \{x_1, a, b, c, \ldots, x_{r-1}, x_r\}$ at some stage r. Either:

    (1) No vertex in column x_r is feasible.

or(2) The path represented by the sequence of vertices in S is of cardinality
       $n - 1$, i.e. it forms a Hamiltonian path.

In case (2) either:

    (i) arc (x_r, x_1) exists in G and a Hamiltonian circuit is therefore found,

or (ii) arc (x_r, x_1) does not exist and no Hamiltonian circuit can be obtained. In cases (1) and (2.(ii)) *backtracking* must occur, whereas in case (2.(i)) the search can either be stopped and the result printed (if only one Hamiltonian circuit is required), or (if all such circuits are required), the output must be followed by backtracking.

Backtracking involves the removal of the last-entered vertex x_r from S to produce the set $S = \{x_1, a, b, c, \ldots, x_{r-1}\}$ and the addition into S of the first feasible vertex following vertex x_r in column x_{r-1} of the $\mathbf{M}$ matrix. If no such feasible vertex exists a further backtracking step is taken and so on.

The end of the search occurs when the set S consists of the vertex x_1 only and no feasible vertex exists for adding into S so that a backtracking step would leave S empty. The Hamiltonian circuits found up to that time are then all the Hamiltonian circuits that exist in the graph.

2.2.1 EXAMPLE. Consider the graph shown in Fig. 10.2. The $\mathbf{M}$ matrix is as shown below, where the vertices in each column are arranged in alphabetical order:

		a	b	c	d	e	f
	1	b	c	a	c	c	a
$\mathbf{M} =$	2	—	e	d	f	d	b
	3	—	—	—	—	—	—

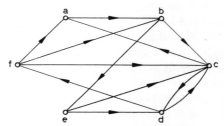

FIG. 10.2. Graph for example 2.2.1

The search to find all Hamiltonian circuits now proceeds as follows:
(Vertex a is taken as the starting vertex.)

Set S		Notes
1. a	:	Add first feasible vertex in column a (i.e. vertex b)
2. a, b	:	Add first feasible vertex in column b (i.e. vertex c)
3. a, b, c	:	First vertex (a) in column c is infeasible ($a \in S$), add next vertex in column c (i.e. vertex d)
4. a, b, c, d	:	Add vertex f
5. a, b, c, d, f	:	No feasible vertex in column f exists. Backtrack
6. a, b, c, d	:	No feasible vertex following vertex f in column d exists. Backtrack
7. a, b, c	:	Similar to case above. Backtrack
8. a, b	:	Add vertex e
9. a, b, e	:	Add vertex c
10. a, b, e, c	:	Add vertex d
11. a, b, e, c, d	:	Add vertex f
12. $\underline{a, b, e, c, d, f}$	:	Hamiltonian path. Hamiltonian circuit closed by arc (f, a). Backtrack
13. a, b, e, c, d	:	Backtrack
14. a, b, e, c	:	Backtrack
15. a, b, e	:	Add vertex d
16. a, b, e, d	:	Add vertex f
17. a, b, e, d, f	:	Add vertex c
18. $\underline{a, b, e, d, f, c}$	:	Hamiltonian path. Circuit closed by arc (c, a). Backtrack
19. a, b, e, d, f	:	Backtrack
20. a, b, e, d	:	Backtrack
21. a, b, e	:	Backtrack
22. a, b	:	Backtrack
23. a	:	Backtrack
24. $\emptyset$	:	End of search

2.2.2 IMPROVEMENTS TO THE BASIC METHOD. Let us say that at some stage during the search, the path being constructed is given by the set $S = \{x_1, x_2, \ldots, x_r\}$, and that the next vertex to be considered for addition into S is $x^* \notin S$. Consider now the following two situations in which a vertex is isolated in the subgraph remaining after deleting from $G = (X, \Gamma)$ all the vertices forming the path constructed so far.

(a) If there exists a vertex $x \in X - S$ such that $x \in \Gamma(x_r)$ and $\Gamma^{-1}(x) \subseteq S$

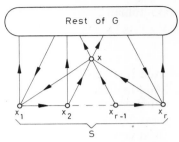

FIG. 10.3 (a). ($x \in \Gamma(x_r)$ and $\Gamma^{-1}(x) \subset S$.) Next arc must be (x_r, x)

(see Fig. 10.3(a)), then if any vertex x^* other than x is added to S, vertex x will not later be reachable by any future end vertex of the path being formed, and the path cannot possibly lead to a Hamiltonian circuit. Thus, under these conditions, x is the only vertex that can be added to S to extend the path.

(b) If there exists a vertex $x \in X - S$ such that $x \notin \Gamma^{-1}(x_1)$ and $\Gamma(x) \subset S \cup \{x^*\}$ for some other vertex x^*, then x^* cannot be added to S since then no possible path between x and x_1 would exist in the remaining subgraph. The path composed of $S \cup \{x^*\}$, therefore, cannot possibly lead to a Hamiltonian circuit and another vertex, different from x^*, should be considered for addition to the set S. (See Fig. 10.3(b).)

In the example given just above, situation (a) arises at step 2 when the set S is $\{a, b\}$. We then see that for vertex e, $\Gamma^{-1}(e) = \{b\} \subset S$ so that e must be the next vertex entered into the set $\{a, b\}$. One can therefore skip steps 3–8 of the above example and go immediately from step 2 to step 9 as shown.

The tests for conditions (a) and (b) will, of course, slow down the iterative procedure and for small graphs (less than about 20 vertices), there is no improvement over the original Roberts and Flores method. For larger graphs, however, these tests cause a marked improvement in the required

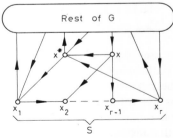

FIG. 10.3 (b). ($x \notin \Gamma^{-1}(x_1)$ and $\Gamma(x) \subset S \cup \{x^*\}$). Next arc must not be (x_r, x^*)

computation times, usually be factors of 2 or more. Detailed computational results for the various methods are shown in Fig. 10.7, Section 3.

2.3 The multi-path method

By examining the operation of the enumeration algorithm of Roberts and Flores, it soon becomes apparent that even with the improvements introduced earlier, insufficient notice is taken of the consequences of extending the path being formed, upon the rest of the graph. In general, the formation of a path S_o by the search (S_0 will be considered both as an ordered sequence of vertices and as an ordinary set) *implies* other paths $S_1, S_2, \ldots$ in other parts of the graph. These implied paths either help to complete a Hamiltonian circuit more quickly, or point to the fact that no Hamiltonian circuit exists which contains the path S_o as a part, in which case backtracking can occur immediately.

The method described in this section was originally developed by Selby [35] for nondirected graphs and a somewhat modified version is described here for directed ones. The method proceeds as follows.

Suppose that at some stage of the search a path S_o has been formed and paths $S_1, S_2, \ldots$ are implied. Considering any "middle" vertex of any of these paths (by "middle" here is meant any vertex other than the beginning and end ones), it is obvious that since this vertex is already linked in a path by two arcs, all other arcs to or from such a vertex can be removed from the graph. For any beginning vertex of the above paths all arcs emanating from such a vertex (except for the arc linking it to the path) can be removed, and for the end vertex of any of the paths all arcs terminating there (except for the arc linking it to the path) can also be removed. Moreover, except for the case where only one path (say S_o) exists and which passes through all the vertices of G (i.e. when S_o is a Hamiltonian path), any existing arcs leading from the end of any path to the beginning vertex of the same path can be removed since such arcs would close non-Hamiltonian cycles.

The removal of all these arcs will leave the graph G with many vertices— all the "middle" vertices of the paths—having only one arc terminating at, and one arc emanating from them. All these "middle" vertices and the arcs incident on them are removed from G and instead a single arc is introduced for each path from its beginning to its end vertex, the result is called a *reduced graph* $G_k = (X_k, \Gamma_k)$; where k is an index indicating the stage of the search.

Consider now the extension of the path S_o being formed by the search, by another vertex x_j which is (by itself) a feasible vertex to add to S_o in the sense mentioned for the Roberts and Flores algorithm, i.e. there is an arc in G_k from the end vertex of S_o—call this vertex $e(S_o)$—to vertex x_j. The consequences of adding x_j to S_o are the following.

(1) First remove from G_k all necessary arcs, i.e.

(i) All arcs terminating at x_j or emanating from $e(S_o)$, except arc $(e(S_o), x_j)$.

(ii) Any arc from x_j to the beginning vertex of S_o.

(iii) If x_j happens to be the beginning vertex of another path S_j, remove also any arc from the end vertex of S_j to the beginning vertex of S_o.

(2) Let the graph after the arc removals be called $G'_k = (X_k, \Gamma'_k)$.

● If there exists a vertex x of G'_k which is not the end of any of the paths $S_o, S_1, \ldots$ etc. and which as a result of the arc deletions now has an indegree of unity, i.e. $|\Gamma_{k'}^{-1}(x)| = 1$, then erase all arcs emanating from vertex $v = \Gamma_{k'}^{-1}(x)$ except for arc (v, x).

● If there exists a vertex x of G'_k which is not the beginning of any path and for which as a result of the arc deletions now has an out-degree of unity, i.e. $|\Gamma_{k'}(x)| = 1$, then erase all arcs emanating from x except for arc $(x, \Gamma_{k'}(x))$.

● Update all paths and remove any arcs from the end to the beginning vertices.

Repeat step 2 until no further arcs can be removed.

(3) Remove from the final graph G'_k any vertices which have both an indegree and outdegree of unity i.e. vertices which have now become "middle" vertices of paths. This removal is carried out in the way mentioned earlier and the result is a new reduced graph G_{k+1} which replaces the previous graph G_k.

If now the addition of vertex x_j to path S_o causes the indegree or outdegree (or both) of some vertex x at the end of step 2 to become zero, then quite obviously, no Hamiltonian circuit exists. Vertex x_j is then rejected and another vertex x_j is chosen from the set of vertices $\Gamma_{k'}[e(S_o)]$ as a possible extension of path S_o, until the set $\Gamma_{k'}[e(S_o)]$ is exhausted and backtracking then occurs—(i.e. $e(S_o)$ is removed from S_o and replaced by another vertex, etc.) Note that backtracking now also implies that enough information must be stored about the arcs being removed by steps 1 and 2 at each stage k, so as to be able to reconstruct G_k from G_{k+1} for any k, when the backtracking step occurs.

If (at some stage) at the end of step 2, it is found that only one path is left passing through all the vertices, then the existence of a Hamiltonian circuit can be immediately checked. If no circuit is found (or if one is found but all are required), backtracking can occur.

A computationally superior way is to check at each iteration of step 2 (instead of at the end) that the indegrees and outdegrees of all vertices in G'_k are non-zero. Thus as soon as one of these becomes zero backtracking can occur immediately and when a Hamiltonian path is found a Hamiltonian circuit is implied without any checking for a return arc.

If none of the above cases occur, i.e. if (at stage k), at the end of step 2, there is more than one path left and all the in and outdegrees are non-zero, no conclusions can yet be drawn. x_j is now added to S_o and another vertex is chosen to extend the new path S_o even further. Steps 1, 2 and 3 are then repeated starting from the new reduced graph.

2.3.1 EXAMPLE. The purpose of this example is to show how the iterative process of step 2 can lead to many paths implied by the forced path S_o, and how these paths can lead to a quick termination of the current search.

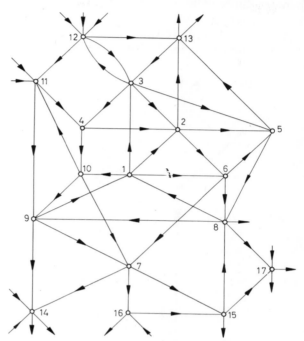

FIG. 10.4. Graph for the example of section 2.3.1

Consider the part of a graph as shown in Fig. 10.4. In the first instance all vertices have indegree and outdegree values greater than unity.

First iteration. Let us begin by taking vertex 1 as the starting vertex of the path to be formed (S_o), and let the first attempted addition to S_o be vertex 6, say, so that S_o becomes $\{1, 6\}$.

Step 1 of the method would then remove from G arcs $(1, 2)$, $(1, 3)$, $(1, 10)$ and $(2, 6)$.

Step 2 would now locate vertex 10 as having indegree one which leads to the implied path $S_1 = \{4, 10\}$ and arc $(4, 2)$ is removed. The second pass through step 2 locates vertex 2 with indegree one which leads to the implied path $S_2 = \{3, 2\}$ and the removal of arcs $(3,12)$, $(3,5)$ and $(3,4)$. The third pass through step 2 would locate vertex 4 with indegree one which leads to an extension of S_1 by the addition of arc $(11, 4)$ so that S_1 is now the path $\{11, 4, 10\}$. Arc $(11, 9)$ is now removed by step 2 and so is arc $(10,11)$ which closes a circuit with path S_1. The fourth pass through step 2 does not produce any further removals and the first reduced graph G_1 at the end of step 3 is then as shown in Fig. 10.5(a), where the paths S_o, S_1 and S_2 are shown by heavy lines. *Second iteration.* Since more than one path exists and since both the in and outdegrees of all vertices are non-zero, we continue to extend path S_o even further. Let us say that of the vertices $\Gamma[e(S_o)] = \{8, 5, 7\}$ vertex 8 is the one chosen so that S_o now becomes the path $\{1, 6, 8\}$.

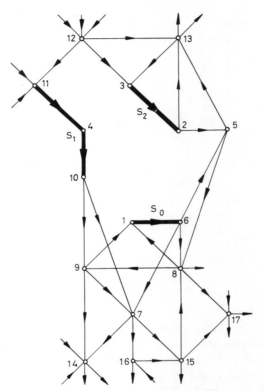

FIG. 10.5 (a). The first reduced graph G_1

Step 1 would now locate vertex 5 with both in and outdegree of one, and vertex 1 with indegree of one. Taking vertex 5 first the second pass through step 2 leads to the addition of arcs $(2, 5)$ and $(5, 13)$ to extend path S_2 which now becomes $\{3, 2, 5, 13\}$. Arcs $(2, 13)$, $(12, 13)$ and the return arc $(13, 1)$ are now removed. The second pass through step 2 would locate vertex 3 with indegree one as well as vertex 1 (with indegree one), and which existed from the previous pass. Taking vertex 3 first, arc $(12, 3)$ is added to S_2 which now becomes $\{12, 3, 2, 5, 13\}$ and arc $(12, 11)$ is removed. The third pass through step 2 would still locate vertex 1 with indegree one but no other vertex, so arc $(9, 1)$ is added to path S_o which now becomes $\{9, 1, 6, 8\}$ and arcs $(9, 7)$ $(9, 14)$ and the return arc $(8, 9)$ are removed. The fourth pass through step 2 would locate two vertices 9 and 7 with indegree one. Taking vertex 9 first, arc $(10, 9)$ is added which now joins paths S_1 and S_o together to become the new path S_o, which now is $\{11, 4, 10, 9, 1, 6, 8\}$. Arc $(10, 7)$ is now removed. The fifth pass through step 2 now locates vertex 7 with indegree zero indicating that the present search cannot possibly lead to any Hamiltonian circuit.

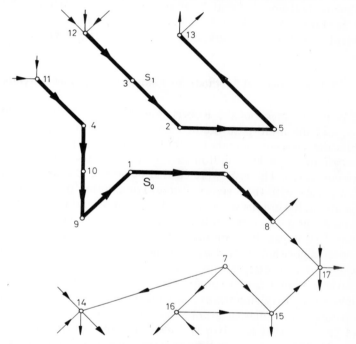

FIG. 10.5 (b). Graph G_1' at the end of the fourth pass through step 2 in the second iteration

The graph G_1' at the end of the fourth pass through step 2 is as shown in Fig. 10.5(b). One must now replace all arcs removed during the second iteration in order to arrive back at the graph G_1 of Fig. 10.5(a) and then back-track, i.e. remove vertex 8 from S_o, replace it by another feasible vertex (either 5 or 7), and continue to apply steps 1, 2 and 3, etc.

From the example above it is seen that this is a very powerful search method; it has deduced in two iterations that no Hamiltonian circuit can contain the path 1, 6, 8 as a part. This, considering the fact that the outdegree of vertex 1 is 4 and that of vertex 6 is 3, is 1/12th of the total search effort. Moreover, this conclusion was arrived at only from the part of the graph shown in Fig. 10.4 which may itself belong to a very much larger graph. For a given average value of vertex degrees one would, therefore, expect this search to be only slightly dependent on the size (number of vertices) of the graph, a fact which is demonstrated by the experimental results of the next section.

By contrast, the method of Roberts and Flores (or its improved version) would require an enormous amount of searching to arrive at the same result since that method would essentially enumerate all paths starting with $1, 6, 8, \ldots$ before backtracking back to $1, 6, \ldots$ and the effort involved is obviously dependent on the size of the graph as well as the vertex degrees.

3. Comparison of Methods for Finding Hamiltonian Circuits

The original version of the Roberts and Flores algorithm, its improved version, and the multi-path method are compared in the present section. The three methods are compared on the basis of the computing times required in order to find a single Hamiltonian circuit if one exists, or indicate that no such circuit exists. The tests were carried out on randomly generated non-directed graphs with their vertex degrees lying within predefined ranges. A total of about 200 graphs were involved in the tests and the results given are averages. All graphs happened to possess Hamiltonian circuits.

Figure 10.6 shows the computing times required by the original Roberts and Flores algorithm plotted against the number of vertices in the graph and where the vertex degrees are in the range 3–5. Because of the very large time variations that can occur for the same size graphs, three curves are shown in this figure giving the average, maximum, and minimum computing times recorded for different graphs having the same number of vertices. It should be noted that Fig. 10.6 is drawn on semi-logarithmic paper which indicates that the computing time requirements increase at an exponential rate with the number of vertices. A formula giving the approximate computation time T as a function of the number of vertices n in the graph with vertex degrees in the range 3–5, is given by:

$$T = 0\cdot85 \times 10^{-4} \times 10^{0\cdot155n} \text{ (seconds—CDC 6600)}$$

The improved version of the Roberts and Flores algorithm is not very much better than the original and the algorithm still requires a computing time more or less exponentially increasing with n. The ratio of the computing times of the two versions for non-directed graphs with vertex degrees in the range 3–5 is shown in Fig. 10.7. From this figure it is noted that for small graphs the "improved" version is actually worse, although for larger graphs (with greater than about 20 vertices), there is more than a 50% saving in computation time by using the improved algorithm.

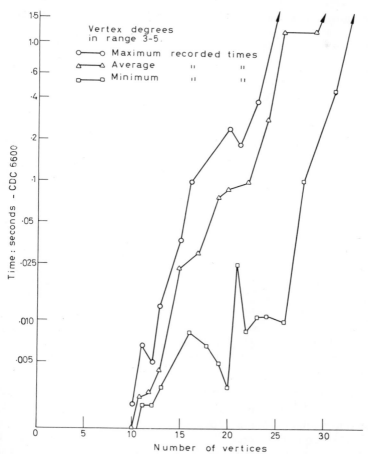

FIG. 10.6. Computational performance of Roberts & Flores' algorithm

With the same set of graphs mentioned above, the multi-path algorithm proved very effective indeed. This is demonstrated by Fig. 10.8 showing the computation times required by this algorithm. It is seen from this figure (which is plotted on linear scale), that the time increases *very* slowly with the number of vertices in the graphs so that very large graphs can be handled by this algorithm.† A further advantage of the method is that there is remarkably little variation in the times taken to solve different graphs of the same size, and therefore one could estimate with reasonable confidence the computational time requirements for various problems. Moreover, experiments with graphs whose vertices have degrees in different ranges from the above 3–5 range, have shown that this method is virtually unaffected by the vertex degrees.

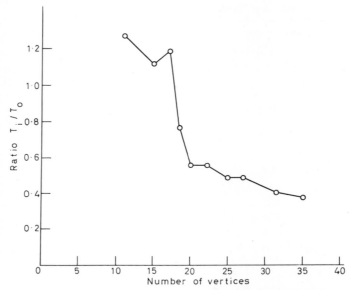

FIG. 10.7. Performance of improved Roberts & Flores' algorithm
T_0: Computing time of original method
T_i: Computing time of improved method

The computational results given in Figs. 10.6 to 10.8 are concerned with finding a single Hamiltonian circuit in a graph. It may, however, be interesting to mention some computational experience with the three types of algorithms when it is required to find all the Hamiltonian circuits. Thus, for a nondirected graph of 20 vertices with vertex degrees in the range 3–5, the original Roberts and Flores algorithm took 2 seconds to find all the Hamiltonian circuits (there were 18 in all); the improved version of the same algorithm took 1·2 seconds and the multi-path algorithm 0·07 seconds, all times being for the CDC 6600 computer.

† This should not be taken to mean that the algorithm is *guaranteed* to terminate in time proportional to n^k for some $k > 0$. Indeed special examples have been constructed which require much larger times to solve than the times shown in Fig. 10.8. Such counterexamples involve graphs which do not contain Hamiltonian circuits.

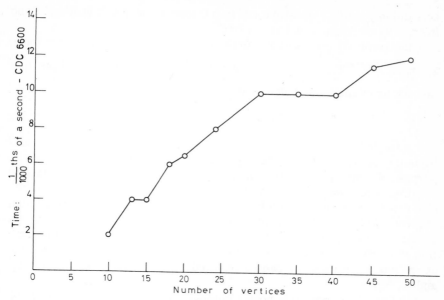

FIG. 10.8. Computational performance of multipath algorithm

4. A Simple Scheduling Problem

In the Introduction we described the case of a manufacturing company producing items $p_1, p_2, \ldots, p_n$ using a single facility on a cyclic rota basis and asked the question what is the production sequence requiring the smallest number of facility resettings that is necessary. A graph $G = (X, A)$ was constructed with vertices x_i representing items p_i $(i = 1, 2, \ldots, n)$ and arcs (x_i, x_j) indicating that item p_j could follow item p_i onto the facility without any resetting.

Now, if the graph $G = (X, A)$ possesses a Hamiltonian circuit say x_{i_1}, $x_{i_2}, x_{i_3}, \ldots, x_{i_n}$ then the corresponding sequence of items $p_{i_1}, p_{i_2}, p_{i_3}, \ldots, p_{i_n}$ can be manufactured by the facility without any resetting, since by definition of the Hamiltonian circuit:

$(x_{i_k}, x_{i_{k+1}}) \in A$ for all $k = 1, 2, \ldots, (x_{i_{n+1}} \equiv x_{i_1})$, and A has been defined above as the set $\{(x_i, x_j) | c_{ij} = 0\}$.

If G possesses no Hamiltonian circuit, we can construct the graph $G_1 = (X_1, A_1)$ where:

$$X_1 = X \cup \{y_1\}$$

and

$$A_1 = A \cup \{(x, y_1) | x \in X\} \cup \{(y_1, x) | x \in X\}$$

i.e. a dummy vertex y_1 is introduced into G together with arcs leading to, and from, it to every other real vertex of G.

If the graph G_1 possesses a Hamiltonian circuit, this will have the form:

$x_{i_1}, x_{i_2}, \ldots, x_{i_r}, \boxed{y_1}, x_{i_{r+1}}, \ldots, x_{i_n}$, which can be interpreted to mean that that the items can be manufactured in a sequence:

$$p_{i_{r+1}}, p_{i_{r+2}}, \ldots, p_{i_n}, p_{i_1}, p_{i_2}, \ldots, p_{i_r}$$

with a single facility resetting operation between the finishing of the last and beginning of the first item in the sequence. Thus the dummy vertex serves the purpose of a marker indicating the position in the sequence where a facility resetting is necessary. In terms of the graph, the dummy vertex y_1 and its associated arcs provide a path between any two real vertices. Thus if an item sequence with one resetting operation exists, i.e. if a Hamiltonian circuit would exist in G provided that one "arc" $(x_i, x_j) \notin A$ could be used, the addition of y_1 to G will always cause a Hamiltonian circuit to exist in G_1 since the extra "arc" (x_i, x_j) needed, can be replaced by the two dummy arcs (x_i, y_1) and (y_1, x_j).

If the graph G_1 possesses no Hamiltonian circuit we construct the graph $G_2 = (X_2, A_2)$ from the graph G_1, where:

$$X_2 = X_1 \cup \{y_2\}$$

and

$$A_2 = A_1 \cup \{(x, y_2) \mid x \in X\} \cup \{(y_2, x) \mid x \in X\}$$

and continue in the same way.

THEOREM 1. *If the graph* $G_m = (X_m, A_m)$ *given by:*

$$X_m = X \cup \left[\bigcup_{j=1}^{m} \{y_j\} \right] \tag{10.2}$$

$$A_m = A \cup \left[\bigcup_{j=1}^{m} \{(x, y_j) \mid x \in X\} \right] \cup \left[\bigcup_{j=1}^{m} \{(y_j, x) \mid x \in X\} \right] \tag{10.3}$$

contains a Hamiltonian circuit, but the graph G_{m-1} *defined in a similar way does not, then m is the minimum number of facility resettings that are necessary and if the Hamiltonian circuit of* G_m *is:*

$$x_{i_1}, \ldots, x_{i_\alpha}, \boxed{y_1}, x_{i_{\alpha+1}}, \ldots, x_{i_\beta}, \boxed{y_2}, x_{i_{\beta+1}}, \ldots, x_{i_\gamma}, \boxed{y_3}, x_{i_{\gamma+1}}, \ldots,$$

$$etc., \ldots, x_{i_\delta}, \boxed{y_m}, x_{i_{\delta+1}}, \ldots, x_{i_n}$$

then the products should be produced in m sequences given by:

$$p_{i_{\alpha+1}}, \ldots, p_{i_{\beta}} \text{ followed by } p_{i_{\beta+1}}, \ldots, p_{i_{\gamma}}, \ldots \text{ etc.} \ldots$$

$$\text{followed by } p_{i_{\delta+1}}, \ldots, p_{i_n}, p_{i_1}, \ldots, p_{i_{\alpha}}$$

Proof. The proof follows immediately by induction from the argument preceding the theorem.

4.1 Computing aspects

In the argument of the preceding section the graph G was increased by a single dummy vertex at a time. If the optimum solution to the problem involves m resettings of the facility, then $m + 1$ attempts have to be made to find Hamiltonian circuits in the graphs G, G_1, ..., G_m, with only the last one of these attempts being successful and leading to a solution of the problem. Obviously m is bounded from above by n and in general for practical problems m will only be a very small fraction of n. Nevertheless, a different procedure for generating and testing the graphs for Hamiltonian circuits is, in fact, necessary from the computational point of view, since it turns out [6] that it is faster to find a Hamiltonian circuit in a graph that possesses one rather than prove that no Hamiltonian circuit exists in a graph that does not possess one. This fact immediately suggests that a better algorithm is one which starts with an upper bound B for the optimal (minimal) number of facility resettings m, and sequentially forms and tests the graphs G_B, G_{B-1}, etc. until a graph G_{m-1} is found which possesses no Hamiltonian circuit.

A short description of such an algorithm is given below.

Step 1. Find an upper bound B to be the optimal (minimal) number of facility resettings that may be necessary. (See Appendix.)

Step 2. Form the graph G_B according to eqns (10.2) and (10.3).

Step 3. Does G_B possess a Hamiltonian circuit? If yes store the circuit in vector H overwriting any previous circuit stored there and go to step 4, else go to step 5.

Step 4. $B \leftarrow B - 1$, go to step 2.

Step 5. Stop. $m = B + 1$ is the minimal number of facility resettings required and the last sequence in H is the required item manufacturing sequence.

Step 1 of the above algorithm requires further explanation. Obviously, the tighter the initial upper bound B is, the fewer will be the number of iterations of the main algorithm. A procedure for calculating a good initial upper bound is given in the Appendix, whereas at step 3 above the existence of a Hamiltonian in graph G_B could be checked by using the multipath algorithm of Section 2.3.

PART II

5. The Travelling Salesman Problem

The travelling salesman problem is closely related to several other problems in graph theory and which are discussed in other parts of this book. Here we will explore two such relationships namely that with the assignment problem (see Chapter 12), and that with the problem of the shortest spanning tree (see Chapter 7). Both of these problems can be solved very much more easily than the travelling salesman problem and one may be able to exploit these relationships in order to produce an efficient method of solution to the latter problem.

5.1 A lower bound from the assignment problem

The linear assignment problem for a graph with a general cost matrix $C = [c_{ij}]$ can be stated as follows (see Chapter 12).

Let ξ_{ij} be an $n \times n$ matrix of 0–1 valued variables so that $\xi_{ij} = 1$ if vertex x_i is "assigned" to vertex x_j and $\xi_{ij} = 0$ if x_i is not assigned to x_j. In the travelling salesman problem we could use a similar scheme where $\xi_{ij} = 1$ would mean that the salesman travels from x_i to x_j directly and $\xi_{ij} = 0$ would mean that he does not. For this last problem we can start by setting all c_{ii} ($i = 1, \ldots, n$) to ∞ thus eliminating non-sensical solutions with $\xi_{ii} = 1$.

The assignment problem now becomes:

Find 0–1 variables ξ_{ij} so as to minimize:

$$z = \sum_{j=1}^{n} \sum_{i=1}^{n} c_{ij} \xi_{ij} \tag{10.4}$$

subject to

$$\sum_i \xi_{ij} = \sum_j \xi_{ij} = 1 \tag{10.5}$$

(for all i and $j = 1, 2, \ldots, n$)

and

$$\xi_{ij} = 0 \text{ or } 1 \tag{10.6}$$

Equations (10.5) simply insure that the solution is cyclic, i.e. one arc enters and one leaves every vertex.

Equations (10.4)–(10.6) together with the additional constraints that the solution must form a *single* (Hamiltonian) circuit and not just a number of disjoint circuits, can also be used to represent a formulation of the travelling salesman problem. (Note that the setting of $c_{ii} = \infty$ at the start could be interpreted as constraints removing the possibility of circuits of cardinality 1 from appearing in the solution of the assignment problem.) Since the addition of any constraint to the assignment problem can only increase or leave unchanged the minimum value of z as calculated from eqns (10.4)–(10.6), this value of z is a valid lower bound to the cost of the solution to the travelling salesman problem for a graph with a cost matrix $[c_{ij}]$.

5.2 A lower bound from the shortest spanning tree

In the case of a graph with a symmetrical cost (distance) matrix **C**, i.e. a nondirected graph, a lower bound to the solution of the travelling salesman problem can be derived from the shortest spanning tree of the graph as follows. Let us suppose that it is specified that link (x_1, x_2) is in the optimal travelling salesman circuit. If this link is removed from the circuit, a path of $n - 1$ links is obtained going through all the vertices starting at x_1 and finishing at x_2. Thus, since the cost of the shortest spanning tree, $L(\text{SST})$ say, is a lower bound to the cost of this path, the length of the shortest spanning tree plus $c(x_1, x_2)$ is a lower bound to the cost of the optimal travelling salesman solution.

In general no link (x_1, x_2) in the optimal cycle will be known, but the longest link in the circuit must be [15] at least as long as $\max_{x_i} [c(x_i, s)]$; where s is the second nearest vertex to vertex x_i. Thus:

$$L(\text{SST}) + \max_{x_i} [c(x_i, s)] \tag{10.7}$$

is a valid lower bound to the cost of the solution to the travelling salesman problem.

5.3 Duality relations

Let us define $G(\text{TSP})$ to be the partial graph of a nondirected graph G, formed by the vertices and those links of G that are used by the optimal travelling salesman circuit. Similarly let us define graphs $G(\text{AP})$ and $G(\text{SST})$ formed by the same vertices but having links which appear in the optimal solutions of the assignment and shortest spanning tree problems respectively.

The graph $G(\text{TSP})$, which in the case of a 5-vertex graph is shown in Fig. 10.9(a), has the following properties:

(i) The graph is connected; i.e. every vertex can reach every other vertex via a path using the links.
and

(ii) The degree of every vertex is 2; i.e. there are two links incident at each vertex.

The graph $G(AP)$ does not necessarily possess property (i) above, (as can be seen from the example of Fig. 10.9(b) but does, by definition, have property (ii). If, however, it so happens that the solution to the assignment problem does have property (i) as well, then this is also the solution to the travelling salesman problem.

The graph $G(SST)$ has property (i), by definition, but does not have property (ii). If, however, it so happens that the shortest spanning tree does have

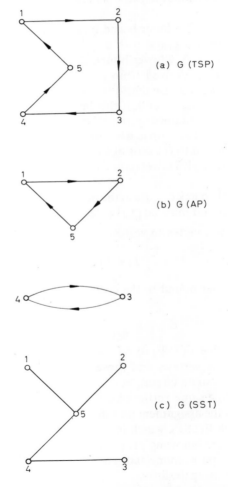

FIG. 10.9. Graphs of travelling salesman, assignment, and shortest spanning tree problems

property (ii)—except for two "end" vertices (x_1 and x_2 say) which by necessity must have a degree of unity—then the shortest spanning tree is also the shortest path passing through all the n vertices. If, moreover, link (x_1, x_2) is in the optimal travelling salesman circuit, then the links of the shortest spanning tree plus link (x_1, x_2) will be the solution to the travelling salesman problem. (If it is not certain that (x_1, x_2) is in the optimal travelling salesman circuit then some small modification is needed. See Section 6.)

Thus, the solutions of the assignment and shortest spanning tree problems are dual in the sense that they possess properties which are complementary with respect to the properties of the travelling salesman problem. Two possible avenues of investigation now reveal themselves which can lead to solutions for this latter problem.

(a) Use the solution of the assignment problem which possesses property (ii) and force this solution to conform with property (i).

or (b) Use the solution of the shortest spanning tree which possesses property (i) and force it to conform with property (ii).

In Section 6 of this Chapter we explore avenues based on method (b) and in Section 7 we consider method (a). It should be remembered, however, that although the assignment problem is defined for graphs with any arbitrary cost structure, the shortest spanning tree is only defined for non-directed graphs—i.e. graphs with symmetrical, $(c_{ij} = c_{ji})$, cost matrices. For asymmetrical cost matrices (i.e. directed graphs), the arborescence—a concept analogous to the spanning tree—was introduced in Chapter 7. Thus, what is being said about the relationship between the travelling salesman problem and the shortest spanning tree for non-directed graphs, has an exact equivalent applying to the relationship between the travelling salesman and the shortest arborescence for the case of directed graphs. In the section that follows, however, we will limit ourselves to the symmetrical problem, the extension to the more general case being quite apparent.

6. The Travelling Salesman and Shortest Spanning Tree Problems

As mentioned in the previous section, the symmetrical travelling salesman problem is closely related to the shortest spanning tree of G and in particular the *open* travelling salesman problem (i.e. that of finding the shortest Hamiltonian *path* of a graph), is equivalent to the problem of finding the shortest spanning tree of G with the restriction that no vertex should have degree greater than 2. Because of this slightly more direct relationship between the open (rather than the *closed*) travelling salesman problem and the shortest spanning tree, we will begin by considering this problem first and postpone till the end of the section the discussion of the few changes needed to deal with the closed (ordinary) travelling salesman problem.

The problem of finding the shortest Hamiltonian path was first investigated by Deo and Hakimi [13] who gave a linear programming formulation of the problem. For a completely connected graph with n vertices, their formulation involves $n(n + 1)$ variables and $n(n + 3)/2 + 1$ constraints which are considered explicitly, together with a very large number of constraints which cannot be included explicitly but which can be considered implicitly by introducing a few of them at a time after each iterative application of the simplex method. Although the linear programming method always gives a solution, it has an inherent bulkiness and inefficiency and will not be discussed here further.

6.1 Definitions

Let $G = (X, A)$ be a link-weighted nondirected graph, and let d_i^G be the degree of vertex x_i with respect to graph G. The number of vertices in G will be denoted by $n = |X|$.

Given a spanning tree $T = (X, A_T)$ of a graph G and denoting the degrees of the vertices with respect to T as d_i^T, we can define the closeness ε_T of this tree to a Hamiltonian path in one of the following ways:

$$\varepsilon_T = \sum_{d_i^T > 2} (d_i^T - 2) \qquad (10.8)\,(a)$$

or

$$\varepsilon_T = \sum_{i=1}^{n} |d_i^T - 2| - 2. \qquad (10.8)\,(b)$$

Equation 10.8(a) bases the closeness only on those vertices whose $d_i^T > 2$, whereas 10.2(b) also includes those vertices which have degree 1. Both of the above definitions give $\varepsilon_T = 0$ for a Hamiltonian path and it is assumed that the larger the value of ε_T the greater is the departure of the tree T from a Hamiltonian path.

Two problems will now be considered as follows.

Problem (a): *Shortest Hamiltonian path.* Find the shortest spanning tree $T^* \equiv (X, A^*)$ of the graph G so that the degree of no vertex exceeds 2.

Problem (b): *Shortest Hamiltonian path with specified end vertices.* Given two vertices x_1 and x_2 $(x_1, x_2 \in X)$, find the shortest spanning tree $T_{1,2}^* \equiv (X, A_{1,2})$, so that the degree of no vertex exceeds 2 and the degree of vertices x_1 and x_2 is one.

The implication in the problems stated above, that a tree T all of whose vertices have degrees less than or equal to 2 is in fact a Hamiltonian path, needs justification. Thus, since T is a tree, the degree of no vertex can be zero and hence $d_i^T = 1$ or 2 for all i. Let q of the vertices have degree 1 and $n - q$

vertices have degree 2. The number of links in the tree is given by

$$m_T = \tfrac{1}{2} \sum_{i=1}^{n} d_i^T \qquad (10.9)$$

since in the summation every link is counted twice, once for each of its end vertices.

Thus eqn (10.9) becomes

$$m_T = \tfrac{1}{2}[q + 2(n - q)] = n - \frac{q}{2}$$

Since the number of links in a tree is $n - 1$, $q = 2$ and hence exactly two vertices have degree 1 and $(n - 2)$ vertices degree 2 i.e. T is a Hamiltonian path.

6.2 A decision-tree search algorithm

6.2.1 SOLUTION OF PROBLEM (A). We will consider Problem (a) first and explain the decision-tree search algorithm with the aid of an example.

Example. The example is a 6-vertex completely connected graph which is the same as that given by Deo and Hakimi [13], and its link-cost matrix is shown below:

	1	2	3	4	5	6
1	0	4	10	18	5	10
2	4	0	12	8	2	6
3	10	12	0	4	18	16
4	18	8	4	0	14	6
5	5	2	18	14	0	16
6	10	6	16	6	16	0

$$\mathbf{C} = [c_{ij}] =$$

The shortest spanning tree of this graph (without any restrictions placed on the d_i^T) is found very simply by the method of Chapter 7, and is as shown in Fig. 10.10 (a). From this figure it is seen that $d_2^T = 3$, and since we want the shortest spanning tree with all $d_i^T \leq 2$, we can say that at least one of the links (2, 1) (2, 6) or (2, 5) must be absent from the final answer. Thus, the solution to the original problem is a solution to at least one of the three subproblems represented by nodes B, C and D in the decision-tree of Fig. 10.11. In this figure node A represents the original problem and nodes B, C and D represent

I

problems whose cost matrices are the same as the cost matrix of the problem represented by node A but with the costs of links (2, 1) (2, 6) or (2, 5), respectively, set to infinity.

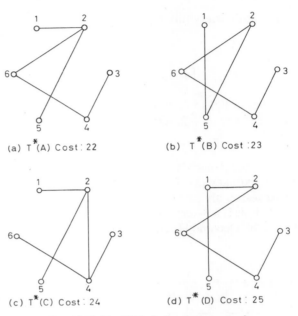

(a) $T^*(A)$ Cost: 22

(b) $T^*(B)$ Cost: 23

(c) $T^*(C)$ Cost: 24

(d) $T^*(D)$ Cost: 25

FIG. 10.10. The SST's during the tree-search

In the decision-tree of Fig. 10.11 the number next to a node is the cost of the shortest spanning tree corresponding to the subproblem defined by that node. It should be noted that at any stage of the decision-tree branching, the smallest of the costs on the free nodes (i.e. the nodes which have not yet been branched from) is a lower bound to the cost of the final answer.

We now find the shortest spanning tree $T(B)$ $T(C)$ and $T(D)$ corresponding to nodes B, C and D; the results are shown in Figs. 10.10(b), 10.10(c) and 10.10(d) respectively, and the costs of these trees are shown next to the nodes.

The trees $T(B)$ and $T(D)$ are Hamiltonian paths (all $d_i^T \leq 2$), whereas tree $T(C)$ is not. The shortest of the two Hamiltonian paths is $T(B)$ with a cost of 23. The lower bound on the cost of the final answer is the smallest of the costs of $T(B)$, $T(C)$ and $T(D)$ which is also 23, and therefore $T(B)$ is the optimal answer, i.e. it is the shortest Hamiltonian path. One should note that no further branching from node C is necessary because any Hamiltonian path that may result from such branching will have a cost of at least 24.

Thus, the decision-tree search algorithm has produced the optimal answer to this particular example by applying the shortest spanning tree algorithm only four times.

The method also has other side advantages in that it often produces additional Hamiltonian paths which are near optimal and may be alternative acceptable solutions to a practical problem.

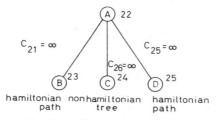

FIG. 10.11. The decision-tree search

6.2.2 SOLUTION OF PROBLEM (B). In Problem (b) the objective is to find the shortest Hamiltonian path whose two end vertices are specified as x_1 and x_2. This problem can be transformed into Problem (a) by the use of the following theorem:

THEOREM 2. *Let* $\mathbf{C} = [c_{ij}]$ *be the link-cost matrix of the original graph G and k be a large positive number greater than the cost of any Hamiltonian path. Then the solution to Problem (a) with the link-cost matrix* $\mathbf{C}'$ *where:*

$$\left.\begin{aligned}
c'_{j1} &= c_{j1} + k \\
c'_{1j} &= c_{1j} + k \\
c'_{2j} &= c_{2j} + k \\
c'_{j2} &= c_{j2} + k
\end{aligned}\right\} \quad (for\ all\ x_j \neq x_1\ or\ x_2),$$

$$(10.10)$$

$$c'_{ij} = c_{ij} + 2k \qquad (for\ x_i\ and\ x_j = x_1\ or\ x_2)$$

$$c'_{ij} = c_{ij} \qquad (for\ all\ x_i, x_j \neq x_1\ or\ x_2)$$

is the solution to Problem (b) with the original cost matrix.

Proof. Each Hamiltonian path belongs to one of the following categories:
  (1) Neither of its end vertices is vertex x_1 or x_2.
  (2) One of its end vertices is either x_1 or x_2.
  (3) One end vertex is x_1 and the other is x_2.
The cost of a Hamiltonian path under cost matrix $\mathbf{C}'$ is greater than the cost

of the same path under matrix **C** by:

If the path belongs to category (1): 4k units;
If the path belongs to category (2): 3k units;
If the path belongs to category (3): 2k units.

Since k is greater than the cost of any Hamiltonian path, the cost (under matrix **C′**) of the longest Hamiltonian path in category (3) is smaller than the cost of the shortest Hamiltonian path in category (2), and the cost of the longest Hamiltonian path in category (2) is smaller than the cost of the longest Hamiltonian path in category (1). Thus, the solution of Problem (a) under cost matric **C′** will produce the shortest Hamiltonian path in category (3), i.e. it will be the solution of Problem (b), under cost matrix **C**. Hence the theorem.

One should note at this point that the relative costs of all Hamiltonian paths within a category remain unchanged under the transformations given by eqn (10.10).

6.3 The vertex penalty algorithm

The spirit of this algorithm is similar to that used in establishing theorem 2 above, i.e. to transform the cost matrix **C** in such a way so that various categories of trees are penalized differently, but at the same time the relative costs of all trees within a category remain unchanged. The objective is to separate the various categories and make the category of Hamiltonian paths the most attractive (least penalized) so that the application of the minimal tree algorithm will automatically produce the shortest Hamiltonian path.

THEOREM 3. *If the cost matrix* **C** *is transformed to a matrix* **C′** *so that*

$$c'_{ij} = c_{ij} + p(i) + p(j) \text{ for all } i, j = 1, 2, \ldots, n \qquad (10.11)$$

where $p(k)$ is an n-dimensional vector of real positive or negative constants, then the relative costs under matrix **C′** *of all Hamiltonian paths having specified end vertices remained unchanged.*

Proof. Let F_C be the cost of an arbitrary Hamiltonian path under cost matrix **C** having end vertices x_1 and x_2. Since every vertex is joined to exactly two other vertices (except for the end vertices x_1 and x_2 which are only joined to one other vertex each), the cost $F_{C'}$, of the same Hamiltonian path under cost matrix **C′** differs from F_C by

$$F_{C'} - F_C = 2 \sum_{j \neq 1, 2} p(j) + p(1) + p(2) \qquad (10.12)$$

This is a constant amount irrespective of the Hamiltonian path chosen. Hence the theorem.

6.3.1 SOLUTION OF PROBLEM (B). An algorithm to find the shortest Hamiltonian path with specified end-vertices (problem (b)) and which is based on theorem 3 will now be explained with the aid of an example.

Example. Consider a completely connected graph G of 10 vertices whose cost matrix C_0 is shown in Table 10.1 and suppose that we want the shortest Hamiltonian path whose end vertices are 8 and 9.

TABLE 10.1

Cost matrix of the example

	1	2	3	4	5	6	7	8	9	10
1	0	28	31	28	22	36	50	67	40	74
2	28	0	31	40	41	64	74	80	63	101
3	31	31	0	14	53	53	53	50	42	83
4	28	40	14	0	50	41	39	41	28	69
5	22	41	53	50	0	40	61	86	53	78
6	36	64	53	41	40	0	24	58	22	39
7	50	74	53	39	61	24	0	37	11	30
8	67	80	50	41	86	58	37	0	36	60
9	40	63	42	28	53	22	11	36	0	41
10	74	101	83	69	78	39	30	60	41	0

In accordance with theorem 2, we add a large number k (say 1000) to rows and columns 8 and 9 of the matrix of Table 10.1 and we will call the resulting cost matrix C.

We now proceed as follows:

Find the shortest spanning tree of G. If this is a Hamiltonian path, the problem is solved. However, in this example the shortest spanning tree is not a Hamiltonian path and is shown in Fig. 10.12(a). The degree of vertices 1 and 7 is 4 instead of 2 as a Hamiltonian path demands—we could therefore "penalize" these vertices in accordance with the spirit of Theorem 3. Suppose we arbitrarily choose a small step size (say 5 units) so that all penalties are in multiples of this. (We could have chosen the step size as 1 unit but this would be unnecessarily fine and would increase the number of iterations required.) Thus, if we decide to penalize only the vertices which have degrees greater than 2, we will put

$$p(i) = 5(d_i^T - 2) \qquad (10.13)$$

Note, however, that this method of penalizing vertices is arbitrary and many other alternatives exist. (See Section 6.3.4.)

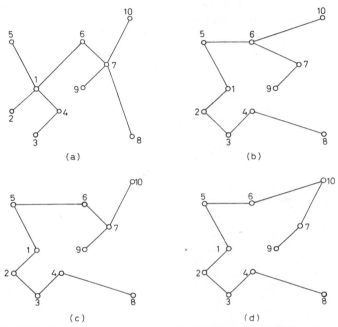

FIG. 10.12. The shortest spanning trees during the vertex penalty iterations

According to eqn (10.13), the penalties on vertices 1 and 7 become $p(1) = p(7) = 10$, all other $p(i)$ being 0. We now calculate the new cost matrix according to eqn (10.11) and solve for the shortest spanning tree. The result is shown in Fig. 10.12(b) and it is seen that this tree is much closer to a Hamiltonian path, having $\varepsilon = 1$ instead of $\varepsilon = 4$ as in the previous case.

In exactly the same way, we continue by penalizing vertex 6 and set $p(6) = 5$. (One should note that this new penalty is added to the previous cost matrix and not to the original matrix **C**.) Thus the total penalty is now $p(1) = p(7) = 10$, $p(6) = 5$ and all other $p(i) = 0$. The resulting shortest spanning tree is shown in Fig. 10.12)c) having an $\varepsilon = 1$ which is the same as that for the previous shortest spanning tree. We now penalize vertex 7, setting $p(7) = 5$, and the resulting shortest spanning tree reverts to that shown in Fig. 10.12(b). We continue by penalizing vertex 6 once more, setting $p(6) = 5$, and again solve for the shortest spanning tree. The result is shown in Fig. 10.12(d) from which it is noted that this is now a Hamiltonian path and hence according to theorem 3, this is the shortest Hamiltonian path. The cost of this path under the original cost matrix C_0 is 258 units.

It is worthwhile to note here that when at any stage a shortest spanning tree is obtained which had also occurred previously, this does not imply that cycling will occur, and any such cycles may be automatically broken as the algorithm contines. (However, as pointed out later, convergence of the shortest spanning tree to the shortest Hamiltonian path is not guaranteed regardless of the method used to penalise vertices.) We have thus obtained the shortest Hamiltonian path having end vertices 8 and 9 in 5 iterations involving 5 calculations of the shortest spanning tree, while complete enumeration would involve cost calculations and comparisons of $8! = 40,320$ paths.

An alternative to the above method of penalizing is the following: Instead of penalizing only the vertices whose $d_i^T > 2$ according to eqn (10.13), we may decide to penalize only the vertices (except for the two specified end vertices) whose $d_i^T = 1 < 2$, by negative amounts $p(i)$, these penalties being in unit step sizes. Alternatively, one could employ a combination of the two penalizing policies or adopt completely different policies as explained in Section 6.3.4. It should be noted here that the choice of penalizing policy can greatly affect the number of iterations needed to achieve convergence.

As an example, suppose that at stage two in the above calculation—when the shortest spanning tree was given by Fig. 10.12(b)—we had decided to adopt the second penalizing policy and instead of setting $p(6) = 5$ we were to set $p(10) = -5$. The resulting shortest spanning tree would then become as shown in Fig. 10.12(d), i.e. it would produce the shortest Hamiltonian path immediately.

It is interesting to note that there is not a unique penalty vector $p(i)$, $i = 1, \ldots, n$, according to which the cost matrix C must be transformed so that the shortest spanning tree becomes a Hamiltonian path and hence the shortest such path. In the above example, for instance, the final value of the penalty vector when the first penalizing policy was adopted was $p(1) = p(6) = 10$, $p(7) = 15$, all other $p(i) = 0$; whereas when the second penalizing policy was adopted from step 2 onward, the final vector was $p(1) = (7) = 10$, $p(10) = -5$, all other $p(i) = 0$.

6.3.2 SOLUTION OF PROBLEM (A). We have illustrated the application of the vertex penalty algorithm to problem (b) first, because this is the problem to which theorem 3 is immediately applicable. Just as problem (b) was made equivalent to problem (a) for solution by the decision-tree search algorithm, problem (a) can be made equivalent to problem (b) for solution by the vertex penalty algorithm with the aid of the following expedient:

To the set X of vertices of the graph G add two more vertices, say x_1 and x_2, to form the new set $X' = X \cup \{x_1, x_2\}$. To the set A of links of G add the two sets of links

$$S_1 = \bigcup_{j=1}^{n} \{(x_1, x_j)\} \quad \text{and} \quad S_2 = \bigcup_{j=1}^{n} \{(x_2, x_j)\}$$

to produce the new set $A' = A \cup S_1 \cup S_2$. Set the cost of links $(x_1, x_j) = a$ and $(x_2, x_j) = b$ for all $j = 1, \ldots, n$, where a and b are any two constants.

THEOREM 4. *The solution of problem* (b) *for graph* $G' = (X', A')$ *with end vertices* x_1 *and* x_2 *is the solution of problem* (a) *for graph* G.

Proof. Any Hamiltonian path of G' with end vertices x_1 and x_2 can be considered as a Hamiltonian path through the vertices of G plus two links (x_1, x_i) and (x_2, x_j); $x_i, x_j \in X$. Thus, the cost of a Hamiltonian path of G' is, say, $F' = F + a + b$, where F is the cost of some Hamiltonian path of G, and hence F' is a minimum only when F is minimum, i.e. when F corresponds to the shortest Hamiltonian path of G. Hence the theorem.

6.3.3 CONVERGENCE OF THE VERTEX-PENALTY METHOD. The vertex penalty method described above was developed by Christofides [7] and in a somewhat different form, at approximately the same time, by Held and Karp [18]. The last authors also show that this iterative method is not necessarily convergent and give as an example the graph of Fig. 10.13. Despite this great shortcoming, the method is nevertheless very valuable for two reasons.

Firstly, it converges in the great majority of cases. (As is mentioned later on, all of 33 randomly generated problems were solved using this method by at least one strategy for placing penalties.)

Secondly, suppose the iterations are stopped at a time when the shortest spanning tree under the modified matrix is T', the vertex penalties $p(i)$ $(i = 1, 2, \ldots, n)$ and the vertex degrees $d_i^{T'}$; and consider the problem when the shortest Hamiltonian path with end vertices x_1 and x_2 is required. The cost of T' under the modified cost matrix is:

$$\bar{F}_{T'} = F_{T'} + \sum_{i \neq 1, 2} p(i) \, d_i^{T'} + p(1) + p(2)$$

where $F_{T'}$ is the cost of T' evaluated under the initial cost matrix.

If H is the shortest Hamiltonian path of the graph, then the cost of H under the modified matrix is:

$$\bar{F}_H = F_H + 2 \sum_{i \neq 1,2} p(i) + p(1) + p(2).$$

Since $\bar{F}_{T'} < \bar{F}_H$ by definition, the difference:

$$f(\boldsymbol{p}) \equiv \bar{F}_H - \bar{F}_{T'} = F_H - \left[F_{T'} + \sum_{i \neq 1,2} p(i)\,(d_i^{T'} - 2) \right] \qquad (10.14)$$

is a measure of how near the tree T' is to the shortest Hamiltonian path H, $f(\boldsymbol{p})$ being 0 when $T' = H$.

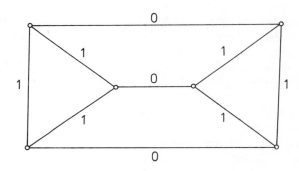

FIG. 10.13.

The quantity $f(\boldsymbol{p})$ can be considered as a function of the penalty vector $\boldsymbol{p} = (p_i | i = 1, \ldots, n)$. Held and Karp gave two methods of finding the penalties $\boldsymbol{p}^*$ which minimize $f(\boldsymbol{p})$, one based on a column-generation technique and the other being a steepest descent method. (An alternative and slightly more general sequential approach based on heuristically guided search was recently used successfully by Camerini et al. [4].) When the vertex-pentalty method converges we will have the minimum value $f(\boldsymbol{p}^*) = 0$ since $T' \to H$ and $d_i^{T'} = 2$ for all $i \neq 1, 2$. On the other hand it was mentioned earlier that the method is not always convergent in which case the minimum value $f(\boldsymbol{p}^*)$ of $f(\boldsymbol{p})$ is not zero but some positive number. In this case it is quite apparent that the quantity:

$$F_{T'} + \sum_{i \neq 1,2} p^*(i)\,(d_i^{T'} - 2) \qquad (10.15)$$

is a lower bound on the value of the shortest Hamiltonian path. In this latter case, the bound derived is very tight and could be used with great effect [19] in a decision-tree search algorithm, similar in spirit to that described in Section 6.2.1.

6.3.4 PENALIZING STRATEGIES. Although, as mentioned in Section 6.3.3, two methods exist [18] for determining the sequence of penalties which will force the shortest spanning tree to become the shortest Hamiltonian path (if convergence is at all possible), both of these methods are elaborate and quite time-consuming. In the present section we will investigate, experimentally, the effect of different strategies for penalizing vertices on the rate of convergence of the method. In all, 33 randomly generated completely connected graphs were used for testing the strategies and in all cases at least one strategy converged to the shortest Hamiltonian path. This leads us to believe that even though problems exist where the vertex penalty method is not convergent (in fact any connected graph without a Hamiltonian circuit is a case in point), the majority of unstructured problems not specifically designed as counter-examples may in fact lead to convergence.

(I) *Fixed penalties*
(a) *Positive penalties only*
    This is the case where all vertices x_i of the tree T, with degrees $d_i^T > 2$ are penalized by $\gamma(d_i^T - 2)$ and all other vertices are not penalized. It was found that the number of iterations to reach a solution (i.e. a Hamiltonian circuit) is approximately inversely proportional to γ—for small γ, but that for a large enough γ it is quite possible for the problem to cycle and not reach a solution. The best value of γ depends on both the number of vertices in the graph and the distribution of link-costs.
(b) *Negative penalties only*
    This is the case where vertices x_i (except for the specified end-vertices) are penalized by $-\gamma$ if $d_i^T = 1$ and not penalized otherwise. The conclusions are roughly the same as case (a) above.
(c) *Combined positive and negative penalties*
    In this case vertices x_i are penalized according to (a) or (b) above, depending on whether $d^T > 2$ or $d_i^T = 1$ respectively. It was found that a value of γ half the size of those used for cases (a) and (b) above produced convergence in approximately the same number of iterations as (a) or (b). The method is in general superior to either (a) or (b).

(II) *Reversible decreasing penalties*
    In using fixed penalties a situation may arise where a vertex x_i with $d_i^T = 3$

(say) and penalized by $+\gamma$ at one instant, is left with $d_i^T = 1$ after the next iteration, i.e. the vertex x_i is "over-penalized". If the combined penalizing strategy (I-c) above is used, then the vertex will be penalized by $-\gamma$ at the next iteration and this may (or may not) cause d_i^T to become 3 again, so that an unnecessarily large number of iterations may result from these oscillations in d_i^T.

It was found that for large graphs a better method in such circumstances is to only remove a proportion of the previously added penalty, so that the next iteration for the above example would involve penalizing x_i by $-\alpha\gamma$, instead of by $-\gamma$, where $0 < \alpha < 1$.

Similar situations to the one above arise when a vertex with degree $d_i^T = 1$ is over-penalized by a negative amount so that after the next iteration it has $d_i^T > 2$. This case is treated in an analogous way to the one above.

In the computational experiments shown in Table 10.2, a value of $\alpha = 0.5$ was used.

(III) Calculated penalties
(a) Positive penalties only

In this case all vertices x_i with $d_i^T > 2$ are penalized by $p(i)$ where $p(i)$ is calculated to be the minimum positive penalty which when applied to vertex x_i (in isolation) causes d_i^T to be reduced by one unit after a single iteration. The $p(i)$ are then applied concurrently to their respective vertices.

The value of $p(i)$ for a given vertex x_i is calculated as follows:

Remove from the tree $T = (X, A_T)$ just one of the links (x_i, x_r) incident at x_i. This separates T into two subtrees T_1 and T_2. Find the least cost link joining these two subtrees, i.e. find the link (x_j^r, x_k^r) with cost:

$$c(x_j^r, x_k^r) = \min_{\substack{x_j \in T_1 \\ \neq x_i}} \min_{\substack{x_k \in T_2 \\ \neq x_i}} [c(x_j, x_k)] \tag{10.16}$$

where x_j^r and x_k^r are the optimizing values of x_j and x_k respectively and T_1 and T_2 are used to represent both the trees and their corresponding vertex sets. The cost $c(x_j^r, x_k^r)$ is therefore the least cost of linking subtrees T_1 and T_2 into a single tree when the original link (x_i, x_r) is removed. Thus, a penalty of $c(x_j^r, x_k^r) - c(x_i, x_r)$ imposed on vertex x_i is the least penalty that would remove link (x_i, x_r) from the shortest spanning tree at the next iteration. If the penalty $p(i)$ is therefore chosen according to:

$$p(i) = \min_{(x_i, x_r) \in A_T} [c(x_j^r, x_k^r) - c(x_i, x_r)] \tag{10.17}$$

then a penalty $p(i)$ imposed on vertex x_i alone would (assuming that $p(i)$ is unique), remove just one link (x_i, x_r^*) from the set of links incident at x_i, i.e. would reduce d_i^T by one. The link (x_i, x_r^*) is that link which minimizes expression (10-17).

It should be noted here that when all penalties $p(i)$, calculated according to (10.17) are imposed on their respective vertices simultaneously, the degrees of some vertices may be reduced by more than 1 whereas the degrees of other vertices may not be reduced at all (or may even increase), due to the combined interaction of the vertex penalties on the link costs at the next iteration. Thus, although the use of penalties according to eqn (10.17) does not guarantee an answer after a given number of iterations, it was found (see Table 10.2), that this is a superior penalizing method than those of (I) and (II) above.

(b) *Negative penalties only*

This is a case where all vertices x_i with $d_i^T = 1$ are penalized by negative numbers $p(i)$, where $p(i)$ is calculated to be the maximum (least negative) penalty which when applied to vertex x_i alone causes d_i^T to become 2 after a single iteration, i.e. a second link is attracted to vertex x_i. The $p(i)$ are, however, applied concurrently to their respective vertices.

The value of $p(i)$ for a given vertex x_i is computed as follows:

Consider the addition of link (x_i, x_r) to the tree T. This addition causes the closing of a cycle composed of link (x_i, x_r) together with the links of T in the path from x_r to x_i. Let the set of links of T in the path x_r to x_i (excluding the last link incident at x_i), be S_{ri}. Then if link (x_i, x_r) is added to the links of T and any one of the links of S_{ri} is removed, another tree results in which the degree of vertex x_i is two. Thus if a single penalty $p(i)$ is imposed on vertex x_i where:

$$p(i) = \min_{x_r \in X} \left[c(x_i, x_r) - \min_{(x_j, x_k) \in S_{ri}} \{c(x_j, x_k)\} \right] \qquad (10.18)$$

(i.e. $p(i)$ is the smallest additional cost of adding a link from x_i to any other vertex x_r and removing the least cost link on the path from x_r to x_i), then the degree of vertex x_i will become 2 after a single iteration.

When all $p(i)$ are imposed simultaneously then, as for case (III-a) above, interactions among the penalties make it impossible to predict the changes to the vertex degrees after one iteration.

(c) *Combined positive and negative penalties*

This strategy involves penalizing vertices x_i according to (III-a) or (III-b) above depending on whether $d_i^T > 2$ or $d_i^T = 1$ respectively.

Computational results and comparisons between the seven penalizing strategies described above are given in Table 10.2. The graphs in this table are derived by distributing n points in a square according to a uniform distribution, and taking the euclidean distance between i and j to be the link cost c_{ij}.

TABLE 10.2. Computational results† of vertex-penalty method

Number of vertices n	Penalising strategy													
	I(a)		I(b)		I(c)		II		III(a)		III(b)		III(c)	
	α	β	α	β	α	β	α	β	α	β	α	β	α	β
10	22	0·15	16	0·13	18	0·11	16	0·18	71	0·18	27	0·16	9	0·12
16	39	0·27	33	0·24	35	0·29	31	0·20	52	0·42	18	0·39	22	0·50
20	83	0·88	56	0·71	74	0·80	39	0·38	55	0·59	24	0·48	27	0·67
26	126	1·83	158	1·41	98	1·66	74	1·02	40	0·47	60	0·71	16	0·46
30	—	—	—	—	—	—	—	—	59	0·91	63	1·11	61	0·95
36									98	1·41	87	1·63	72	1·37
40									172	2·23	159	1·99	57	1·00
46									—	—	—	—	81	3·09
50													92	4·43
56													146	9·70
60													111	13·6

α: Number of iterations.

β: Computing time (CDC 6600).

†: Each entry is the average value for 3 graphs of the same size.

The table shows the number of iterations and computing time, (CDC 6600 sec.), needed to reach the final answer. It is seen from this table that the

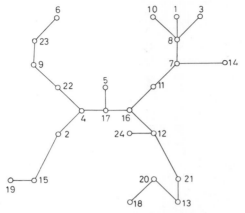

(a) The initial shortest spanning tree.

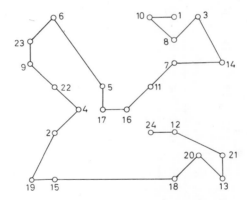

(b) The shortest hamiltonian circuit obtained after 40 iterations using penalty III c

FIG. 10.14. A 24-vertex problem.

performance of all penalizing strategies is problem-dependent although it can be seen that the best strategy is (III-c), and with this strategy one can find the shortest Hamiltonian path in graphs of about 60 vertices in less than 15 seconds.

Figure 10.14 shows the initial shortest spanning tree and the final shortest Hamiltonian path of a 24-vertex graph. The lowest and highest numbered vertices are the specified end vertices.

6.4. The "closed" travelling salesman problem

So far in Section 6, we dealt with the "open" travelling salesman problem, i.e. with the shortest Hamiltonian path (rather than circuit). This was justified at the beginning as a means of dealing with the more direct relation of the "open" problem with the shortest spanning tree. On the other hand, only a very minor modification is needed to deal with the "closed" travelling salesman problem. Held and Karp [18], for example, introduce the notion of a shortest 1-tree of G, this being defined as a shortest spanning tree of the subgraph of G with vertex 1 removed, plus the two shortest links from vertex 1 to two other vertices of the tree. Obviously, the same relationship exists between the shortest 1-tree and the "closed" travelling salesman problem as between the shortest spanning tree and the "open" problem. Thus, penalizing vertices changes the relative costs of 1-trees but leaves the relative ordering of Hamiltonian circuits invariant. It is also quite obvious that, just as for the "open" problem the shortest spanning tree with all vertex degrees of value 2 (except for the two end vertices) became the shortest Hamiltonian path between those end vertices; similarly for the "closed" problem the shortest 1-tree with all vertex degrees of value 2 is the shortest Hamiltonian circuit of the graph. The vertex-penalty method discussed earlier on in the present section can, therefore, be used with virtually no change to solve the "closed" travelling salesman problem as well.

7. The Travelling Salesman and Assignment Problems

In Section 5 it was pointed out that the assignment problem defined by eqns (10.4), (10.5) and (10.6) may have solutions composed of a number of disjoint circuits, and one method of solving the travelling salesman problem would be to impose restrictions until the solution becomes a single (Hamiltonian) circuit. In the present section we will investigate procedures of imposing these restrictions within the framework of a decision-tree search algorithm, conceptually in much the same way as was done in Section 6.2 for the decision-tree search based on the shortest spanning tree.

7.1 A decision-tree search algorithm

Let the solution of the assignment problem with a cost matrix $[c_{ij}]$ (and $c_{ii} = \infty, \forall i$), be composed of a number of disjoint circuits. For example if the solution of an 8-vertex problem is given by $\xi_{1,2} = \xi_{2,6} = \xi_{35} = \xi_{47} = \xi_{54} = \xi_{61} = \xi_{78} = \xi_{83} = 1$ and all other $\xi_{ij} = 0$, then the solution corresponds to the two circuits shown in Fig. 10.15(a). What is now needed is to eliminate this solution together with as many other solutions as possible without eliminating the solution to the travelling salesman problem under the same cost matrix. Since the travelling salesman solution is a Hamiltonian circuit we will attempt to eliminate any solutions which correspond to more than one circuit.

(A) A SIMPLE BRANCHING RULE. In general, let the solution to the assignment problem contain the (non-Hamiltonian) circuit $(x_1, x_2, \ldots, x_k, x_1)$ with cardinality k. This circuit (and all solutions containing it), can be removed from further consideration by insisting that at least one of the arcs (x_1, x_2), $(x_2, x_3) \ldots (x_k, x_1)$ must not be in the solution. This can be done quite simply by subdividing the original problem P_o with the cost matrix $[c_{ij}]$, into k subproblems $P_1, P_2, \ldots, P_k$. In problem P_1, $c(x_1, x_2)$ is set to ∞ all other c_{ij} remaining unchanged (i.e. as for problem P_o); in P_2, $c(x_2, x_3) = \infty$, etc., and for problem P_k, $c(x_k, x_1) = \infty$. Obviously any solution to problem P_o not containing the circuit $(x_1, x_2, \ldots, x_k, x_1)$ is a solution to at least one of the problems $P_1, \ldots, P_k$ and hence the optimal travelling salesman solution is the solution to one or more of these subproblems.

For the example of Fig. 10.15(a), choosing to eliminate the circuit of cardinality 3 results in the decision-tree of Fig. 10.16 in which the problems

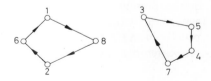

(a) Solution to problem P_0

(b) Solution to problem P_1

10.15. Solutions to assignment problems

P_1, P_2 and P_3 are represented by nodes of the tree derived from the initial problem P_o. Let the problems P_1, P_2 and P_3 be solved as assignment problems and let the corresponding costs be C_1, C_2 and C_3. Since C_1 is a lower bound on the travelling salesman solution to problem P_1 and similarly for P_2 and P_3 the number $L = \min [C_1, C_2, C_3]$ is a lower bound to the value of the solution of the initial travelling salesman problem.

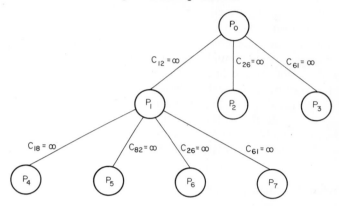

FIG. 10.16. Decision-tree with simple branching rule A

Let us now say that $L = C_1$ (i.e. $C_1 \leqslant C_2, C_3$). If the solution to P_1 is a Hamiltonian circuit, then this is the solution to the initial travelling salesman problem. If not, let us say that the solution is as shown in Fig. 10.15(b). Choosing to eliminate circuit (1, 8, 2, 6, 1) we can again form and solve the subproblems P_4, P_5, P_6 and P_7 as shown in Fig. 10.16. It is seen from this figure that P_4 corresponds to a problem whose cost matrix has the entry $c_{1,8}$ set to ∞ (in addition to $c_{1,2}$ which was set to ∞ earlier), and similarly for P_5, P_6 and P_7. The lower bound is now redefined as $L = \min [C_2, C_3, C_4, C_5, C_6, C_7]$ and let us say that P_3 is the problem corresponding to the cost L (i.e. $L = C_3$). If the solution to P_3 is a Hamiltonian circuit then it is the solution to the initial travelling salesman problem. Otherwise, further branching must be resumed from node P_3 in exactly the same way as was done for P_1, and one continues in this way until the solution to the problem whose cost is the current value of L, becomes a Hamiltonian circuit. When this occurs, that circuit is the final solution, since its cost is (by the definition of L) less than or equal to the lower bounds on the cost of any other Hamiltonian circuit that may result by branching further from the remaining nodes in the tree.

From the above description of the algorithm it should be quite apparent that the decision-tree search employed is of the "breadth-first" type as explained in Appendix I.

(B) A DISJOINT BRANCHING RULE. As explained in Appendix I all that is required for a valid branching from a problem P_o to subproblems P_1, P_2 and P_3, is that every feasible solution of P_o (except the ones being eliminated) should be a solution of at least one subproblem. As mentioned in that Appendix, however, an obviously desirable characteristic for a branching method to possess, is for the sub-problems created to be disjoint as far as the feasible solutions of P_o are concerned, i.e. that every feasible solution of P_k should be a solution to one and only one of these subproblems.

The previously described branching rule was based on the fact that a circuit such as $(x_1, x_2, \ldots, x_k, x_1)$ could be removed by excluding one of its arcs. This, however, does not lead to branching into mutually exclusive subproblems. Thus, in the example given earlier, the solution of P_o corresponding to circuits (1, 3, 6, 1) and (2, 5, 4, 7, 8, 2) is a feasible solution to both subproblems P_1 (with $c_{1,2} = \infty$) and P_2 (with $c_{2,6} = \infty$).

A different branching rule which removes a circuit $(x_1, x_2, \ldots, x_k)$ but produces disjoint subproblems is as follows.

For problem P_1 set $c(x_1, x_2) = \infty$

For problem P_2 set $c(x_1, x_2) = -M$ and $c(x_2, x_3) = \infty$

For problems P_3 set $c(x_1, x_2) = c(x_2, x_3) = -M$ and $c(x_3, x_4) = \infty$

.
.
.

For problem P_k set $c(x_1, x_2) = c(x_2, x_3) = \ldots = c(x_{k-1}, x_k) = -M$

$$\text{and } c(x_k, x_1) = \infty$$

where $-M$ is a large negative number to ensure that the arc whose cost is $-M$ is in the optimal solution.

With this branching rule the subproblems are certainly disjoint since for any two subproblems there is at least one arc excluded from the solution in one, and which is definitely included in the solution in the other subproblem. It is also easy to see that no feasible solution of P_o is lost, i.e. that any solution to the initial problem P_o must also be represented as a solution to one of the subproblems. This is obvious since any solution of P_o has some sequence of arcs leading from x_1, such as $(x_1, x_\alpha), (x_\alpha, x_\beta)$ etc. and these must coincide in the first r arcs with the arcs of the path $(x_1, x_2, x_3, \ldots, x_k)$ for some value of $r = 0, 1, \ldots, k$; $r = 0$ corresponding to the case where there is no coincidence at all, $r = 1$ to the case where $x_\alpha = x_2$ but $x_\beta \neq x_3$ etc.

In the example given earlier, the initial problem P_o would be partitioned into the three subproblems as shown in Fig. 10.17, compared to the first-level partition of Fig. 10.16 which resulted from the previous branching rule.

(C) A BETTER BRANCHING RULE. Both of the previous two branching rules, eliminated (at each branching), all solutions containing a given circuit such as $(x_1, x_2, \ldots, x_k, x_1)$. However, not only must this circuit not exist in a travelling salesman solution but, obviously, there must be at least one arc leading from the set of vertices $S = \{x_1, \ldots, x_k\}$ to the set of vertices $\bar{S} = X - S$. In fact the existence of an arc from S to $\bar{S}$ not only guarantees that solutions containing the circuit based on S are eliminated, but also that solutions in which the subset of vertices in S are joined to form several circuits (instead of just one) are also removed. Thus, a branching rule, based on the insistence that some arc from S to $\bar{S}$ must exist could be expected to be uniformly better than the previous two branching rules [2].

Since an arc from S to $\bar{S}$ must start from some vertex in S, a problem could be split up into k subproblems $P_1, P_2, \ldots, P_k$ where for subproblem P_i we would insist that the initial vertex of the arc is $x_i \in S$ and the final vertex is some vertex in $\bar{S}$. This can be done by setting $c(x_i, x_j) = \infty \ \forall \ x_j \in S$ and leaving all other costs unchanged. In the solution of the resulting assignment problem we would then certainly have the arc from x_i leading into $\bar{S}$ since all other alternatives have had their costs set to ∞.

In the example given above, the initial problem P_o would—according to the present branching rule—be partitioned into the three subproblems defined and shown in Fig. 10.18, and similarly for the following branchings.

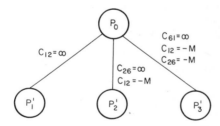

FIG. 10.17. Decision-tree with disjoint branching rule B

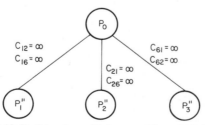

FIG. 10.18. Decision-tree with improved branching rule C

7.2 Example

Find the solution to the travelling salesman problem whose cost matrix is:

$$\mathbf{C}_A =$$

	1	2	3	4	5	6	7	8
1	∞	76	43	38	51	42	19	80
2	42	∞	49	26	78	52	39	87
3	48	28	∞	36	53	44	68	61
4	72	31	29	∞	42	49	50	38
5	30	52	38	47	∞	64	75	82
6	66	51	83	51	22	∞	37	71
7	77	62	93	54	69	38	∞	26
8	42	58	66	76	41	52	83	∞

We will now use the circuit elimination algorithm just described, taking the least cardinality circuit at each stage and using branching rule C.

The solution of the assignment problem under matrix $\mathbf{C}_A$ is composed of the two circuits (2, 4, 3) and (1, 7, 8, 6, 5) with cost 232. This corresponds to node A of the decision tree shown in Fig. 10.19. Eliminating circuit (2, 4, 3) according to branching rule C leads to the three subproblems represented by

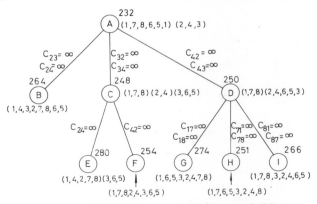

FIG. 10.19. Decision-tree search for example 7.2

nodes B, C and D in Fig. 10.19. The cost matrices of these subproblems, their assignment problem solutions, and the value of these solutions are: (respectively)

$C_B = $

	1	2	3	4	5	6	7	8
1	∞	76	43	38	51	42	19	80
2	42	∞	∞	∞	78	52	39	87
3	48	28	∞	36	53	44	68	61
4	72	31	29	∞	42	49	50	38
5	30	52	38	47	∞	64	75	82
6	66	51	83	51	22	∞	37	71
7	77	62	93	54	69	38	∞	26
8	42	58	66	76	41	52	83	∞

Solution:
(1, 4, 3, 2, 7, 8, 6, 5)
Cost: 264

$C_C = $

	1	2	3	4	5	6	7	8
1	∞	76	43	38	51	42	19	80
2	42	∞	49	26	78	52	39	87
3	48	∞	∞	∞	53	44	68	61
4	72	31	29	∞	42	49	50	38
5	30	52	38	47	∞	64	75	82
6	66	51	83	51	22	∞	37	71
7	77	62	93	54	69	38	∞	26
8	42	58	66	76	41	52	83	∞

Solution:
(1, 7, 8), (2, 4), (3, 6, 5)
Cost: 248

$$C_D =$$

	1	2	3	4	5	6	7	8
1	∞	76	43	38	51	42	19	80
2	42	∞	49	26	78	52	39	87
3	48	28	∞	36	53	44	68	61
4	72	∞	∞	∞	42	49	50	38
5	30	52	38	47	∞	64	75	82
6	66	51	83	51	22	∞	37	71
7	77	62	93	54	69	38	∞	26
8	42	58	66	76	41	52	83	∞

Solution:
(1, 7, 8), (2, 4, 6, 5, 3)
Cost: 250

The solution to subproblem B is a Hamiltonian circuit of value 264. However, the lower bounds in both nodes C and D are less than this value so there is a possibility of a better solution by continuing the branching from there. The lowest of the lower bounds is $L = \min [248, 250] = 248$ corresponding to node C so branching is continued from this node by eliminating circuit (2, 4). Two further subproblems now result as shown by nodes E and F in Fig. 10.19. The cost matrices, assignment problem solutions and their costs for these two subproblems are:

$$C_E =$$

	1	2	3	4	5	6	7	8
1	∞	76	43	38	51	42	19	80
2	42	∞	49	∞	78	52	39	87
3	48	∞	∞	∞	53	44	68	61
4	72	31	29	∞	42	49	50	38
5	30	52	38	47	∞	64	75	82
6	66	51	83	51	22	∞	37	71
7	77	62	93	54	69	38	∞	26
8	42	58	66	76	41	52	83	∞

Solution:
(3, 6, 5), (1, 4, 2, 7, 8)
Cost: 280

$$C_F =$$

	1	2	3	4	5	6	7	8
1	∞	76	43	38	51	42	19	80
2	42	∞	49	26	78	52	39	87
3	48	∞	∞	∞	53	44	68	61
4	72	∞	29	∞	42	49	50	38
5	30	52	38	47	∞	64	75	82
6	66	51	83	51	22	∞	37	71
7	77	62	93	54	69	38	∞	26
8	42	58	66	76	41	52	83	∞

Solution:
(1, 7, 8, 2, 4, 3, 6, 5)
Cost: 254

At this stage it should be noted that the solution to subproblem F is a Hamiltonian circuit with cost 254. This cost is less than that of the previously best solution (cost 264) and hence replaces it as the current best. Two nodes (E and D) still have solutions with non-Hamiltonian circuits and are therefore candidates for continuing the branching. However, the lower bound on node E is 280 > 254 (the value of the current best solution) and hence no improvement to the solution can result by branching from there. Node D, with lower

$$C_G =$$

	1	2	3	4	5	6	7	8
1	∞	76	43	38	51	42	∞	∞
2	42	∞	49	26	78	52	39	87
3	48	28	∞	36	53	44	68	61
4	72	∞	∞	∞	42	49	50	38
5	30	52	38	47	∞	64	75	82
6	66	51	83	51	22	∞	37	71
7	77	62	93	54	69	38	∞	26
8	42	58	66	76	41	52	83	∞

Solution:
(1, 6, 5, 3, 2, 4, 7, 8)
Cost: 274

bound $250 < 254$ is therefore the only node from which branching can continue with any possibility of an improvement. Eliminating circuit $(1, 7, 8)$ from the solution to node D then leads to the three subproblems shown as G, H and I in Fig. 10.19. The corresponding cost matrices and assignment problem solutions to these subproblems are:

$$C_H =$$

	1	2	3	4	5	6	7	8
1	∞	76	43	38	51	42	19	80
2	42	∞	49	26	78	52	39	87
3	48	28	∞	36	53	44	68	61
4	72	∞	∞	∞	42	49	50	38
5	30	52	38	47	∞	64	75	82
6	66	51	83	51	22	∞	37	71
7	∞	62	93	54	69	38	∞	∞
8	42	58	66	76	41	52	83	∞

Solution:
   $(1, 7, 6, 5, 3, 2, 4, 8)$
Cost: 251

$$C_I =$$

	1	2	3	4	5	6	7	8
1	∞	76	43	38	51	42	19	80
2	42	∞	49	26	78	52	39	87
3	48	28	∞	36	53	44	68	61
4	72	∞	∞	∞	42	49	50	38
5	30	52	38	47	∞	64	75	82
6	66	51	83	51	22	∞	37	71
7	77	62	93	54	69	38	∞	26
8	∞	58	66	76	41	52	∞	∞

Solution:
   $(1, 7, 8, 3, 2, 4, 6, 5)$
Cost: 266

At this stage we note that the solution to subproblem H is a Hamiltonian circuit with cost 251 which is less than the previously best value of 254. The solution to subproblem H, i.e. (1, 7, 6, 5, 3, 2, 4, 8) therefore replaces the previously best solution. Moreover, the value of 251 is less than the lower bounds on any terminal node of the tree and hence it is the optimal solution to the complete problem. (The solutions to all terminal nodes—except node E from which branching was stopped earlier—are in fact Hamiltonian circuits and no further branching would be needed from there in any case regardless of their bounds.)

7.3 Computational comments and performance

The decision-tree search algorithm based on circuit elimination and which has just been described, depends upon finding solutions to assignment problems with only minor modification of the cost matrix. It should be noted here that each of these problems can be solved very easily by storing the solution to the previous problem from which it was derived. In particular, the modifications involved in the case of all three branching rules mentioned above involve the setting of some entry $c(x_i, x_j)$ to ∞ for an arc (x_i, x_j) which is in the current assignment problem solution.

In the hungarian algorithm (see Chapter 12), for the solution of the assignment problem, the entry $c(x_i, x_j)$ in the final relative cost matrix would have had the value 0 and an associated marker indicating that the assignment was in the solution. The change of value of $c(x_i, x_j)$ would obviously necessitate the reallocation of that assignment, but would leave the other $(n - 1)$ assignments still valid. Thus, starting from the solution (assignments) of the problem before the modifications to the $[c_{ij}]$ were made, and removing the affected assignment, the solution of the new modified problem can be derived by reentering the hungarian algorithm at the last step since a single "breakthrough", i.e. a single increase in the number of zero-assignments from $(n - 1)$ to n would in fact produce the optimal solution to the new assignment problem. (See Chapter 12 for details.)

The performance of the above tree-search algorithm, with circuit elimination done according to the branching rule C, was investigated by Bellmore and Malone [2]. At any one stage branching from a node was continued by choosing to eliminate the lowest cardinality circuit in the solution of the problem corresponding to that node. Travelling salesman problems with random asymmetric cost matrices could be solved in T seconds on the IBM 7094 II, where

$$T \approx 0{\cdot}55 \times 10^{-4} \times n^{3{\cdot}46}$$

n being the number of vertices in the problem. (The other two branching rules were shown to be inferior, especially in problems where the graph formed clusters with arcs between vertices in the same cluster having small costs and arcs between vertices in different clusters having large costs.)

The above algorithm, however, does not perform at all well in symmetrical problems, because the solutions of assignment problems are almost always found to consist of large numbers of circuits of cardinality 2 which require very many branchings to be completely eliminated. An even worse performance results in cases where the travelling salesman problem is defined on a graph which does not contain a Hamiltonian circuit of finite cost.

7.4 A better bound for the tree search

In Section 7.1 we described a tree-search algorithm in which the lower bound at a node was taken as the value of the solution to the corresponding assignment problem. In fact it is quite obvious that the method of circuit elimination remains unchanged whatever lower bound is used for limiting the search. However, since branching from a node is stopped only when, either the solution to the subproblem corresponding to that node becomes a Hamiltonian circuit, or which the lower bound on the node exceeds the value of the best solution so far; the quality of the bound obviously has a very marked influence on the number of branchings in the decision-tree and hence on the computational efficiency of the method. The purpose of the present section is to describe a lower bound which can be calculated from the solution to the assignment problem with little extra effort, and which is tighter than the previously used bound. Let the solution to the assignment problem contain a number of circuits as shown for example in Fig. 10.20(a). Let the ith of the circuits be called $S_{1,i}$ and let their number be n_1. (We will use the same symbol $S_{1,i}$ to also represent the set of vertices in circuit i.)

A *contraction* is defined as a replacement of a circuit by a single vertex, thus forming a contracted graph containing n_1 vertices $S_{1,i}$ ($i = 1, 2, \ldots, n_1$). The cost matrix $\mathbf{C}_1 \equiv [c_1(S_{1,i}, S_{1,j})]$ of the contracted graph is taken as:

$$c_1(S_{1,i}, S_{1,j}) = \min_{\substack{k_i \in S_{1,i} \\ k_j \in S_{1,j}}} [f_1(k_i, k_j)] \tag{10.19}$$

where $F_1 = [f_1(k_i, k_j)]$ is the resulting relative cost matrix at the end of the solution to the assignment problem by (say) the hungarian method (see Chapter 12).

In Fig. 10.20(a), for example,

$$c_1(S_{1,5}, S_{1,6}) = \min_{\substack{k_5 \in \{11, 12, 13\} \\ k_6 \in \{6, 14, 15, 16\}}} [f_1(k_5, k_6)]$$

Now a second solution to the assignment problem of this contracted problem under the matrix C_1 may still produce circuits having the previous circuits as vertices. Fig. 10.20(b) shows one possible formation of the new circuits $S_{2,i}$ ($i = 1, 2, \ldots, n_2$) where n_2 is their total number. ($n_2 = 3$ in Fig. 10.20(b)). These circuits may again be contracted into vertices to form a new problem where the new cost matrix $C_2 \equiv [c_2(S_{2,i}, S_{2,j})]$ is calculated from an equation similar to equation (10.19), i.e.

$$c_2(S_{2,i}, S_{2,j}) = \min_{\substack{k_i \in S_{2,i} \\ k_j \in S_{2,j}}} [f_2(k_i, k_j)] \qquad (10.20)$$

where the various S_2 are the union of all the sets S_1 forming the particular circuit, and where $F_2 = [f_2(k_i, k_j)]$ is the relative cost matrix at the end of the second solution to the assignment problem.

A solution to the assignment problem of the new doubly contracted graph may still produce circuits and the iterative process of solution–contraction can be continued until the problem is reduced to a single vertex.

Compression is defined as the transformation of a matrix which does not satisfy the triangularity condition of metric space into one that does. Thus, to compress a matrix M what is necessary is to replace every element m_{ij} for which

$$m_{ij} > m_{ik} + m_{kj} \qquad \text{(for some } k)$$

by the value of $\min_{k} [m_{ik} + m_{kj}]$, and to continue this replacement until all $m_{ij} \leqslant m_{ik} + m_{kj}$ for any k.

THEOREM 5. *The sum of the values of the solutions to the assignment problems obtained during the "solution-contraction-compression" process (up to the stage when the contracted problem becomes a single vertex) is a valid lower bound to the travelling salesman problem.*

Proof. It is shown in Chapter 12 that entry (i, j) in the relative cost matrix resulting at the end of the solution to an assignment problem represents a

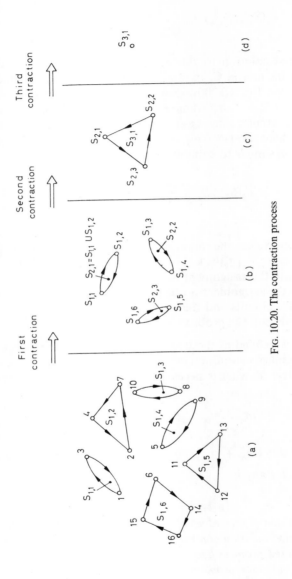

Fig. 10.20. The contraction process

lower bound on the extra cost that would result from including arc (i, j) into the solution.

Consider circuit $S_{1,i}$ formed at the end of the solution to the first assignment problem. Any Hamiltonian circuit passing through all the n vertices will have at least two arcs incident (one directed inwards and the other outwards) from vertices not in $S_{1,i}$ to one or more of the vertices in $S_{1,i}$, and the actual number of such arcs is obviously even. In Fig. 10.21(a) a Hamiltonian circuit is shown in heavy lines whereas the solution to the assignment problem is shown in light lines. After contraction, the Hamiltonian circuit would become the graph (assignment) of Fig. 10.21(b) with two arcs incident on $S_{1,3}$ and $S_{1,4}$ but four arcs incident on $S_{1,1}$ and $S_{1,2}$.

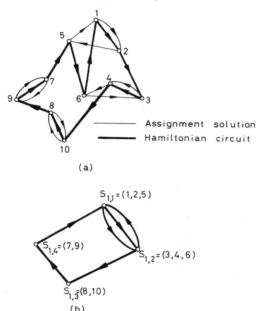

(a)

(b)

FIG. 10.21. Transformation of a Hamiltonian circuit under contraction

Now if the triangularity condition applies, then

$$c_1(S_{1,4}, S_{1,3}) \leqslant c_1(S_{1,4}, S_{1,1}) + c_1(S_{1,1}, S_{1,2}) + c_1(S_{1,2}, S_{1,3}) \quad (10.21)$$

and hence the assignment of Fig. 10.22—(in which links $(S_{1,4}, S_{1,1}), (S_{1,1}, S_{1,2})$ and $(S_{1,2}, S_{1,3})$ have been replaced by link $(S_{1,4}, S_{1,3})$) has a value less than or equal to the value of the assignment of Fig. 10.21(b). Since the assignment of Fig. 10.22 has two incident arcs per vertex, and the solution to the assignment problem of the contracted graph is the least-cost assignment, the value of the solution to the assignment problem, $V(AP_1)$ say, is a lower bound on the value of the aissignment of Fig. 10.21(b), i.e. on the value of the increment in cost that would be necessary in order to join the various circuits together.

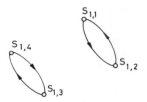

FIG. 10.22. Assignment corresponding to Fig. 10.21 (b) with two arcs per vertex

It is apparent that, since the graph of the assignment obtained by any Hamiltonian circuit after the first contraction (i.e. the assignment corresponding to Fig. 10.21(b)) contains an Eulerian circuit, it is always possible to transform it into an assignment with only two incident arcs per vertex, and having lower (or equal) cost, by replacing a path of arcs from one vertex to another by a single arc between the vertices as illustrated above.

If, on the other hand, the relative cost matrix C_1 does not satisfy the triangularity condition, then eqn (10.21) may not apply. In that case the matrix C_1 may be compressed first and the value of the solution to the assignment problem under the compressed matrix will then be the lower bound on the value of the increment in cost that would be necessary in order to join the circuits together. This is so, because compressing a matrix can only reduce or leave unchanged the cost of any assignment under the original matrix.

Similarly, $V(AP_2)$, the cost of the solution to the assignment problem after the second contraction, is a lower bound on the cost of linking the circuits resulting from this contraction, and so on for the third, fourth, etc. contractions. Therefore,

$$L = \sum_{i=0}^{k} V(AP_i) \qquad (10.22)$$

—where $V(AP_0)$ is the value of the initial solution to the assignment problem under the initial matrix C, and k is the number of contractions necessary to reduce the graph of the problem to a single vertex—is a valid lower bound to the cost of the solution of the travelling salesman problem. Hence the theorem.

One should perhaps note here that even if the initial cost matrix satisfies the triangularity condition, the consequent relative cost matrices may not, and their compression may be necessary at any one stage.

Example for the calculation of the bound

Consider the 10-vertex travelling salesman problem whose cost matrix is symmetrical and is given in Table 10.3.

The solution to the initial assignment problem gives the value $V(AP_o) = 184$. The resulting relative cost matrix is given in Table 10.4, and the solution is given by the graph of Fig. 10.23.

The contraction of the graph of Fig. 10.23 produces a 4-vertex graph whose cost matrix can be calculated according to eqn (10.19) to be as shown in Table 10.5(a). This cost matrix does not satisfy the triangularity condition and is therefore compressed into the matrix shown in Table 10.5(b).

The solution to the assignment problem under the matrix of Table 10.5(b) gives the value $V(AP_1) = 20$. The resulting relative cost matrix is given in Table 10.6, and the solution to this assignment problem is given by the graph of Fig. 10.24.

The contraction of the graph of Fig. 10.24 produces a 2-vertex graph whose cost matrix is calculated (from eqn 10.20) to be as shown in Table 10.7. (This trivial 2×2 matrix needs no compression since it satisfies the triangularity condition.)

The solution to the trivial assignment problem under the matrix of Table 10.7 has a value $V(AP_2) = 10$, and the solution is given by the graph of Fig. 10.25, which becomes a single vertex after the next contraction.

Thus, a lower bound on the value of the travelling salesman problem under the matrix of Table 10.3 is given by

$$L = V(AP_o) + V(AP_1) + V(AP_2) = 214$$

compared with the value of the optimal solution to the travelling salesman problem of 216, an error of 0·93 %, as compared with an error of 14·8 % when $V(AP_o)$ only is used as the lower bound. Although the present lower bound is not, in general, as tight as for the case of the above example, the results given in Reference [8] show that the bound is, on average, significantly better than $V(AP_0)$.

The computation time to obtain the bound on the travelling salesman problem is, on average, only 9% greater than the time required to solve the assignment problem under the same cost matrix. The computation time for the solution to the assignment problem using the hungarian method varies as kn^3 (see Chapter 12), where k is a constant and n is the size of the matrix.

The worst possible case for the computation of the bound (as far as the computation times are concerned), appears when all the circuits at each contraction contain only two vertices each, in which case the total computing

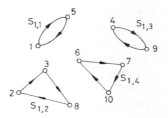

FIG. 10.23. First solution of the assignment problem

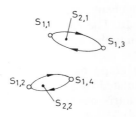

FIG. 10.24. Solution after first contraction

FIG. 10.25. Solution after second contraction

time spent on solving assignment problems would be:

$$kn^3 + k\left(\frac{n}{2}\right)^3 + k\left(\frac{n}{4}\right)^3 + \ldots = \frac{8}{7}kn^3 = 1\cdot143\,kn^3 \qquad (10.23)$$

Hence from eqn (10.23), it can be seen that, at worst, the time required to calculate the suggested bound to the travelling salesman problem is only 14·3% greater than the time required to solve an assignment problem of the

TABLE 10.3. Initial cost Matrix

	1	2	3	4	5	6	7	8	9
2	32								
3	41	22							
4	22	50	63						
5	20	42	41	36					
6	57	51	30	78	45				
7	54	61	45	72	36	22			
8	32	20	10	54	32	32	41		
9	22	54	60	20	22	67	57	50	
10	45	51	36	64	28	20	10	32	50

Solution to
assignment problem →

TABLE 10.4. Relative cost matrix

	1	2	3	4	5	6	7	8	9	10
1	∞	12	31	2	0	37	44	24	2	37
2	0	∞	0	18	10	19	39	0	22	31
3	19	0	∞	41	19	8	33	0	38	26
4	2	30	53	∞	16	58	62	46	0	56
5	0	22	31	16	∞	25	26	24	2	20
6	25	19	8	46	13	∞	0	12	35	0
7	32	39	33	50	14	0	∞	31	35	0
8	12	0	0	34	12	12	31	∞	30	24
9	2	34	50	0	2	47	47	42	∞	42
10	25	31	26	44	8	0	0	24	30	∞

K

TABLE 10.5(a) Matrix of contracted graph

	1	2	3	4
1	∞	12	2	20
2	0	∞	18	∞
3	2	30	∞	42
4	∞	8	30	∞

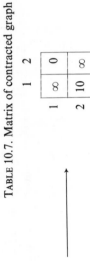

TABLE 10.7. Matrix of contracted graph

	1	2
1	∞	0
2	10	∞

TABLE 10.5(b) Compressed matrix

	1	2	3	4
1	∞	12	2	20
2	0	∞	2	∞
3	2	14	∞	22
4	∞	8	10	∞

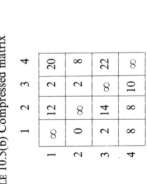

Solution to the assignment problem

TABLE 10.6. Relative cost matrix

	1	2	3	4
1	∞	10	0	10
2	0	∞	2	0
3	0	12	∞	12
4	∞	0	2	∞

same size. (It should be noted here that the computing times required for the "contraction" and "compression" parts of the process vary as n^2).

7.5 The example of Section 7.2 with the improved bound

Consider once more the example of Section 7.2 with the improved bound described above being used instead of simply using the value of the assignment problem solution as the bound. The bound obtained for the initial problem with cost matrix $\mathbf{C}_A$ (Section 7.2) would then be $V(AP_o) + V(AP_1) = 232 + 13 = 245$. (A second contraction is needed and the solution to the contracted assignment problem is 13.) Subproblems B, C and D are generated as before, but the new bound on subproblem C is $V(AP_o) + V(AP_1) = 248 + 6 = 254$ instead of the previous value of 248. The new bound on subproblem D is $V(AP_o) + V(AP_1) = 250 + 0 = 250$ unchanged from its previous value. (The bound on B would naturally remain unchanged since the solution of B

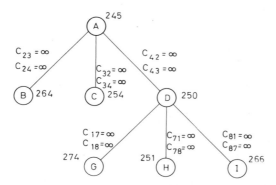

FIG. 10.26. Decision-tree search for example 7.2 with the improved AP bound

is in fact a Hamiltonian circuit.) The branching would now continue from D (since $250 < 254$) instead of from C which was the case previously. This branching leads to the subproblem G, H and I as before with the best solution (subproblem H) having value 251. Further branching from C—whose solution is not a Hamiltonian circuit—is now unnecessary since its lower bound of 254 is greater than 251. Previously, the bound on C—which was $248 < 251$—was insufficient to stop further branching from there. Therefore, the improved bound caused a saving of 2 nodes in the decision-tree search as shown in Fig. 10.26. This represent a saving of only 2/9ths for this small problem, but as the problem size increases, so does the percentage saving in the nodes of the tree, i.e. in the number of (n by n) assignment problems that have to be solved.

8. Problems P10

1. Show that if a nondirected graph G satisfies the conditions: (i) For every positive integer $k < \frac{1}{2}(n - 1)$, the number of vertices with degree not exceeding k is less than k; and (ii) The number of vertices with degree not exceeding $\frac{1}{2}(n - 1)$ is less than or equal to $\frac{1}{2}(n - 1)$; then G possesses a Hamiltonian circuit. (See Ref. [25, 29]. Also note that a graph G consisting of a single Hamiltonian circuit does not satisfy the above conditions, i.e. the conditions are sufficient but not necessary).

2. Prove that if for a nondirected graph G and any pair of non-adjacent vertices x_i and x_j we have the degrees satisfying:

$$d(x_i) + d(x_j) \geqslant n,$$

then G has a Hamiltonian circuit. (See Ref. [26]).

3. Consider the lattice graph G formed by p horizontal and q vertical grid lines where each intersection point is considered as a vertex and each grid line as a link. For what values of p and q does G possess a Hamiltonian circuit?

4. Show that the graph whose vertices and links correspond to the vertices and edges of the n-dimentional hypercube has a Hamiltonian circuit.

5. Show that the graph of problem 4 above has a Hamiltonian path between two diametrically opposite vertices if and only if n is odd.

6. Prove that in every complete antisymmetric directed graph there exists a Hamiltonian path.

7. Use the methods of Sections 2.1, 2.2 and 2.3 to either find all Hamiltonian circuits in the graph of Fig. 10.27 (if such circuits exist), or to indicate that no such circuit can be found. Compare the computational effort involved.

8. Use the method of Section 2.3 to verify that the graph shown in Fig. 10.28 contains no Hamiltonian circuit. Prove the same result by an argument.

9. For the travelling salesman problem defined by the link-cost matrix $\mathbf{C}$ given below, calculate lower bounds to the optimal solution using the assignment and shortest spanning tree problems.

	1					
2	41	2				
3	12	30	3			
$\mathbf{C} = $ 4	27	40	19	4		
5	52	14	39	43	5	
6	62	25	53	65	31	6
7	47	17	38	52	28	15

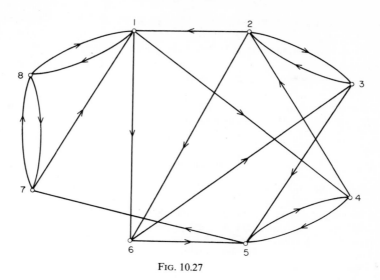

FIG. 10.27

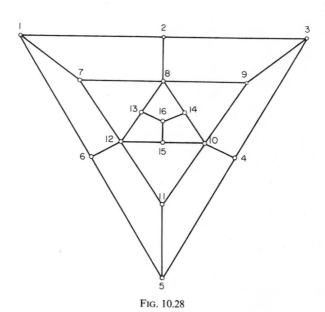

FIG. 10.28

10. Find the shortest Hamiltonian path between vertices 4 and 7 for the graph of problem 9.

11. Find the absolute shortest Hamiltonian path of the graph of problem 9.

12. Solve the travelling salesman problem with the link-cost matrix given in problem 9 by using the decision-tree search algorithm of Section 7.1 with branching rule B.

13. Repeat problem 12 above using the same method but calculating the lower bounds according to the results given in Section 7.4.

9. References

1. Bellmore, M. and Nemhauser, G. L. (1968). The travelling salesman problem—a survey, *Ops. Res.*, **16**, p. 538.
2. Bellmore, M. and Malone, J. C. (1971). Pathology of travelling salesman subtour-elimination algorithms, *Ops. Res.*, **19**, p. 278.
3. Burstall, R. M. (1967). Tree-searching methods with an application to a network design problem, *In*: "Machine Intelligence, Vol. 1", Collins and Michie, Eds., Oliver and Boyd, London.
4. Camerini, P. M., Fratta, L. and Maffioli, F. (In press). The travelling salesman problem: heuristically guided search and modified gradient techniques.
5. Charlton, J. M. and Death, C. C. (1970). A method of solution for general machine-scheduling problems, *Ops. Res.*, **18**, p. 689.
6. Christofides, N. (1973). Large scheduling problems with bivalent costs, *The Computer Jl.*, **16**, p. 263.
7. Christofides, N. (1970). The shortest Hamiltonian chain of a graph, *Jl. of SIAM (Appl. Math.)*, **19**, p. 689.
8. Christofides, N. (1972). Bounds for the travelling salesman problem, *Ops. Res.*, **20**, p. 1044.
9. Christofides, N. and Eilon, S. (1969). An algorithm for the vehicle dispatching problem, *Opl. Res. Quart.*, **20**, p. 309.
10. Christofides, N. and Eilon, S. (1972). Algorithms for large-scale travelling salesman problems, *Opl. Res. Quart.*, **23**, p. 511.
11. Danielson, G. H. (1968). On finding the simple paths and circuits in a graph, *IEEE Trans.*, CT-**15**, p. 294.
12. Demoucron, G., Malgrange, Y. and Petruiset, R. (1964). Graphes planaires; reconnaissance at construction de representations planaires topologiques, *Rev. Fran. de Resh. Oper.*, **8**, p. 33.
13. Deo, N. and Hakimi, S. L. (1965). The shortest generalized Hamiltonian tree, Proc. 3rd Allerton Conference on Circuit and Systems Theory, University of Illinois, Urbana, p. 879.
14. Dhawan, V. (1969). Hamiltonian circuits and related problems in graph theory, M.Sc. Report, Imperial College, London.
15. Eilon, S., Watson-Gandy, C. D. T. and Christofides, N. (1971). "Distribution Management: Mathematical modelling and practical analysis", Griffin, London.
16. Gilmore, P. C. and Gomory, R. E. (1964). Sequencing a one-state variable machine; a solvable case of the travelling salesman problem, *Ops. Res.*, **12**, p. 655.

17. Haring, D. R. (1966). Sequential-circuit synthesis, MIT Press, Research Monograph 31, Cambridge, Massachusetts.
18. Held, M. and Karp, R. M. (1970). The travelling salesman problem and minimum spanning trees, *Ops. Res.*, **18**, p. 1138.
19. Held, M. and Karp, R. M. (1971). The travelling salesman problem and minimum spanning trees. Part II, *Math. Prog.*, **1**, p. 6.
20. Krolak, P., Felts, W. and Marble, G. (1971). A man-machine approach toward solving the travelling salesman problem, *Comm. of ACM*, **14**, p. 327.
21. Lawler, E. L. (1971). A solvable case of the travelling salesman problem, *Math. Prog.*, **1**, p. 267.
22. Lin, S. (1965). Computer solutions of the travelling salesman problem, *Bell Syst. Tech. Jl.*, **44**, p. 2245.
23. Lin, S. and Kernighan, B. W. (1971). A heuristic technique for solving a class of combinatorial optimization problems, Princeton Conf. on System Science.
24. Maffioli, F. (1973). The travelling saleman problem and its implications, Report, Instituto di Elettrotecnica ed Elettronica, Politecnico di Milano.
25. Nash-Williams, C. St. J. A. (1966). On Hamiltonian circuits in finite graphs, *Proc. American Mathematical Soc.*, **17**, p. 466.
26. Ore, O. (1962). "Theory of Graphs", American Mathematical Society, New York.
27. Pevepeliche, V. A. and Gimadi, E. X. (1969). On the problem of finding the minimum Hamiltonian circuit in a graph with weighted arcs, *Diskret Analyz.*, (In Russian), **15**, p. 57.
28. Pohl, J. (1970). Heuristic search viewed as path finding in a graph, *Artificial Intelligence*, **1**, p. 193.
29. Pósa, L. (1962). A theorem concerning Hamiltonian lines, *Magyar Tnd. Akad. Mat. Kutató Int. Közl*, **7**, p. 225.
30. Roberts, S. M. and Flores, B. (1967). An engineering approach to the travelling salesman problem, *Man. Sci.*, **13**, p. 269.
31. Roberts, S. M. and Flores, B. (1966). Systematic generation of Hamiltonian circuits, *Comm. of ACM*, **9**, p. 690.
32. Roy, B. (1959). Recherche des circuits elementaires et des circuits Hamiltoniens dans un graphe quelconque, Mimeographe, Soc. de Math. Appl., Paris.
33. Roy, B. (1969, 1970). "Algebre modern et theorie des graphes", Vols 1 and 2, Dunod, Paris.
34. Rubinshtein, M. I. (1971). On the symmetric travelling salesman problem, *Automatica i Telemekanika*, **9**, p. 126.
35. Selby, G. R. (1970). The use of topological methods in computer-aided circuit layout, Ph.D. Thesis, London University.
36. Syslo, M. M. (1973). A new solvable case of the travelling salesman problem, *Math. Prog.*, **4**, p. 347.
37. Yau, S. S. (1967). Generation of all Hamiltonian circuits, paths and centres of a graph and related problems, *IEEE Trans.*, **CT-14**, p. 79.

10. Appendix

The following is an algorithm for calculating an upper bound B on the minimum number of facility resettings required in the scheduling problem of Section 4, and is used for initializing the algorithm of that section.

Step 0. index $\leftarrow$ 0, $B \leftarrow$ 0.

Step 1. Set labels $l(x_i) = 0 \; \forall \; x_i \in X$. Set $p = 0$.

Step 2. $G = (X, A)$, choose any $x_o \in X$, set $S = \{x_o\}$.

Step 3. If index $= 0$ form $\bar{S} = S \cup \Gamma(S)$; else form $\bar{S} = S \cup \Gamma^{-1}(S)$

Where: $\Gamma(x_i) = \{x_j | (x_i, x_j) \in A\}$ and $\Gamma(S) = \bigcup_{x_i \in S} \Gamma(x_i)$

and also: $\Gamma^{-1}(x_i) = \{x_j | (x_j, x_i) \in A\}$ and $\Gamma^{-1}(S) = \bigcup_{x_i \in S} \Gamma^{-1}(x_i)$

Step 4. If $\bar{S} = S$ go to step 5, else $p \leftarrow p + 1$, $l(x_i) \leftarrow p \; \forall \; x_i \in \bar{S} - S$, $\bar{S} \leftarrow S$ and return to step 3.

Step 5. Find $x \in \{x_i | l(x_i) = p\}$.

Step 6. If index $= 0$ find an $x' \in \{x_i | l(x_i) = p - 1$ and $(x', x) \in A\}$.
else: find an $x' \in \{x_i | l(x_i) = p - 1$ and $(x, x') \in A\}$.

Step 7. $X \leftarrow X - \{x\}$, $A \leftarrow A - \{(x, x_i) | x_i \in X\} - \{(x_i, x) | x_i \in X\}$.
If $x = \{x_o\}$, $B \leftarrow B + 1$ and go to step 12; else go to step 8.

Step 8. $x \leftarrow x'$, $p \leftarrow p - 1$. If $x' = x_o$ go to step 9; else go to step 5.

Step 9. If index $= 0$, index $\leftarrow 1$ and go to step 1; else go to step 10.

Step 10. If $X = \{x_o\}$, $B \leftarrow B + 1$ and go to step 12; else go to step 11.

Step 11. index $\leftarrow$ 0, $B \leftarrow B + 1$.
$X \leftarrow X - \{x_o\}$, $A \leftarrow A - \{(x_o, x_i) | x_i \in X\} - \{(x_i, x_o) | x_i \in X\}$.

Step 12. Stop. B is the required upper bound.

The algorithm requires some explanation. When index $= 0$ *forward* paths are traced through the vertices G starting with vertex x_o. These paths are traced by labelling with p those vertices of G that require p arcs to be reached from x_o. (Steps 1, 2, 3 and 4.) When none of these paths can be extended the algorithm proceeds to steps 5, 6 and 7 which trace the longest of these paths backwards to the vertex x_o erasing from the graph vertices (and their associated arcs), which lie on the longest path. Step 8 terminates the erasing process. Step 9 returns the algorithm to the beginning with index $= 1$ in order to start forming *backward* paths, i.e. paths terminating at x_o. Again the longest of these is found and erased.

The longest of the forward and backward paths (to or from x_o), can be considered together as a single long path which contains x_o. The number B (of path sequences necessary), is then increased by unity at step 11 and the process continued by choosing another vertex x_o from the remaining graph to form new longest forward and backward sequences e.t.c. until the graph is exhausted.

The final value of B, which is the number of path sequences used to erase the whole graph (i.e. cover all the vertices) is then, obviously, an upper bound on the required minimum number of such covering sequences (i.e. on the minimum number of facility resettings required).

Chapter 11

Network Flows

1. Introduction

One of the most interesting and important problems to do with graphs is that of determining the value of the maximum flow that can be transmitted from a specified source vertex s of a graph, to a specified terminal (sink) vertex t. In this context a capacity q_{ij} is associated with every arc (x_i, x_j) of the graph G, and this capacity represents the largest amount of flow that can be transmitted along the arc. This problem and its variants can arise in a large number of practical applications, for example in determining the maximum rate of traffic flow between two locations on a road map represented by a graph. In this example, a solution to the maximum flow problem would also indicate the parts of the road network which are "saturated" and form a "bottleneck" as far as the flow between the two specified end locations is concerned.

A method of solution of this (s to t) *maximum flow* problem, was developed by Ford and Fulkerson [12] and their "labelling technique" forms the basis of other solution algorithms for a number of problems that are simple generalization or extensions of the above problem. Some possible variations of the (s to t) maximum flow problem, which have been discussed in the literature are:

(i) Let us suppose that with each arc of the graph is associated not only a capacity q_{ij} giving an upper bound for the flow in arc (x_i, x_j), but also a "capacity" r_{ij} giving a lower bound on the arc flows. In such a case, it is not even obvious that a feasible set of flows exists which satisfies both the lower and upper arc capacities. However, if—in general—many such flows are possible, and if in addition to the capacities there are also costs c_{ij} associated with a unit flow along the arcs, then the problem becomes one of the finding the *minimum-cost feasible flow* from vertex s to vertex t.

(ii) Consider the case where the *maximum flow between every pair of vertices* is required. Although this problem can be solved as $n(n-1)/2$ separate (s to t) problems, this is a very laborious process. Just as in the case of

282

determining the shortest path between every pair of vertices of a graph, it was not found necessary to consider every (s to t) shortest path problem individually (see Chapter 8), so with the present problems a method exists which—for nondirected graphs—does not involve the solution of an (s to t) maximum flow problem for every pair of vertices s and t.

(iii) If, instead of a single-source single-sink problem one considers a number of specified source and sink vertices, with different commodities flowing between different specified source–sink combinations, then the problem of maximizing the sum of all the flows from the sources to the sinks is called the *multi-commodity* flow problem. In this problem the capacity q_{ij} of an arc (x_i, x_j) is a limit on the sum of the flows of all the commodities along that arc.

(iv) The implicit assumption has been made in all of the above cases that the flow leaving an arc is the same as the flow that enters it. If we abandon this assumption and consider the case of a graph where the output flow of an arc is its input flow multiplied by some non-negative number, then the (s to t) maximum flow problem is referred to as the problem of *flows in graphs with gains*. In this last type of problem flows can be both "generated" and "absorbed" by the graph itself, so that the flow entering s and that leaving t can be varied quite independently.

The general area of graph theory dealing with flows in graphs has been extensively studied by a large number of workers. The purpose of this Chapter is to give an overview of the general problems encountered, show the relations between them and give a description of some important algorithms used to solve flow problems. This is necessary, firstly as an end in its own right, secondly because the algorithms themselves reveal relationships between problems which would not otherwise be apparent, and thirdly because the algorithms which are described here can be used (either directly or as elementary steps in larger algorithms), to analyse a wide variety of other problems.

In this Chapter we will be discussing the basic (s to t) maximum flow problem and its generalization (i), (ii) and (iv) above. Because the algorithms for the multi-commodity flow problem are of a different nature, and not at all as efficient as the labelling methods discussed in this Chapter, we will not be dealing with this problem here and the interested reader is referred to [18, 27, 26, 25, 24, 15].

2. The Basic (s to t) Maximal Flow Problem

Consider the graph $G = (X, A)$ with arc capacities q_{ij}, a source vertex s and a terminal vertex t; (s and $t \in X$). A set of numbers ξ_{ij} defined on the arcs

$(x_i, x_j) \in A$ are called flows in the arcs if they satisfy the following conditions: †

$$\sum_{x_j \in \Gamma(x_i)} \xi_{ij} - \sum_{x_k \in \Gamma^{-1}(x_i)} \xi_{ki} = \begin{cases} v \text{ if } x_i = s \\ -v \text{ if } x_i = t \\ 0 \text{ if } x_i \neq s \text{ or } t \end{cases} \quad (11.1)$$

and $\xi_{ij} \leqslant q_{ij}$ for all $(x_i, x_j) \in A$ (11.2)

Equation (11.1) is an equation of conservation of flow and states that the flow into a vertex x_i is equal to the flow out of the same vertex except for the source and sink vertices s and t for which there is a net outflow and inflow of *value v* respectively. Equation (11.2) simply states the capacity constraint for each arc of the graph G. The objective is to find a set of arc flows so that

$$v = \sum_{x_j \in \Gamma(s)} \xi_{sj} = \sum_{x_k \in \Gamma^{-1}(t)} \xi_{kt} \quad (11.3)$$

is maximized where ξ_{sj} and ξ_{kt} are written for the flows from vertex s to x_j and from x_k to t respectively.

The labelling algorithm developed by Ford and Fulkerson [12] for the solution of this problem, is based on the following theorem (see Chapter 9 for the definition of a cut-set).

THEOREM 1. (Maximum-flow minimum-cut theorem) [8, 10, 11]. *The value of the maximum flow from s to t is equal to the value of the minimum cut-set* $(X_m \to \tilde{X}_m)$ *separating s from t.*

A cut-set $(X_o \to \tilde{X}_o)$ *separates s from t if* $s \in X_o$ *and* $t \in \tilde{X}_o$. *The value of such a cut-set is the sum of the capacities of all arcs of G whose initial vertices are in* X_o *and the final vertices in* $\tilde{X}_o$; *i.e.*

$$v(X_o \to \tilde{X}_o) = \sum_{(x_i, x_j) \in (X_o \to \tilde{X}_o)} q_{ij}$$

The minimum cut-set $(X_m \to \tilde{X}_m)$ *is then the cut-set with the smallest such value.*

Proof. A constructive proof of the maximum-flow minimum-cut theorem is given here and the method of construction immediately suggests the labelling algorithm which follows.

It is quite apparent that the maximum flow from s to t cannot be greater than $v(X_m \to \tilde{X}_m)$ since all paths leading from s to t use one of the arcs of this cut-set. The aim of the proof is, therefore, to show that a flow exists which attains this value. Let us now assume a flow given by the m-dimensional vector ξ, and define a cut set $(X_o, \tilde{X}_o)$ by recursively applying step (b) below:

(a) Start by setting $X_o \leftarrow \{s\}$

† When ambiguity arises we will also write $\xi(x_i, x_j)$, $q(x_i, x_j)$, $r(x_i, x_j)$ and $c(x_i, x_j)$ instead of the shorthand expressions ξ_{ij}, q_{ij}, r_{ij} and c_{ij} respectively.

(b) If $x_i \in X_o$, and either $\xi_{ij} < q_{ij}$, or $\xi_{ji} > 0$ place x_j in the set X_o and repeat the step until X_o can not be increased further.
Then two cases can occur, either $t \in X_o$ or $t \notin X_o$.

CASE (i) $t \in X_o$ According to step (b) above $t \in X_o$, implies that a chain of arcs from vertex s to vertex t exists so that for every arc (x_i, x_j) used by the chain in the forward direction (forward arcs), $\xi_{ij} < q_{ij}$; and for every arc (x_k, x_l) used by the chain in the backward direction i.e. in the direction from x_l to x_k (backward arcs), $\xi_{kl} > 0$. (This chain of arcs from s to t will be called a *flow-augmenting chain*.)
Let:

$$\delta_f = \min_{(x_i, x_j)} [q_{ij} - \xi_{ij}]; \; (x_i, x_j) \text{ forward} \tag{11.5}$$

$$\delta_b = \min_{(x_k, x_l)} [\xi_{kl}]; \qquad (x_k, x_l) \text{ backward} \tag{11.6}$$

and $$\delta = \min [\delta_f, \delta_b] \tag{11.7}$$

If now δ is added to the flow in all forward arcs and subtracted from the flow in all backward arcs of the chain, the net result is a new feasible flow with a value δ units greater than the previous one. This is apparent since the addition of δ to the flow in the forward arcs cannot violate any of the arc capacities of these arcs (since $\delta \leqslant \delta_f$) and the subtraction of δ from the flow in the backward arcs cannot make the flow in these arcs negative (since $\delta \leqslant \delta_b$).

Using the new improved flow one can then reapply steps (a) and (b) above to define a new cut set $(X_o, \tilde{X}_o)$ and repeat the argument.

CASE (ii) $t \notin X_o$ (i.e. $t \in \tilde{X}_o$). According to step (b) $\xi_{ij} = q_{ij}$ for all $(x_i, x_j) \in (X_o \to \tilde{X}_o)$, and $\xi_{kl} = 0$ for all $(x_k, x_l) \in (\tilde{X}_o \to X_o)$.
Hence:

$$\sum_{(x_i, x_j) \in (X_o \to \tilde{X}_o)} \xi_{ij} = \sum_{(x_i, x_j) \in (X_o \to \tilde{X}_o)} q_{ij}$$

and:
$$\sum_{(x_k, x_l) \in (\tilde{X}_o \to X_o)} \xi_{kl} = 0$$

i.e. the value of the flow which is

$$\sum_{(x_i, x_j) \in (X_o \to \tilde{X}_o)} \xi_{ij} - \sum_{(x_k, x_l) \in (\tilde{X}_o \to X_o)} \xi_{kl}$$

is equal to the value of the cut $(X_o \to \tilde{X}_o)$.

Since in case (i) the flow is continuously increased by a least one unit, then assuming all q_{ij} are in integers, the maximum flow must be obtained in a finite number of steps when case (ii) occurs. That flow then equals the value of the current cut $(X_o \to \tilde{X}_o)$ which must, therefore, then be the minimum cut. Hence the theorem.

The constructive method used for the proof of the maximum-flow minimum-cut theorem immediately suggests an algorithm for the calculation of the maximum flow from a given vertex s to a given vertex t in a capacitated graph. Such an algorithm will now be given.

The algorithm starts with an arbitrary feasible flow (zero flow may be used) and then tries to increase the flow value by systematically searching all possible flow-augmenting chains from s to t. The search for a flow-augmenting chain is carried out by attaching labels to vertices indicating the arc along which the flow may be increased and by how much. Once such a chain is found, the flow along it is increased to its maximum value, all vertex labels are erased and the new flow is used as a basis for relabelling. When no flow-augmenting chain can be found the algorithm terminates with the maximal flow. The algorithm proceeds as follows:

2.1 Labelling algorithm for the (s to t) maximum flow problem

A. THE LABELLING PROCESS

A vertex can only be in one of three possible states; labelled and scanned (i.e. it has a label and all adjacent vertices have been "processed"), labelled and unscanned (i.e. it has a label but not all its adjacent vertices have been processed) and unlabelled (i.e. it has no label). A label on a vertex x_i is composed of two parts and takes one of the two forms $(+x_j, \delta)$ or $(-x_j, \delta)$. The part $+x_j$ of the first type of label implies that the flow along arc (x_j, x_i) can be increased. The part $-x_j$ of the alternative type of label implies that the flow along arc (x_i, x_j) can be decreased. δ represents in both cases the maximum amount of extra flow that can be sent from s to x_i along the augmenting chain being constructed. The labelling of a vertex x_i corresponds to finding a flow-augmenting chain from s to x_i.

Initially all vertices are unlabelled.

Step 1. Label s by $(+s, \delta(s) = \infty)$. s is now labelled and unscanned and all other vertices are unlabelled.

Step 2. Choose any labelled unscanned vertex x_i and suppose its label is $(\pm x_k, \delta(x_i))$.

(i) To all vertices $x_j \in \Gamma(x_i)$ that are unlabelled and for which $\xi_{ij} < q_{ij}$ attach the label $(-x_i, \delta(x_j))$ where:

$$\delta(x_j) = \min\left[\delta(x_i), q_{ij} - \xi_{ij}\right]$$

and

(ii) To all vertices $x_j \in \Gamma^{-1}(x_i)$ that are unlabelled and for which $\xi_{ji} > 0$ attach the label $(-x_i, \delta(x_j))$ where:

$$\delta(x_j) = \min\left[\delta(x_i), \xi_{ji}\right].$$

(The vertex x_i is now labelled and scanned and the vertices x_j labelled by (i) and (ii) above are labelled and unscanned.) Indicate that x_i is now scanned by marking it in some way.

Step 3. Repeat step 2 until either t is labelled in which case proceed to step 4 or t is unlabelled and no more labels can be placed in which case the algorithm terminates with ξ as the maximum flow vector. It should be noted here that if X_o is the set of labelled vertices and $\tilde{X}_o$ the set of unlabelled ones then $(X_o \to \tilde{X}_o)$ is the minimum cut.

B. FLOW AUGMENTING PROCESS
Step 4. Let $x = t$ and go to step 5.

Step 5. (i) If the label on x is of the form $(+z, \delta(x))$, change the flow along the arc (z, x) from $\xi(z, x)$ to $\xi(z, x) + \delta(t)$.
(ii) If the label on x is of the form $(-z, \delta(x))$, change the flow along the arc (x, z) from $\xi(x, z)$ to $\xi(x, z) - \delta(t)$.

Step 6. If $z = s$ erase all labels and return to step 1 to repeat the labelling process starting from the new improved flow calculated in step 5 above.
If $z \neq s$ set $x = z$ and return to step 5.

2.2 Example
Consider the graph of Fig. 11.1 and take vertex x_1 to be the source vertex s and vertex x_9 to be the sink vertex t. The capacity of each arc is given by the number next to the arc, and what is required is to find the maximum flow from x_1 to x_9.

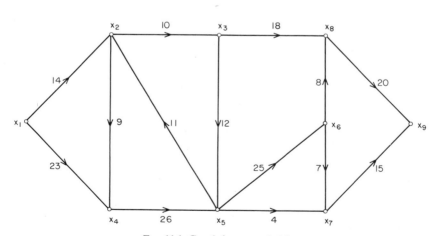

FIG. 11.1. Graph for example 2.2

Let us start from an initial flow of value zero in all arcs. The algorithm then proceeds as follows:

Step 1. Label x_1 by $(+x_1, \infty)$.

Step 2.

(i) The set of vertices $\{x_j | x_j \in \Gamma(x_1), \xi_{ij} < q_{ij} \text{ and } x_j \text{ unlabelled}\}$ is $\{x_2, x_4\}$

   label x_2 by $(+x_1, \min[\infty, 14 - 0])$, i.e. by $(+x_1, 14)$

   label x_4 by $(+x_1, \min[\infty, 23 - 0])$, i.e. by $(+x_1, 23)$.

(ii) The set of vertices $\{x_j | x_j \in \Gamma^-(x_1), \xi_{j1} > 0 \text{ and } x_j \text{ unlabelled}\}$ is empty. Hence x_1 is labelled and scanned x_2 and x_4 labelled and unscanned and all other vertices are unlabelled.

Repeating step 2 and taking x_2 to scan first:

(i) The set $\{x_j | x_j \in \Gamma(x_2), \xi_{2j} < q_{2j} \text{ and } x_j \text{ unlabelled}\}$ is $\{x_3\}$

   label x_3 by $(+x_2, \min[14, 10 - 0]) = (+x_2, 10)$.

(ii) The set $\{x_j | x_j \in \Gamma^{-1}(x_2), \xi_{j2} < 0 \text{ and } x_j \text{ unlabelled}\}$ is empty. Vertices x_1 and x_2 are now labelled and scanned, and x_3 and x_4 labelled and unscanned.

Taking x_3 to scan next and repeating step 2 we get the following labellings:

   label x_5 by $(+x_3, \min[10, 12 - 0]) = (+x_3, 10)$

   label x_8 by $(+x_3, \min[10, 18 - 0]) = (+x_3, 10)$.

Taking x_4 to scan next we find that no labels can be placed, so continuing the scanning with x_5, etc., we obtain labellings in the following order:

   label x_6 by $(+x_5, \min[\bar{1}0, 25 - 0]) = (+x_5, 10)$

   label x_7 by $(+x_5, \min[10, 4 - 0]) = (+x_5, 4)$

   label x_9 by $(+x_7, \min[4, 15 - 0]) = (+x_7, 4)$

Going on to steps 4 and 5 we get:

$$x = x_9; \quad \xi_{7,9} = 0 + 4 = 4$$
$$x = x_7; \quad \xi_{5,7} = 4$$
$$x = x_5; \quad \xi_{3,5} = 4$$
$$x = x_3; \quad \xi_{2,3} = 4$$
$$x = x_2; \quad \xi_{1,2} = 4$$

The flow pattern at the end of step 5 and the vertex labels before being erased by step 6 are as shown in Fig. 11.2(a). All flows are shown underlined.

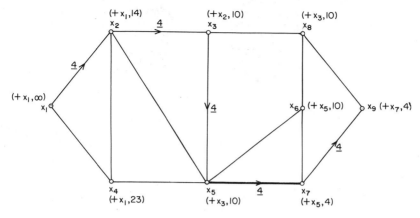

FIG. 11.2 (a). Flows and labels after 1st iteration
———————: Saturated arcs

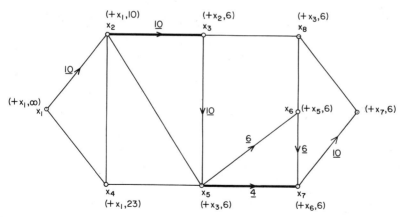

FIG. 11.2 (b). Flows and labels after 2nd iteration
———————: Saturated arcs

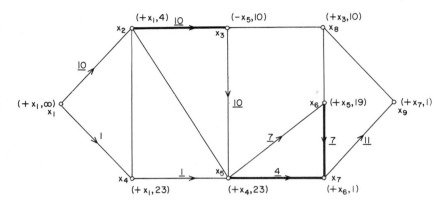

FIG. 11.2 (c). Flows and labels after 3rd iteration
———————— Saturated arcs

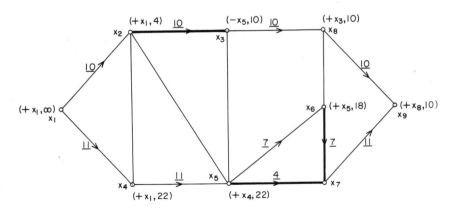

FIG. 11.2 (d). Flows and labels after 4th iteration
———————— Saturated arcs

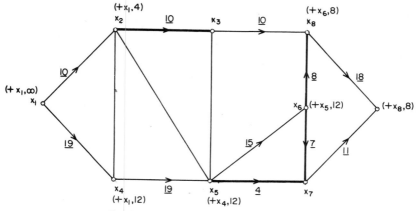

FIG. 11.2 (e). Flows and labels after 5th iteration
——————— Saturated arcs

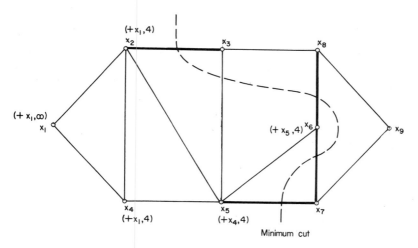

FIG. 11.2 (f). Labels at end of 6th iteration
——————— Saturated arcs

Erasing the labels on the vertices and returning to step 1 for a second pass we obtain the following new vertex labels: (labelled and unscanned vertices are scanned in ascending order of their suffices).

Step 3.

$$\text{label } x_1 \text{ by } (+x_1, \infty)$$

$$\text{label } x_2 \text{ by } (+x_1, \min [\infty, 14 - 4]) = (+x_1, 10)$$

$$\text{label } x_4 \text{ by } (+x_1, \min [\infty, 23 - 0]) = (+x_1, 23)$$

vertex x_1 is now labelled and scanned.

$$\text{label } x_3 \text{ by } (+x_2, \min [10, 10 - 4]) = (+x_2, 6)$$

vertex x_2 is now labelled and scanned.

$$\text{label } x_5 \text{ by } (+x_3, \min [6, 12 - 4]) = (+x_3, 6)$$

$$\text{label } x_8 \text{ by } (+x_3, \min [6, 18 - 0]) = (+x_3, 6)$$

vertex x_3 is now labelled and scanned

vertex x_4 is also labelled and scanned

$$\text{label } x_6 \text{ by } (+x_5, \min [6, 25 - 0]) = (+x_5, 6)$$

vertex x_5 is now labelled and scanned

$$\text{label } x_7 \text{ by } (+x_6, \min [6, 7 - 0]) = (+x_6, 6)$$

vertex x_6 is now labelled and scanned

$$\text{label } x_9 \text{ by } (+x_7, \min [6, 15 - 4]) = (+x_7, 6)$$

Steps 4 and 5.

The new flows are increased as follows:

$$\xi_{7,9} = 4 + 6 = 10; \quad \xi_{6,7} = 0 + 6 = 6; \quad \xi_{5,6} = 0 + 6 = 6;$$

$$\xi_{3,5} = 4 + 6 = 10; \quad \xi_{2,3} = 4 + 6 = 10; \quad \xi_{1,2} = 4 + 6 = 10;$$

all other values of flow remaining the same.

The new flow pattern and vertex labels before erasing are shown in Fig. 11.2(b).

Proceeding in the same way, the flows and labellings at the end of each pass through the algorithm are shown successively in Figs. 11.2(c) to 11.2(e). The algorithm terminates with the labelling shown in Fig. 11.2(f) when x_9 cannot be labelled.

The flow in Fig. 11.2(e) is therefore the maximal flow of value 29 and the corresponding minimum cut is shown dotted in Fig. 11.2(f).

2.3 The incremental graph

The process of finding a flow-augmenting chain in a graph $G = (X, A)$ when the arc flows are given by the vector ξ, can be considered as an (s to t) path finding process in an *incremental* graph $G^\mu(\xi) = (X^\mu, A^\mu)$ defined as follows:

$$X^\mu = X$$

and

$$A^\mu = A^\mu_1 \cup A^\mu_2$$

where:

$$A^\mu_1 = \{(x^\mu_i, x^\mu_j) \mid \xi_{ij} < q_{ij}\}$$

with the capacity of an arc $(x^\mu_i, x^\mu_j) \in A^\mu_1$ being $q^\mu_{ij} = q_{ij} - \xi_{ij}$,

and

$$A^\mu_2 = \{(x^\mu_j, x^\mu_i) \mid \xi_{ij} > 0\}$$

with the capacity of an arc $(x^\mu_j, x^\mu_i) \in A^\mu_2$ being $q^\mu_{ji} = \xi_{ij}$.

The labelling procedure of the algorithm described earlier in Section 2.1 is then no more than a method of calculating the reachable set $R(s)$ in the incremental graph $G^\mu(\xi)$. If $t \in R(s)$, i.e. if vertex t receives a label, then a path P from s to t in the graph $G^\mu(\xi)$ has been found. The flow-augmenting chain of G is then the path P where those arcs of P in A^μ_1 are the forward arcs and those arcs of P in A^μ_2 are the backward arcs.

Figure 11.3(a) shows a flow pattern ξ in a graph G, where the underlined numbers are the flows, and the numbers next to the arcs are the arc capacities. Fig. 11.3(b) shows the corresponding incremental graph $G^\mu(\xi)$.

2.4 Decomposition of a flow pattern

It is sometimes desirable to represent a complex flow pattern as the sum of simpler flow patterns. This is useful not so much because such a decomposition is required in practice but because it contributes to a better understanding of the nature of network flows, and serves as a convenient means of justifying many of the network flow algorithms.

Let us denote by $h \circ (S)$ the flow pattern in a graph G in which arcs $(x_i, x_j) \in S$ have $\xi_{ij} = h$ and arcs $(x_i, x_j) \notin S$ have $\xi_{ij} = 0$. Obviously, $h \circ (S)$ is not a flow pattern for any arbitrary arc set S since a flow pattern must satisfy the continuity eqns (11.1) for some value of v. It is quite apparent that for $h \circ (S)$ to represent a flow pattern, the set S of arcs must form either a path in G from s to t, or a circuit of G.

Given two flow patterns ξ and ψ, let us denote by $\xi + \psi$ the flow pattern for which the net flow in arc (x_i, x_j) is $\xi_{ij} + \psi_{ij}$.

THEOREM 2. *If ξ is any $(s$ to $t)$ flow pattern of (integer) value v in a graph G, then ξ can be decomposed as:*

$$\xi = 1 \circ (P_1) + 1 \circ (P_2) + \ldots + 1 \circ (P_v) + 1 \circ (\Phi_1) + 1 \circ (\Phi_2) + \ldots + 1 \circ (\Phi_k),$$

where $P_1, \ldots, P_v$ are elementary $(s$ to $t)$ paths of G and $\Phi_1, \ldots, \Phi_k$ are elementary circuits of G.

(The P_i and Φ_i need not necessarily be distinct.)

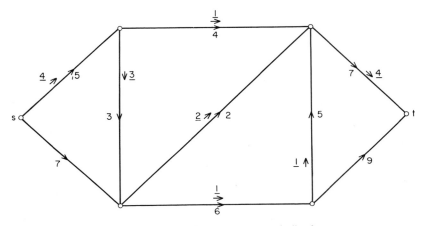

(a) Graph G with flows ξ_{ij} shown underlined

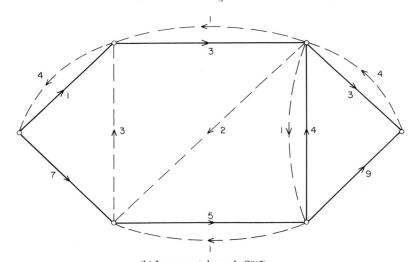

(b) Incremental graph $G^\mu(\xi)$

FIG. 11.3

Proof. From the graph $G = (X, A)$ with the flow pattern ξ construct the *unitary* graph $G^e = (X^e, A^e)$ as follows: The set X^e of vertices of G^e is the same as the set X of vertices of G. If ξ_{ij} is the flow in arc (x_i, x_j) of G, then place ξ_{ij} arcs in parallel between the corresponding vertices x_i^e and x_j^e of G^e. If $\xi_{ij} = 0$, then no arc is placed between x_i^e and x_j^e. The graph G^e is then an s-graph and since each arc in G^e corresponds to a unit arc flow in the graph G, G^e represents the flow ξ in G.

In the graph G^e the vertex degrees must (by the continuity conditions 11.1) satisfy:

$$d_o(x_i^e) = d_t(x_i^e) \qquad \forall x_i^e \neq s^e \text{ or } t^e$$

$$d_o(s^e) = d_t(t^e) = v$$

Now if v return arcs are added to G^e from vertex t^e to vertex s^e, the graph G^e will then possess an Eulerian circuit. (See Chapter 9). The removal of these v arcs from the Eulerian circuit then leaves v paths from s^e to t^e which in totality traverse each arc of G^e exactly once. Let these paths be $P_1', P_2', \ldots, P_v'$. The paths P_i' are not necessarily elementary although (by the definition of an Eulerian circuit), they must be simple. However, since any non elementary path can be considered as the sum of an elementary (s^e to t^e) path and a number of elementary arc disjoint circuits, we obtain:

$$\xi = 1 \circ (P_1) + 1 \circ (P_2) + \ldots + 1 \circ (P_v) + 1 \circ (\Phi_1) + \ldots + 1 \circ (\Phi_k),$$

where the P_i are the elementary (s^e to t^e) paths and the Φ_i are the elementary circuits. Hence the theorem.

In general not all the paths and circuits will be distinct. If only the first v' paths and k' circuits are distinct, with path P_i appearing h_i times in the list $P_1, \ldots, P_v$ and circuit Φ_i appearing l_i times in the list $\Phi_1, \ldots, \Phi_k$, then ξ can be written as:

$$\xi = \sum_{i=1}^{v'} h_i \circ (P_i) + \sum_{i=1}^{k'} l_i \circ (\Phi_i)$$

where the summation is meant to imply the flow pattern addition as explained earlier.

In the course of the proof of theorem 2 another important result is obtained namely that the unit flows into which ξ has been decomposed are *conformal*, i.e. if ξ_{ij} is the net flow in arc (x_i, x_j) of G, then there is a total number of ξ_{ij} unit flows all using this arc in the direction from x_i to x_j.†

Figure 11.4 shows an example of a flow pattern ξ and its decomposition into elementary (s to t) paths and circuits.

---

† The flows would not be conformal, for example, if there was a total number of $\xi_{ij} + p$ (say) unit flows using the arc in the direction x_i to x_j and p unit flows using the arc in the opposite direction thus producing a net flow of ξ_{ij} from x_i to x_j.

In general two flow patterns ξ and ψ are conformal if $\xi_{ij} \cdot \psi_{ji} = 0$ for any arc (x_i, x_j).

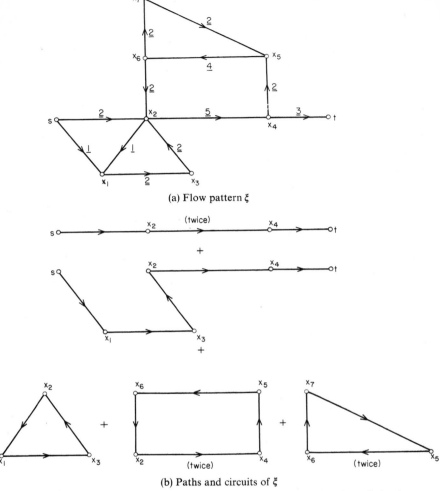

(a) Flow pattern ξ

(b) Paths and circuits of ξ

FIG. 11.4. Decomposition of a flow pattern into elementary (s to t) paths and circuits

3. Simple Variations of the (s to t) Maximum Flow Problem

We will now give some problems concerned with flow which can be immediately recast into the (s to t) maximum flow problem discussed in the above section.

3.1. Graphs with many sources and sinks

Consider a graph with n_s source vertices and n_t sink vertices and assume that flow can go from any source to any sink. The problem of finding the maximum total flow from all the sources to all the sinks can be converted to the simple (s to t) maximum flow problem by adding a new artificial source vertex s and a new artificial sink vertex t with added arcs leading from s to each of the real source vertices and from every real sink to t.

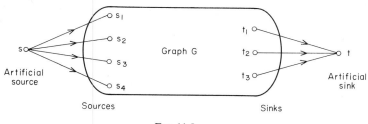

FIG. 11.5

Figure 11.5 shows how multiple sources and sinks can be reduced to a single source and a single sink. The capacities of the arcs leading from s to the source can be set to infinity, or, in the case where the supply at a source s_k is limited, the capacity of the corresponding arc (s, s_k) can be set to this limit. Similarly the capacities of the arcs leading from the sinks to t can be set to the limited demand at the sinks or to infinity if there is no such limit.

If the problem is one where certain sinks can be supplied only by certain sources and *vice versa*, then the problem ceases to be a simple variation of the (s to t) maximum flow problem and becomes the multi-commodity flow problem as mentioned in the Introduction (see problem (iii)).

3.2. Graphs with arc and vertex capacities

For a graph G let the arcs have capacities q_{ij}, and in addition let the vertices have capacities w_j, say ($j = 1, 2, ..., n$) so that the total flow entering vertex x_j must have a value less than w_j i.e.

$$\sum_{x_i \in \Gamma^{-1}(x_j)} \xi_{ij} \leqslant w_j \quad \text{for all } x_j.$$

Let the maximum flow between vertices s and t for such a graph be required.

Let us define a graph G_o so that every vertex x_j of graph G corresponds to two vertices x_j^+ and x_j^- in the graph G_o, in such a way so that for every arc (x_i, x_j) of G incident at x_j corresponds an arc (x_i^-, x_j^+) of G_o incident at x_j^+, and for every arc (x_j, x_k) of G emanating from x_j corresponds an arc (x_j^-, x_k^+) of G_o emanating from x_j^-. Moreover, we introduce an arc between x_j^+ and x_j^- of capacity w_j, i.e. equal to the capacity of vertex x_j.

Figure 11.6(a) shows an example of a graph with vertex capacities and

Fig. 11.6(b) shows the graph G_o constructed as indicated above. Since the total flow entering a vertex x_j^+ must, by necessity, travel along the arc (x_j^+, x_j^-) whose capacity is w_j, the maximum flow in the graph G with vertex capacities, is equal to the maximum flow in the graph G_o which has only arc capacities. One should note that if the minimum cut of G_o does not contain arcs of the form (x_j^+, x_j^-), then the vertex capacities in G are inactive and superfluous; if on the other hand the minimum cut of G_o contains such arcs then the corresponding vertices of G are saturated by the maximum flow solution.

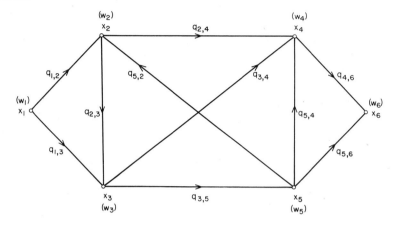

(a) Graph with both vertex and arc capacities

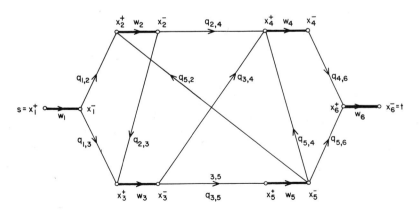

(b) Equivalent graph with arc capacities only

FIG. 11.6

3.3 Graphs with arcs having both upper and lower bounds on the flow

Let us consider a graph $G = (X, A)$ whose arcs (x_i, x_j) have capacities (upper bounds on the flow) of, say, q_{ij} and also lower bounds on the flow of, say, r_{ij}. Suppose we want to know if a feasible flow exists between s and t, i.e. a flow ξ_{ij} satisfying condition (11.1) and also $r_{ij} \leqslant \xi_{ij} \leqslant q_{ij}$ for all arcs (x_i, x_j) of G.

Start by introducing an artificial source s_a and an artificial sink t_a into G to produce a new graph G_a. For every arc (x_i, x_j) for which $r_{ij} \neq 0$, introduce an arc (s_a, x_j) of capacity r_{ij} and lower bound zero, and also a second arc (x_i, t_a) of capacity r_{ij} and lower bound zero. Decrease q_{ij} to $q_{ij} - r_{ij}$ and

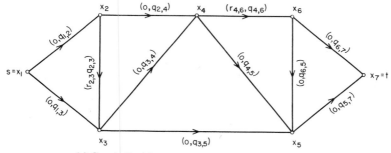

(a) Graph G with upper and lower bounds on arc flows

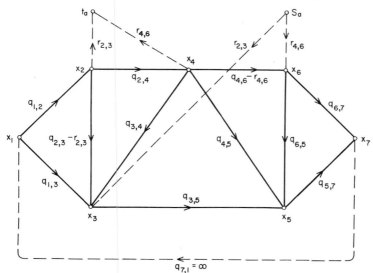

(b) Transformed graph with upper bounds on flow only
(Values shown are for upper bounds, lower bounds are all zero)

FIG. 11.7

decrease r_{ij} to zero. In addition introduce the arc (t, s) with $q_{ts} = \infty$ and $r_{ts} = 0$.

For the example of Fig. 11.7(a) where only two arcs have non-zero lower bounds, these transformations produce the graph of Fig. 11.7(b). Let us now find the maximum flow between the artificial vertices s_a and t_a of the transformed graph G_a. If the value of the maximum flow $(s_a$ to $t_a)$ is $\sum\limits_{r_{ij} \neq 0} r_{ij}$ (i.e. if all arcs leading from s_a, and all arcs leading to t_a are saturated) and, say, the flow along arc (t, s) is ξ_{ts}; then a feasible flow of value ξ_{ts} exists in the original graph. This can be seen as follows: If we subtract r_{ij} from the flow in arcs (x_i, t_a) and (s_a, x_j) and add r_{ij} to the flow in arc (x_i, x_j) then the value of the overall flow from s_a to t_a is reduced by the amount r_{ij}, the flow in (x_i, t_a) and (s_a, x_j) reduced to zero and the flow in the arc (x_i, x_j) will take a value $r_{ij} \leqslant \xi_{ij} \leqslant q_{ij}$. (The final value of ξ_{ij} is equal to r_{ij} if the original value of ξ_{ij} corresponding to the maximum flow is zero, and the final value of ξ_{ij} is equal to q_{ij} if the original value of ξ_{ij} is the maximum possible, i.e. $q_{ij} - r_{ij}$.) The step of substracting flow from the maximum flow is the opposite of the flow augmentation step in the maximum flow algorithm. Since we have assumed the maximum flow from s_a to t_a to be equal to $\sum\limits_{r_{ij} \neq 0} r_{ij}$, the process of flow subtraction will eventually reduce the flow from s_a to t_a to zero (thus making these two artificial vertices and their incident arcs superfluous), and the flow in all arcs with $r_{ij} \neq 0$ will be in the range $r_{ij} \leqslant \xi_{ij} \leqslant q_{ij}$. The final result is therefore a "circulation" of flow in the graph, the value of this flow being ξ_{ts}.

On the other hand:

THEOREM 3. *If the value of the maximum $(s_a$ to $t_a)$ flow in graph G_a is not* $\sum\limits_{r_{ij} \neq 0} r_{ij}$, *then no feasible flow exists in G.*

The proof of this theorem will be left as an exercise for the reader. (See problem 5).

4. Maximum Flow Between Every Pair of Vertices

In some cases, for example in the case where the graph represents a telephone network with the vertices corresponding to stations and the arcs corresponding to the telephone cables, what may be required is the maximum rate of communication between *any* two stations s and t, assuming that only a single pair of stations can communicate with each other at any one time. In this example the capacity of a cable is the number of independent calls that can be routed through it. A similar situation exists in the case where the graph represents a road network and where, say, a system of one-way streets is proposed. What is then required is to determine how the one-way streets affect the maximum rate of traffic flow between any two areas, when those areas are considered in isolation.

As was mentioned in the Introduction, this problem could be solved by simply considering the (s to t) maximum flow problem for every pair of vertices (s, t). However, this is an unnecessarily laborious method and here we give an algorithm due to Gomory and Hu [14] which is much more efficient for the case of *nondirected* graphs. Since this algorithm is founded on two basic concepts these will be discussed first so that the description of the algorithm which follows becomes clearer.

4.1. Flow equivalence

THEOREM 4. *Let* f_{ij} *be the value of the maximum flow from vertex* x_i *to vertex* x_j *of* G. (*Since* G *is undirected* $f_{ij} = f_{ji}$.) *These flows then satisfy the relation:*

$$f_{ij} \geqslant \min [f_{ik}, f_{kj}] \quad \text{for all } i, j \text{ and } k. \tag{11.8}$$

Proof. If (X_o, $\tilde{X}_o$) is the minimum cut-set giving f_{ij}, i.e.

$$f_{ij} = v(X_o, \tilde{X}_o), \qquad x_i \in X_o, \quad x_j \in \tilde{X}_o;$$

then if $x_k \in X_o$, $f_{ij} \geqslant f_{kj}$, since (X_o, $\tilde{X}_o$) is also a cut-set separating x_k from x_j; and if $x_k \in \tilde{X}_o$, $f_{ij} \geqslant f_{ik}$, since (X_o, $\tilde{X}_o$) is also a cut-set separating x_i from x_k. The theorem then flows immediately.

Repeated application of inequality (11.8) to the flows in the parenthesis of (11.8) then gives:

$$f_{ij} \geqslant \min [f_{ik_1}, f_{k_1k_2}, f_{k_2k_3}, \ldots, f_{k_pj}] \tag{11.9}$$

for any sequence of vertices $x_i, x_{k_1}, x_{k_2}, x_{k_3}, \ldots, x_{k_p}, x_j$.

If we now consider a hypothetical complete graph of G' on n vertices, and take the "length" of a link (x'_k, x'_j) to be the maximum flow f_{ij} between the corresponding vertices x_i and x_j of G, we can then use these "lengths" to construct a "longest spanning tree" T^* of G' according to one of the methods for constructing shortest (or longest) spanning trees given in Chapter 7. We can now say, according to eqn (11.9), that the flow f_{ij} is greater than or equal to the "length" of the shortest link which lies on the unique path in T^* from x'_i to x'_j. However, if f_{ij} is greater than the "length" of this shortest link, the removal of this link and the addition of (x'_i, x'_j) would produce a new tree of G' longer than T^*, which is a contradiction of the assumption that T^* is maximal. Thus, f_{ij} must be *equal* to the "length" of the shortest link on the unique path from x'_i to x'_j in the tree T^*. Hence, the tree graph T^* with link capacities equal to the corresponding link "lengths" of G', would then be *flow-equivalent* to the original graph G. By flow-equivalent is meant that if the graphs G and T are considered as "black boxes" with the vertices as "terminals" then the two graphs would be indistinguishable as far as the maximum flow between terminals is concerned. The above construction of T^*

is quite general and shows that for any graph G there is always a flow-equivalent tree graph T^*.

4.2. Vertex condensation

Suppose that the (s to t) maximum flow problem has been solved for a graph $G = (X, \Gamma)$, with s and t being two vertices of G chosen at random. Let us say that $(X_o, \tilde{X}_o)$ is the minimum cut-set corresponding to this maximum flow and consider two vertices x_i and x_j which are both in X_o (or both in $\tilde{X}_o$). If we now wish to find f_{ij}, the maximum flow from x_i to x_j, then all vertices of $\tilde{X}_o$ (or X_o if x_i and $x_j \in \tilde{X}_o$) may be "condensed" into a single vertex $\tilde{x}_o$, say, as far as this flow calculation is concerned. The condensation is such that links (x_a, x_b), $x_a \in X_o$ and $x_b \in \tilde{X}_o$ are replaced by links $(x_a, \tilde{x}_o)$, and any parallel links between the same pair of vertices (which may result) are replaced by a single link of capacity equal to the sum of the capacities of the parallel links. Figures 11.8(a), (b) and (c) illustrate the condensation process. The fact that such a condensation of $\tilde{X}_o$ is possible has been shown by Gomory and Hu [14, 16] who have developed the method we are about to describe. The proof involves the demonstration that all vertices of $\tilde{X}_o$ must lie on the same side of the minimum cut-set $(Y_o, \tilde{Y}_o)$ separating x_i and x_j (i.e. either $\tilde{X}_o \subseteq Y_o$ or $\tilde{X}_o \subseteq \tilde{Y}_o$) so that the internal properties of the subgraph (X_o, Γ) of G do not enter into the calculation of the minimum cut-set between x_i and x_j.

4.3. Algorithm for maximum flow between all pairs of vertices

The algorithm that is given in this section generates a tree T^* which is flow-equivalent to the nondirected graph G. The maximum flow f_{ij} between two vertices x_i and x_j of the graph G can then be found from this tree as:

$$f_{ij} = \min \left[q'_{ik_1}, q'_{k_1 k_2}, q'_{k_2 k_3}, \ldots, q'_{k_p j} \right] \qquad (11.10)$$

where $(x'_i, x'_{k_1}, x'_{k_2}, \ldots, x'_j)$ is the unique path along links of T^* which leads from x'_i to x'_j. Every vertex x'_k of T^* corresponds to a vertex x_k of G and q'_{kl} is the capacity of link (x'_k, x'_l) of T^*.

The algorithm uses the fact that since a tree T^* flow-equivalent to the graph G exists and, since T^* has only $(n - 1)$ links in it, all that is necessary is to calculate the capacities of these links. The algorithm described in this section calculates the capacity of these links of T^* by $(n - 1)$ flow calculations on a graph which is continuously condensed as mentioned in Section 4.2 above.

4.3.1. DESCRIPTION OF THE ALGORITHM Since the algorithm generates T^* gradually, and since at any one of the $(n - 1)$ stages of the algorithm the "vertices" of T^* may in fact be sets composed of vertices of G, we will, in

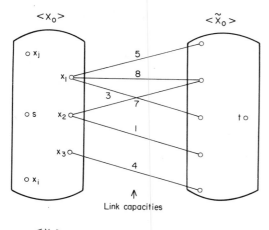

(a) Minimum (s to t) cut-set

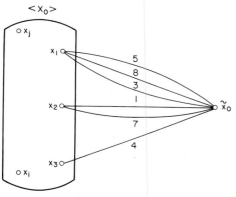

(b) Vertex condensation for x_i to x_j flow calculation

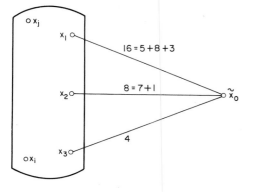

(c) Condensed graph

Fig. 11.8

order to avoid confusion, refer to the vertices of G as G-vertices and to the vertices of T^* as T^*-vertices. Enough comments are given in the following description to make the algorithm self-explanatory.

Step 1: Set $S_1 = X$, $N = 1$.
At any stage, T^* is the graph defined by N T^*-vertices $S_1, S_2, \ldots, S_N$ each one of which corresponds to a set of G-vertices. T^* is initialized to be the single vertex S_1.

Step 2: Find a set $S^* \in \{S_1, S_2, \ldots, S_N\}$ which contains more than one G-vertex in it. If none exist go to step 6, otherwise go to step 3.

Step 3: If S^* were to be removed from T^*, the tree would in general be reduced to a number of subtrees (connected components). Condense the T^*-vertices in each subtree into single vertices to form the condensed graph. Take any two vertices x_i and $x_j \in S^*$ and calculate the minimum cut-set $(X_o, \tilde{X}_o)$ of G separating x_i from x_j by doing an $(x_i$ to $x_j)$ maximum flow calculation.

Step 4: Remove the T^*-vertex S^* together with its incident links from T^*, and replace it by two T^*-vertices composed of the G-vertex sets $S^* \cap X_o$ and $S^* \cap \tilde{X}_o$, and a link of capacity $v(X_o, \tilde{X}_o)$ between them. Also, for all T^*-vertices S_i which were incident on S^* (say with a link of capacity c_i^o) before the replacement, add the link $(S_i, S^* \cap X_o)$ to T^* if $S_i \subset X_o$; or add the link $(S_i, S^* \cap \tilde{X}_o)$ to T^* if $S_i \subset \tilde{X}_o$. The capacities of the links being taken as c_i^o, regardless of which one is added.

Note: As mentioned earlier in Section 4.2 it was shown by Gomory and Hu [14] that S_i is either wholly in X_0 or wholly in $\tilde{X}_o$, so that the only two cases possible are those mentioned above.

Step 5: Set $N = N + 1$. The vertices of T^* are now the sets of G-vertices $S_1, S_2, \ldots, S^* \cap X_o, S^* \cap \tilde{X}_o, \ldots, S_N$ where S^* has been replaced by the two T^*-vertices $S^* \cap X_o$ and $S^* \cap \tilde{X}_o$ as explained earlier. Goto step 2.

Step 6: Stop. T^* is the required flow-equivalent tree of G and its T^*-vertices are now single G-vertices. The link capacities of T^* corresponds to the values of the $(n - 1)$ independent cut-sets of G. Equation (11.10) can then be used to calculate the f_{ij} (for any $x_i, x_j \in X$), directly from T^*.

The fact that the above algorithm produces the optimal answer follows immediately from the properties of the minimal cut-sets and flow-equivalent tree of a graph G given in Sections 4.1 and 4.2 earlier, and a formal proof can be found in Hu [16].

4.4. Example
Consider the nondirected graph G shown in Fig. 11.9 where the link

capacities are shown next to the links. It is required to find the maximum flow between every pair of vertices of G.

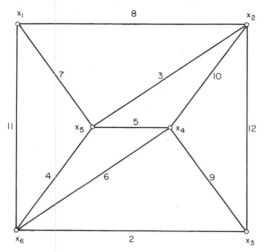

FIG. 11.9. Graph for example 4.4

From the above algorithm we get.

Step 1: $S_1 = \{x_1, x_2, x_3, x_4, x_5, x_6\}$; $N = 1$

Step 2: $S^* = S_1$.

Step 3: The graph cannot be condensed. Take $x_i = x_1$ and $x_j = x_2$ arbitrarily. From an $(x_1$ to $x_2)$ maximum flow calculation we find the minimum cut-set to be $(X_o, \tilde{X}_o)$ where $X_o = \{x_1, x_5, x_6\}$ and $\tilde{X}_o = \{x_2, x_3, x_4\}$, the value of the cut-set being 24.

Step 4: The tree T^* and its link capacities are now shown in Fig. 11.10(a).

Step 5: $N = 2$ (i.e. T^* now has 2 vertices S_1 and S_2 as shown in Fig. 11.10(a)).

Step 2: Take $S^* = S_2$.

Step 3: Choose $x_i = x_3$ and $x_j = x_4$ arbitrarily. The condensed graph now becomes as shown in Fig. 11.10(b). From this graph a maximum flow calculation produces the minimum cut-set $(X_o, \tilde{X}_o)$ where $X_o = \{x_3\}$ and $\tilde{X}_o = \underbrace{\{x_1, x_5, x_6, x_2, x_4\}}_{S_1}$. The value of this cut-set is 23.

Step 4: The T^*-vertex $S_2 = \{x_2, x_3, x_4\}$ is now replaced by the two new T^*-vertices $\{x_3\}$ and $\{x_2, x_4\}$ with a link of value 23 between them, and since

L

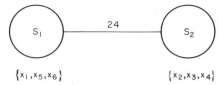

FIG. 11.10 (a). T^* after the first stage

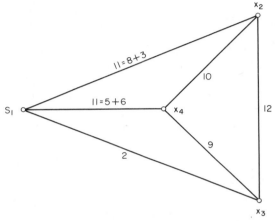

FIG. 11.10 (b and d). The condensed graph after the 2nd (b) and 3rd stages (d)

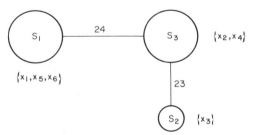

FIG. 11.10 (c). T^* after the 2nd stage

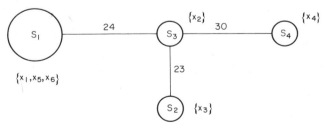

FIG. 11.10 (e). T^* after the 3rd stage

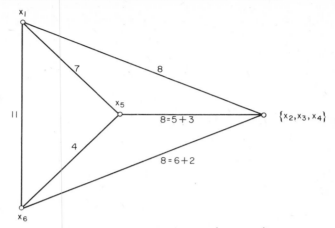

FIG. 11.10 (f and h). The condensed graph after the 4th (f) and 5th (h) stages

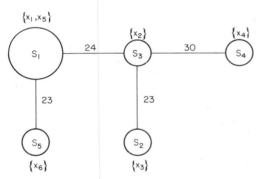

FIG. 11.10 (g). T^* after the 4th stage

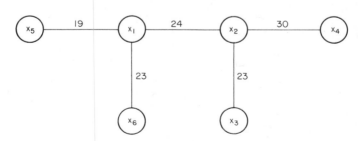

FIG. 11.10 (i). Final flow-equivalent tree T^*

$S_1 \subset \tilde{X}_0$, link (S_1, S_2) is removed and replaced by the link (S_1, S_3) where $S_3 = \{x_2, x_4\}$.

Step 5: $N = 3$ (i.e. T^* has now 3 vertices S_1, S_2 and S_3 as shown in Fig. 11.10(c).

Step 2: Take $S^* = S_3$.

Step 3: Choose $x_i = x_2$ and $x_j = x_3$. The condensed graph becomes as shown in Fig. 11.10(d). The minimum cut-set $(X_o, \tilde{X}_o)$ with value 30, is given by $X_o = \{x_4\}$ and $\tilde{X}_o = \{x_2, x_3, x_1, x_5, x_6\}$.

Step 4: The T^*-vertex $S_3 = \{x_2, x_4\}$ is now replaced by the two new T^*-vertices $\{x_2\}$ and $\{x_4\}$, and the new tree is shown in Fig. 11.10(e).

Continuing in the same way and taking $S^* = S_1$ next, the development of the flow-equivalent tree is illustrated by Figs. 11.10(f) to 11.10(i). From this last figure, the maximum flow matrix $[f_{ij}]$ of the original graph can be calculated immediately by applying eqn (11.10), and is shown below:

		x_1	x_2	x_3	x_4	x_5	x_6
	x_1	—	24	23	24	19	23
	x_2	24	—	23	30	19	23
Maximum	x_3	23	23	—	23	19	23
flow matrix:	x_4	24	30	23	—	19	23
	x_5	19	19	19	19	—	19
	x_6	23	23	23	23	19	—

The 5 (in general $n - 1$) cut-sets of G corresponding to the links of the tree T^* are shown dotted in Fig. 11.11.

It should be noted that, in general, the flow-equivalent tree T^* is not unique and that other trees exist which are also flow-equivalent to G. One such tree for the present example—and which is in fact a Hamiltonian path—is shown in Fig. 11.12.

5. Minimum Cost Flow from s to t

In Section 2 we considered the problem of maximizing the flow from s to t without reference to any costs. We will now consider the problem of finding

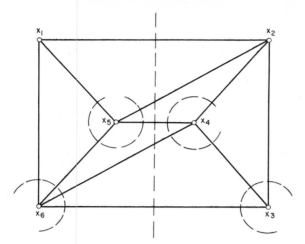

FIG. 11.11. Cut-sets forming tree T^*

FIG. 11.12. Another flow-equivalent tree

a flow of a given value v from s to t so that the cost of the flow is minimized. In this problem, each arc (x_i, x_j) has two numbers associated with it, a capacity q_{ij} and a cost per unit flow along that arc of value c_{ij}.

Obviously, if v is set to be greater than the value of the maximum flow from s to t no solution would exist. However, if v is set to be smaller or equal to this value a number of different flow patterns will, in general, be possible and what is then required is to find that flow of value v and which incurs the minimum cost. The best-known procedure for the minimum cost flow problem is the so-called "out-of-kilter" algorithm of Ford and Fukerson and the interested reader can find a description of this efficient algorithm in References [12], [4], and [7]. The algorithms we will describe here are due to Klein [20], and Busacker and Gowen [6]. These algorithms are conceptually simpler than the out-of-kilter method and use techniques which have already been introduced. Computationally the methods are comparable [2].

5.1. An algorithm based on negative circuit determination

Let us suppose that a feasible flow ξ of value v exists in the graph and that this flow pattern is known. Such a flow pattern can be obtained by applying the (s to t) maximum-flow algorithm of Section 2 and sending along the flow-augmentation chains (steps 4 to 6 of that algorithm), an amount of flow $\delta(t)$, not until the maximum is reached but until the flow f_{st} reaches the value v.

With this feasible flow, define a new incremental graph $G^\mu(\xi) = (X^\mu, A^\mu)$, in exactly the same way as explained earlier in Section 2.3, and with arc costs specified as follows:

For each arc $(x_i^\mu, x_j^\mu) \in A_1^\mu$ set the cost $c_{ij}^\mu = c_{ij}$.

For each arc $(x_j^\mu, x_i^\mu) \in A_2^\mu$ set the cost $c_{ji}^\mu = -c_{ij}$.

The new graph $G^\mu(\xi)$ now represents incremental capacities and costs (relative to the initial flow pattern ξ) of any extra flow pattern to be introduced into G. The algorithm is then based on the following theorem:

THEOREM 5. ξ *is a minimum cost flow of value* v *if and only if there is no circuit* Φ *in* $G^\mu(\xi)$ *such that the sum of the costs of the arcs in* Φ *is negative.*

Proof. Let $c[\xi]$ be the cost of flow pattern ξ in the graph G and $c[\Phi \,|\, G^\mu(\xi)]$ be the sum of the costs of the arcs in circuit Φ with respect to graph $G^\mu(\xi)$.

Necessity. Let $c[\Phi \,|\, G^\mu(\xi)] < 0$ for some circuit Φ of $G^\mu(\xi)$. The circulation of an additional unit flow around circuit Φ produces the new flow pattern $\xi + 1 \circ (\Phi)$ and leaves the value v of the flow from s to t unchanged. The cost of flow $\xi + 1 \circ (\Phi)$ is $c[\xi] + c[\Phi \,|\, G^\mu(\xi)] < c[\xi]$ which contradicts the assumption that ξ is the minimum cost flow of value v.

Sufficiency. Assume that $c[\Phi \,|\, G^\mu(\xi)] \geq 0$ for every circuit Φ in the graph $G^\mu(\xi)$ and that $\xi^* (\neq \xi)$ is the minimum cost flow of value v.

Now let us write $\xi^* - \xi$ for the flow pattern for which the flow in arc (x_i, x_j) is $\xi_{ij}^* - \xi_{ij}$.

Since ξ^* and ξ can be decomposed into the sum of flows along (s to t) paths in G, the construction of the unitary graph G^e (see Section 2.4) for the flow pattern $\xi^* - \xi$ leads to vertex degrees:

$$d_0^{G^e}(x_i) = d_t^{G^e}(x_i) \qquad \forall \, x_i \in X.$$

Thus, according to Section 2.4, G^e consists of a collection of conformal unit circuit flows around circuits $\Phi_1, \Phi_2, \ldots, \Phi_k$, (say), and we can then write:

$$\xi^* - \xi = 1 \circ (\Phi_1) + 1 \circ (\Phi_2) + \ldots + 1 \circ (\Phi_k).$$

Since the flows $1 \circ (\Phi_i)$, $i = 1, \ldots, k$ are pairwise conformal and we know that flow $\xi^* = \xi + 1 \circ (\Phi_1) + \ldots + 1 \circ (\Phi_k)$ is feasible, any sum $\xi + 1 \circ (\Phi_1) + \ldots + 1 \circ (\Phi_l)$ is feasible for any $1 \leq l \leq k$. Thus, considering the flow pattern

$\xi + 1 \circ (\Phi_1)$ we have:

$$c[\xi + 1 \circ (\Phi_1)] = c[\xi] + c[\Phi_1 | G^\mu(\xi)]$$
$$\geqslant c[\xi]$$

Consider now the incremental graph $G^\mu(\xi + 1 \circ (\Phi_1))$. The only arcs of this graph which have their costs reduced as compared with their corresponding costs in the graph $G^\mu(\xi)$ are those which are the "reverse" of arcs in Φ_1. However, since the flows $1 \circ (\Phi_1), 1 \circ (\Phi_2), \ldots$ are conformal, such arcs can never be used by any of the remaining circuits $\Phi_2, \ldots, \Phi_k$ and hence become irrelevant to the following argument.

We then have:

$$c[\Phi_l | G^\mu(\xi + 1 \circ (\Phi_1))] \geqslant c[\Phi_l | G^\mu(\xi)]$$

for any $l = 2, \ldots, k$.

The cost of the flow pattern $\xi + 1 \circ (\Phi_1) + 1 \circ (\Phi_2)$ is then:

$$
\begin{aligned}
c[\xi + 1 \circ (\Phi_1) + 1 \circ (\Phi_2)] &= c[\xi + 1 \circ (\Phi_1)] + c[\Phi_2 | G^\mu(\xi + 1 \circ (\Phi_1))] \\
&\geqslant c[\xi + 1 \circ (\Phi_1)] + c[\Phi_2 | G^\mu(\xi)] \\
&\geqslant c[\xi + 1 \circ (\Phi_1)] \\
&\geqslant c[\xi]
\end{aligned}
$$

Continuing in the way we finally get $c[\xi^*] \geqslant c[\xi]$ which contradicts the assumption that ξ^* is a minimum cost flow. Hence the theorem.

Thus, according to the theorem 5, all that is required in order to find a minimum cost flow of value v in G, is to start with a feasible flow ξ of value v, form the graph $G^\mu(\xi)$ and check it, for negative cost circuits using any of the algorithms for finding shortest paths which were given in Sections 3 and 4 of Chapter 9. If no negative cost circuit exists the flow is a minimum cost one. If a negative cost circuit Φ exists, find this circuit and send the maximum possible flow δ around the circuit. The overall flow from s to t then remains unchanged at the value v although its cost is reduced by $\delta \cdot c(\Phi)$ where $c(\Phi)$ is the cost of the negative cost circuit Φ. Obviously δ must be chosen so that the capacities of the arcs in $G^\mu(\xi)$ are not violated, i.e.

$$\delta = \min_{(x_i^\mu, x_j^\mu) \, in \, \Phi} [q_{ij}^\mu] \tag{11.11}$$

Because of the original choice of capacities for the arcs of $G^\mu(\xi)$; when this flow δ is imposed on the flow ξ already in G, the resulting new flow is still feasible. The process can then be repeated by starting with this new flow ξ

forming a new graph $G^\mu(\xi)$ relative to the new flow and checking for negative cost circuits again.

A description of the algorithm is given below.

5.1.1 DESCRIPTION OF THE ALGORITHM

Step 1. Use the (s to t) maximum flow algorithm of Section 2 to find a feasible flow ξ of value v in the graph G.

Step 2. Relative to this flow ξ form the graph $G^\mu(\xi)$ according to the rules given earlier.

Step 3. With the cost matrix of the graph $G^\mu(\xi)$ as a starting point, use a shortest path algorithm (see Sections 3.1 and 4 of Chapter 8) to find whether any negative cost circuit exists in $G^\mu(\xi)$. If such a circuit exists, identify the negative cost circuit Φ and go to step 5. If no such circuit can be found stop. The current flow pattern ξ is the minimum cost one.

Step 4. Calculate δ according to eqn (11.11).
(i) For all (x_i^μ, x_j^μ) in Φ with $c_{ij}^\mu < 0$ change the flow ξ_{ji} in the corresponding arc (x_j, x_i) of G from ξ_{ji} to $\xi_{ji} - \delta$.
and (ii) For all (x_i^μ, x_j^μ) in Φ with $c_{ij}^\mu > 0$ change the flow ξ_{ij} in the corresponding arc (x_i, x_j) of G from ξ_{ij} to $\xi_{ij} + \delta$.

Step 5. With this new flow pattern ξ return to step 2.

5.1.2 EXAMPLE Consider the graph G shown in Fig. 11.13, where the first number of the label on each arc is its capacity and the second number its cost. We want to find the minimum cost flow of value 20 from s to t.

We will use Floyd's shortest path algorithm (Section 3.1 Chapter 8) to determine negative cost circuits at step 3 of the above method.

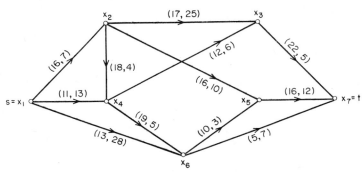

FIG. 11. 13. Graph for example 5.1.2. First label is arc capacity and second label is arc cost

Step 1. The maximum flow algorithm produces the first feasible flow pattern as shown in Fig. 11.14(a). The cost of this flow is:

$$16(7 + 25 + 5) + 4(28 + 7) = 732.$$

Step 2. Relative to this flow we form the graph $G^{\mu}(\xi)$ as shown in Fig. 11.14(b), where negative cost arcs are shown dotted.

Step 3. Starting with the cost matrix:

	x_1	x_2	x_3	x_4	x_5	x_6	x_7
x_1	0			13		28	
x_2	-7	0	25	4	10		
x_3		-25	0				5
x_4			6	0		5	
x_5					0		12
x_6	-28				3	0	7
x_7			-5			-7	0

(where blank entries are taken as infinities) and applying Floyd's algorithm, the following least-cost matrices together with their associated path matrices are obtained.

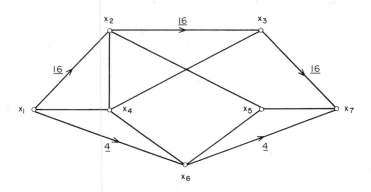

FIG. 11.14 (a). Initial flow pattern ξ

Least cost matrix

	x_1	x_2	x_3	x_4	x_5	x_6	x_7
x_1	0			13		28	
x_2	−7	0	25	4	10	21	
x_3		−25	0				5
x_4			6	0		5	
x_5					0		12
x_6	−28			−15	3	0	7
x_7			−5			−7	0

Path matrix

	x_1	x_2	x_3	x_4	x_5	x_6	x_7
x_1	1	1	1	1	1	1	1
x_2	2	2	2	2	2	1	2
x_3	3	3	3	3	3	3	3
x_4	4	4	4	4	4	4	4
x_5	5	5	5	5	5	5	5
x_6	6	6	6	1	6	6	6
x_7	7	7	7	7	7	7	7

Matrices at end of first iteration ($k = 1$).

Least cost matrix

	x_1	x_2	x_3	x_4	x_5	x_6	x_7
x_1	0			13		28	
x_2	−7	0	25	4	10	21	
x_3	−32	−25	0	−21	−15	−4	5
x_4			6	0		5	
x_5					0		12
x_6	−28			−15	3	0	7
x_7			−5			−7	0

Path matrix

	x_1	x_2	x_3	x_4	x_5	x_6	x_7
x_1	1	1	1	1	1	1	1
x_2	2	2	2	2	2	1	2
x_3	2	3	3	2	2	1	3
x_4	4	4	4	4	4	4	4
x_5	5	5	5	5	5	5	5
x_6	6	6	6	1	6	6	6
x_7	7	7	7	7	7	7	7

Matrices at end of second iteraction ($k = 2$)

Least cost matrix

	x_1	x_2	x_3	x_4	x_5	x_6	x_7
x_1	0			13		28	
x_2	−7	0	25	4	10	21	30
x_3	−32	−25	0	−21	−15	−4	5
x_4	−26	−19	6	−15	−9	2	11
x_5					0		12
x_6	−28			−15	3	0	7
x_7	−37	−30	−5	−26	−20	−9	0

Path matrix

	x_1	x_2	x_3	x_4	x_5	x_6	x_7
x_1	1	1	1	1	1	1	1
x_2	2	2	2	2	2	1	3
x_3	2	3	3	2	2	1	3
x_4	2	3	4	2	2	1	3
x_5	5	5	5	5	5	5	5
x_6	6	6	6	6	6	6	6
x_7	2	3	7	2	2	1	7

Matrices at end of third iteration ($k = 3$)

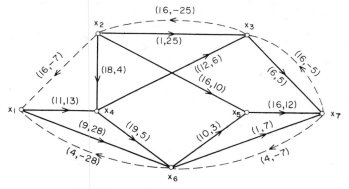

FIG. 11.14 (b). $G^\mu(\xi)$ relative to flow in fig. (a)

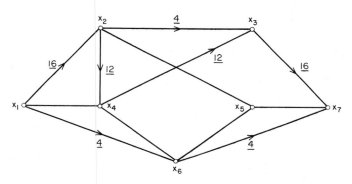

FIG. 11.14 (c). Improved flow pattern ξ

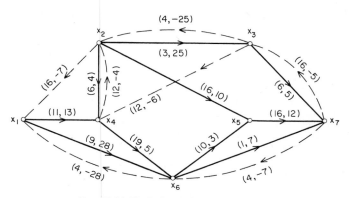

FIG. 11.14 (d). $G^\mu(\xi)$ relative to flow in fig. (c)

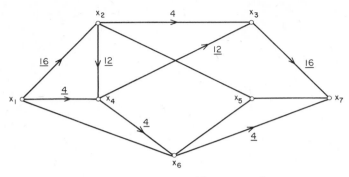

Fig. 11.14 (e). Improved flow pattern ξ

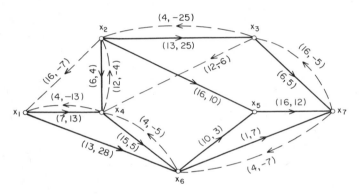

Fig. 11.14 (f). $G^{\mu}(\xi)$ relative to flow in fig. (e)

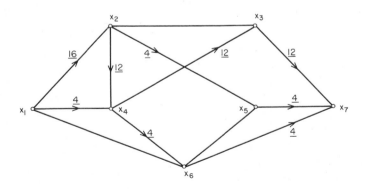

Fig. 11.14 (g). Improved flow pattern ξ

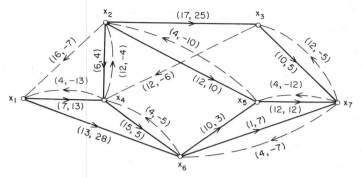

Fig. 11.14. (h). $G^\mu(\xi)$ relative to flow in fig. (g)

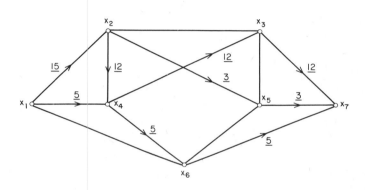

Fig. 11.14 (i). Improved flow pattern ξ

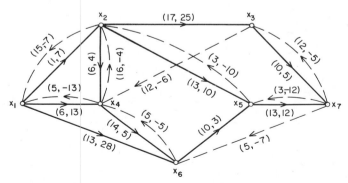

Fig. 11.14 (j). $G^\mu(\xi)$ relative to flow in fig. (i)

During this last iteration $c_{4,4}^k$ becomes negative with value -15 indicating the existence of a negative cost circuit involving vertex x_4. This circuit is found from the path matrix to be (x_4, x_3, x_2, x_4).

Step 4. Equation (11.11) gives the value of δ as:

$$\delta = \min\, [12\quad,\quad 16\quad,\quad 18] = 12$$
$$\quad\quad (x_4,\!\uparrow\!x_3)\quad (x_3,\!\uparrow\!x_2)\quad (x_2,\!\uparrow\!x_4)$$

The new flow pattern after a circulation flow of δ is introduced into the circuit is, therefore, as shown in Fig. 11.14(c). The cost of this new flow is 552.

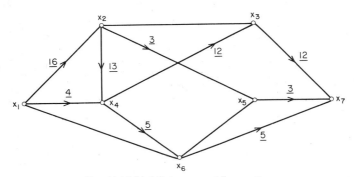

FIG. 11.14 (k). Minimum cost flow pattern

Returning to step 2, the new graph $G^\mu(\xi)$ is found and is as shown in Fig. 11.14(d).

Proceeding in the same way as previously we get:

Step 3. Negative cost circuit (x_6, x_1, x_4, x_6) of cost -10 detected.

Step 4. $\delta = \min\,[4, 11, 19] = 4$.
The new flow pattern with cost 512 is shown in Fig. 11.14(e).

Step 2. The graph $G^\mu(\xi)$ relative to the new flow is now as shown in Fig. 11.14(f).

Step 3. Negative cost circuit $(x_7, x_3, x_2, x_5, x_7)$ of cost -8 detected.

Step 4. $\delta = \min\,[16, 4, 16, 16] = 4$.
The new flow pattern with cost 480 is shown in Fig. 11.14(g).

Step 2. The graph $G^\mu(\xi)$ now becomes that of Fig. 11.14(h).

Step 3. Negative cost circuit $(x_1, x_4, x_6, x_7, x_5, x_2, x_1)$ of cost -4 detected.

Step 4. $\delta = \min\,(7, 15, 1, 4, 4, 16) = 1$.
The new flow pattern with cost 476 is shown in Fig. 11.14(i).

Step 2. The graph $G^\mu(\xi)$ is now that shown in Fig. 11.14(j).

Step 3. Negative cost circuit (x_1, x_2, x_4, x_1) of cost -2 detected.

Step 4. $\delta = (1, 6, 5) = 1$.
The new flow pattern with cost 474 is shown in Fig. 11.14(k).

No negative cost circuits can now be detected in the incremental graph so that the flow pattern shown in Fig. 11.14(k) is the least cost flow pattern of value 20 and its cost is 474.

5.1.3 STEP 3 OF THE ALGORITHM OF 5.1.1 In the above example Floyd's method was used to find negative cost circuits in the incremental cost graph $G^\mu(\xi)$ at step 3 of the minimum cost flow algorithm given in 5.1.1. However, as mentioned in Section 4 of Chapter 8, this is not the best method of determining negative cost circuits in a graph G in which a vertex x_0 reaching all other vertices of G exists. A better way is to use an algorithm which finds shortest paths from x_0 to all other vertices, rather than use Floyd's algorithm which finds shortest paths between every pair of vertices [2]. In this respect the shortest path algorithm of Section 2.2.1 of Chapter 8 could be used to advantage as explained in Section 4 of that Chapter.

For the incremental graph $G^\mu(\xi)$ it is quite easy to show that at least one vertex always exists which can reach every other vertex of $G^\mu(\xi)$, and that in fact the terminal vertex t has this property. This can be shown as follows.

Since in the incremental graph $G^\mu(\xi)$ there is an arc (x_j^μ, x_i^μ) for every arc (x_i, x_j) of G along which there is some non-zero flow, all vertices x_i^μ of $G^\mu(\xi)$ corresponding to vertices x_i of G which lie on the flow paths from s to t, can be reached from t using the "reverse" paths formed by the arcs (x_j^μ, x_i^μ). In this way the source vertex s will also be reached. Now every vertex x_i of G must be reachable from s; (otherwise x_i could be removed from G without any effect on the flow). Therefore, since those arcs (x_i, x_j) not carrying any flow in G also appear as arcs (x_i^μ, x_j^μ) in $G_\mu(\xi)$ it is apparent that those vertices x_i^μ of $G^\mu(\xi)$ corresponding to vertices x_i of G not lying on the flow paths, could then be reached from t via s; i.e. by simply reaching s first as mentioned earlier and then following a path from s to x_i^μ.

In the incremental graph $G^\mu(\xi)$, therefore, all vertices can be reached from t and if this vertex is used as the source vertex in a shortest path calculation from t to all other vertices of $G^\mu(\xi)$, (for example using the algorithm of Section 2.2.1, Chapter 8), the negative cost circuits of $G^\mu(\xi)$ could be detected.

Using such an approach for performing step 3 of the main minimum cost flow algorithm of Section 5.1, Bennington [2] published computational experience indicating that this algorithm is at least as good as the "out-of-kilter" algorithm [12, 4, 7]. The best least squares fit for the computing time t

in tests performed on randomly generated graphs with n vertices and m arcs was found to be:

$$t = -2{\cdot}3 + 0{\cdot}0113n + 0{\cdot}00166m \quad (\text{sec IBM 360/75}).$$

A graph with 200 vertices and 5000 arcs requiring approximately 8 seconds.

5.1.4 MINIMUM COST FLOWS IN GRAPHS WITH CONVEX ARC COSTS. We have, up to now, assumed that the cost of a unit of flow ξ_{ij} in an arc (x_i, x_j) is c_{ij}, a constant number. However, this is not a necessary assumption and as shown by Hu [17] and Klein [20], the algorithm given in 5.1.1 can easily be adapted to find minimum cost flows in graphs whose arc costs are convex piecewise linear functions of flow.

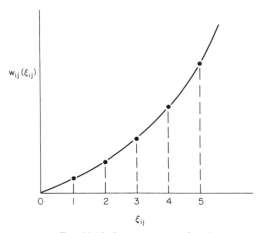

FIG. 11.15. Convex arc cost function

If we assume the cost of a flow ξ_{ij} in arc (x_i, x_j) to be given by the convex function $w_{ij}(\xi_{ij})$ as shown in Fig. 11.15, then for any flow pattern ξ we can define the incremental cost graph $G^\mu(\xi) = (X^\mu, A_1^\mu \cup A_2^\mu)$ in exactly the same way as explained in Section 2.3, except that now the arc capacities q_{ij}^μ and q_{ji}^μ need never be calculated, and where costs are allocated to the arcs as follows:

For an arc $(x_i^\mu, x_j^\mu) \in A_1^\mu$ set:

$$c_{ij}^\mu = w_{ij}(\xi_{ij} + 1) - w_{ij}(\xi_{ij})$$

For an arc $(x_j^\mu, x_i^\mu) \in A_2^\mu$ set:

$$c_{ji}^\mu = - [w_{ij}(\xi_{ij}) - w_{ij}(\xi_{ij} - 1)],$$

assuming $w_{ij}(0) = 0$.

We could now use the algorithm of Section 5.1.1 directly except that δ—the flow to be sent round any negative cost circuit Φ that may be found in $G^\mu(\xi)$—is always taken as 1 unit. Notice that all incremental costs have been defined so that they apply to a single unit increase or decrease in the value of the flow. Also note that with only unit increments in flows, the incremental capacities of the arcs in $G^\mu(\xi)$ need not be calculated since these are only involved in the computation of δ.

If a lower accuracy in the representation of the cost function is acceptable, then $w_{ij}(\xi_{ij})$ can be approximated by peicewise linear sections. The restriction of choosing δ as a unit of flow can then be lifted and faster convergence to the minimum cost flow pattern can be achieved. The incremental cost of each arc can now be defined as the slope of the linear section of the cost function on which the current flow through this arc lies. Capacities can then be calculated to ensure that any increase or decrease of flow δ in this arc will not produce a net flow beyond the limits of applicability of the current linear section approximating the cost function.

5.2 Least cost path flow algorithm

The algorithm of Section 5.1 started with an arbitrary flow pattern ξ of value v and continuously improved the flow pattern until the minimum cost flow ξ^* was obtained. In mathematical programming terminology, such an algorithm would be called *primal*, since its first aim is to obtain a feasible flow ξ and then improve the flow whilst maintaining feasibility. An alternative approach is the *dual* method which for this problem would involve the building up of the minimum cost flow pattern ξ^* with value v by starting with a minimum cost flow pattern having some value $v_0 < v$ and adding extra flow in the network to obtain a new minimum cost flow having value $v_1 > v_0$ etc. until the minimum cost flow pattern ξ^* with value v is reached. In this respect one should note that a starting minimum cost flow pattern is always available since the zero flow is the minimum cost flow of value 0.

The algorithm we are about to describe is based on the calculation of shortest paths and on the following theorem.

THEOREM 6. *Given that ξ is a minimum cost flow pattern in a graph G having value v and P^* is the shortest (minimum cost) path from s to t in the incremental graph $G^\mu(\xi)$, then $\xi + 1 \circ (P^*)$ is the minimum cost flow of value $v + 1$.*

The proof of the theorem is very simple and involves the demonstration that if $G^\mu(\xi)$ contains no negative cost circuit, then $G^\mu(\xi + 1 \circ (P^*))$ also does not contain such circuits. The result then follows directly from theorem 5.

5.2.1 DESCRIPTION OF THE ALGORITHM

Step 1. Start with all arc flows equal to zero and the flow value equals zero.

Step 2. Construct the incremental graph $G^\mu(\xi) = (X^\mu, A_1^\mu \cup A_2^\mu)$ as explained in Section 2.3, and allocate arc costs as follows:

For any arc $(x_i^\mu, x_j^\mu) \in A_1^\mu$, set the cost $c_{ij}^\mu = c_{ij}$.

For any arc $(x_j^\mu, x_i^\mu) \in A_2^\mu$, set the cost $c_{ji}^\mu = -c_{ij}$.

Step 3. Find the shortest path P^* in $G^\mu(\xi)$ from vertex s to vertex t.

Step 4. Calculate the largest amount δ of flow that can be sent along P^* before the structure of $G^\mu(\xi)$ changes. δ is given by:

$$\delta = \min_{\substack{(x_i^\mu, x_j^\mu) \\ \text{in } P^*}} [q_{ij}^\mu]$$

Step 5. (i) If the value of flow $\xi + \delta \circ (P^*)$ is less than the required value v, update $\xi \leftarrow \xi + \delta \circ (P^*)$ and return to step 2.

(ii) If the value of flow $\xi + \delta \circ (P^*)$ is equal to v, stop. $\xi + \delta \circ (P^*)$ is the required minimum cost flow.

(iii) If the value of $\xi + \delta \circ (P^*) = h > v$, stop. $\xi + (\delta - h + v) \circ (P^*)$ is the required minimum cost flow.

It is important to note here that the graph $G^\mu(\xi)$ at step 2 contains both positive and negative cost arcs, and hence the shortest path algorithm used at step 3 cannot be Dijkstra's method but could be the (less efficient) algorithm described in Section 2.2.1 of Chapter 8. However, Edmonds and Karp [9] have recently described a technique which circumvents the above difficulty and allows the more efficient shortest path algorithms to be used.

5.2.2 EXAMPLE

Let us recalculate the minimum cost flow of value 20 for the graph shown in Fig. 11.13.

Initially take $\xi = 0$. The incremental graph $G^\mu(\xi)$ is then as shown in Fig. 11.13. The shortest path in this graph is $P_1^* = (x_1, x_2, x_4, x_3, x_7)$ with cost 22, and δ is calculated to be 12.

The flow ξ now becomes $12 \circ (P_1^*)$. The shortest path in the corresponding incremental graph is $P_2^* = (x_1, x_2, x_4, x_6, x_7)$ with cost 23 and δ is calculated to be 4.

The flow ξ now becomes $12 \circ (P_1^*) + 4 \circ (P_2^*)$. The shortest path in $G^\mu(\xi)$ is $P_3^* = (x_1, x_4, x_6, x_7)$ with cost 25 and δ is calculated to be 1.

The flow becomes $12 \circ (P_1^*) + 4 \circ (P_2^*) + 1 \circ (P_3^*)$. The shortest path in $G^\mu(\xi)$ is $P_4^* = (x_1, x_4, x_2, x_5, x_7)$ with cost 31 and along which the remaining 3 units of flow can be sent to produce the minimum cost flow pattern as shown in Fig. 11.14(k).

6. Flows in Graphs with Gains

Up to now in this Chapter the assumption was made that the flow that leaves an arc is the same as the flow that has entered it. In a number of practical situations, however, this is not a valid assumption. For example in pipeline networks carrying liquids or wire networks carrying electricity, leakage in the system implies loss of flow along an arc. In transportation systems items may be damaged en route which also implies loss of flow along the arc representing that route. In a manufacturing process that can be represented by a graph with arcs corresponding to operations, the value of materials leaving an arc is greater than the value of materials entering the arc i.e. there is a gain associated with an arc representing the "valued added" by the operation. In the problem of exchanging commodities (e.g. the buying and selling of currencies), the arc gains represent exchange rates which convert the input flow (measured in one currency) to the output flow (measured in a different currency).

In this section we will study the problem of finding the maximum flow from (s to t) in a graph with arbitrary nonnegative gains g_{ij} and capacities q_{ij} associated with the arcs (x_i, x_j) of a graph G. The problem is analogous to the (s to t) maximum flow problem discussed in Section 2 although now the input and output flows are unrelated except through the graph G which is capable of both "generating" and "absorbing" flow.

If we denote ξ_{ij}^e as the entering (input) flow into an arc (x_i, x_j) and ξ_{ij}^o as the leaving (output) flow from that arc we have

$$\xi_{ij}^o = g_{ij}\xi_{ij}^e \tag{11.12}$$

We will further assume that the arc capacities refer to the input flows so that for any arc we have

$$\xi_{ij}^e \leqslant q_{ij} \tag{11.13}$$

regardless of the value of ξ_{ij}^o. If we denote the net input flow into s as v_s and the net flow out of t as v_t we can then say that a flow pattern $\Xi = (\xi^e, \xi^o)$ is feasible if it satisfies the flow continuity conditions at the vertices, i.e.

$$\sum_{x_j \in \Gamma(x_i)} \xi_{ij}^e - \sum_{x_j \in \Gamma^{-1}(x_i)} \xi_{ji}^o = \begin{cases} v_s \text{ for } x_i = s \\ -v_t \text{ for } x_i = t \\ 0 \text{ for } x_i \neq s \text{ or } t \end{cases} \tag{11.14}$$

and also conditions (11.12) and (11.13) for every arc (x_i, x_j) of G.

We will now introduce two definitions.

Definition. A feasible flow pattern $\hat{\Xi} = (\hat{\xi}^e, \hat{\xi}^o)$ is referred to as *maximum* if it produces the largest value of v_t (say $\hat{v}_t$) over all feasible flow patterns.

Definition. A feasible pattern $\tilde{\Xi} = (\tilde{\xi}^e, \tilde{\xi}^o)$ is referred to as *optimum* if for any other feasible flow pattern Ξ.

$$\text{either: } v_t \leqslant \tilde{v}_t \quad \text{when} \quad v_s = \tilde{v}_s$$

$$\text{or: } v_s \geqslant \tilde{v}_s \quad \text{when} \quad v_t = \tilde{v}_t$$

where $\tilde{v}_s$ and $\tilde{v}_t$ are respectively the net source and sink flows corresponding to $\tilde{\Xi}$.

These last conditions correspond to the intuitive concept of optimality that: given $\tilde{v}_s$, $\tilde{v}_t$ is the largest possible; or given $\tilde{v}_t$, $\tilde{v}_s$ is the smallest possible.

Definition. A feasible flow pattern $\hat{\tilde{\Xi}}$ is referred to as *optimum–maximum* if it is an optimum and also a maximum flow pattern.

We will illustrate the above definitions by an example. Consider the graph of Fig. 11.16(a) where the first label on an arc refers to its capacity and the second to its gain. Then:

(i) The flow shown in Fig. 11.16(b), is a maximum flow pattern of value $\hat{v}_t = 18$ and $\hat{v}_s = 6$.

(ii) The flow shown in Fig. 11.16(c) is an optimum flow pattern of value $\tilde{v}_t = 3$ and $\tilde{v}_s = 0$. (Note that flow is generated by circulation round a circuit whose total gain is greater than 1. Also note that $\tilde{v}_t < \hat{v}_t$ and hence this is not a maximum flow, however, $\tilde{v}_t$ is the maximum that can be obtained with $v_s = 0$.)

(iii) The flow shown in Fig. 11.16(d) is optimum maximum having $\hat{\tilde{v}}_t = \hat{v}_t$, but $\hat{\tilde{v}}_s = 5 < \hat{v}_s = 6$, and in fact, as we shall show later, 5 is the smallest possible value of v_s to produce an output flow of 18.

6.1 Augmenting chains

Consider a graph G with gains and with an existing feasible flow pattern $\Xi = (\xi^e, \xi^o)$, flowing in it. We will denote this by $G(\Xi)$. A chain (non necessarily simple), of arcs from vertex x_{i_1} to vertex x_{i_t} is called an *augmenting* chain from x_{i_1} to x_{i_t} if flow can be sent in $G(\Xi)$ along the chain from x_{i_1} to x_{i_t}. (This is analogous to the definition of an (s to t) flow augmenting chain defined for the (s to t) maximum flow problem.) If the chain is given by the vertex sequence $x_{i_1}, x_{i_2}, \ldots, x_{i_t}$, then let F be the set of all "forward" arcs in this chain—i.e. arcs $(x_{i_p}, x_{i_{p+1}})$ with $x_{i_{p+1}} \in \Gamma(x_{i_p})$—and let B be the set of all "backward" arcs— i.e. those with $x_{i_p} \in \Gamma(x_{i_{p+1}})$. A chain is then an augmenting chain if for every forward arc $(x_i, x_j) \in F$ we have $\xi_{ij}^e < q_{ij}$ and for every backward arc $(x_j, x_i) \in B$ we have $\xi_{ji}^e > 0$.

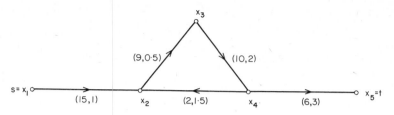

(a) Graph G. First label is arc capacity, second label is arc gain

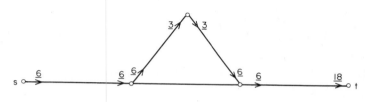

(b) Maximum flow pattern

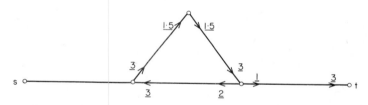

(c) Optimum flow pattern

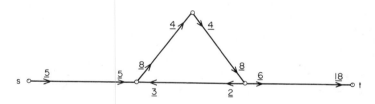

(d) Optimum-maximum flow pattern

FIG. 11.16

A. *Chain gain*

The *gain* of a chain $S = x_{i_1}, x_{i_2}, \ldots, x_{il}$ is given by:

$$g(S) = \prod_{(x_i,\, x_j) \in F} g_{ij} \cdot \prod_{(x_j,\, x_i) \in B} 1/g_{ji} \qquad (11.15)$$

B. *Chain capacity*

The *incremental capacity* of an augmenting chain S is the maximum input flow into x_{i_1} which can be sent along the chain to x_{i_t}, up to the level when either the flow in some forward arc of S is saturated, or the flow in some backward arc becomes zero.

If S is a simple chain, and a flow of δ is entered into x_{i_1}, then the flow entering x_{i_p} in arc $(x_{i_p}, x_{i_{p+1}})$ of S is

$$\delta_{i_p i_{p+1}} = \delta \cdot \prod_{(x_i, x_j) \in F_p} g_{ij} \cdot \prod_{(x_j, x_i) \in B_p} 1/g_{ji} \qquad (11.16)$$

$$\equiv \delta \cdot g(S_p)$$

where F_p and B_p are respectively the forward and backward arcs of the sub-chain $S_p = x_{i_1}, x_{i_2}, \ldots, x_{i_p}$ and $g(S_p)$ is the gain of S_p. If $(x_{i_p}, x_{i_{p+1}})$ is a forward arc of S carrying flow $\xi^e_{i_p i_{p+1}}$, the arc will become saturated when $\xi^e_{i_p i_{p+1}} + \delta_{i_p i_{p+1}} = q_{i_p i_{p+1}}$. If it is a backward arc carrying $\xi^e_{i_{p+1} i_p}$, it will have its flow reduced to 0 when $\delta_{i_p i_{p+1}} = \xi^o_{i_{p+1} i_p}$, i.e. when $\delta_{i_p i_{p+1}} = g_{i_{p+1} i_p} \xi^e_{i_{p+1} i_p}$. The incremental capacity of chain S is then given by:

$$q(S) = \min \left[\min_{(x_{i_p}, x_{i_{p+1}}) \in F} \left\{ \frac{q_{i_p i_{p+1}} - \xi^e_{i_p i_{p+1}}}{g(S_p)} \right\}, \right.$$

$$\left. \min_{(x_{i_{p+1}}, x_{i_p}) \in B} \left\{ \frac{g_{i_{p+1} i_p} \xi^e_{i_{p+1} i_p}}{g(S_p)} \right\} \right] \qquad (11.17)$$

If S is not a simple chain and some arc appears more than once, an expression similar to eqn (11.17) above can be easily derived.

For the chain S of Fig. 11.17(a) and the initial flow values given, the addition of a flow δ into x_1 results in the flow pattern shown in Fig. 11.17(b). The maximum value of δ which leaves the flow pattern feasible is then $q(S) = 0.6$. For this value the flow in arc (x_4, x_3) is reduced to zero, i.e. S ceases to be an augmenting chain in the new $G(\Xi)$. Had the initial flow entering arc (x_4, x_3) been 8 (say) instead of 2, the maximum feasible value of δ would then have been 2.1, the point at which arc (x_4, x_5) of S becomes saturated.

For the non-simple chain S with the initial flow pattern shown in Fig. 11.17(c), the addition of a flow δ into x_1 results in the flow pattern shown in Fig. 11.17(d). The incremental capacity of this chain is then $\frac{1}{2}$, the point at which arc (x_3, x_4) becomes saturated.

6.2 Active cycles

Since a cycle can be considered as a chain with identical initial and final vertices, the gain of a cycle (and its capacity), may be defined in the same way as that of a chain. A cycle Φ is said to be *active* in $G(\Xi)$ with respect to the

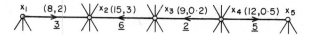

(a) Chain S with initial input flows shown underlined. First label in parenthesis is arc capacity and second is arc gain

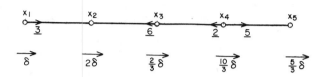

(b) Chain S with additional flow introduced

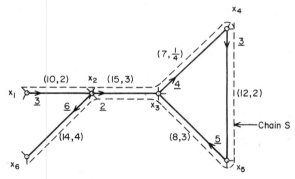

(c) Non-simple chain S with initial input flows underlined. First label in parenthesis is arc capacity and second is arc gain.

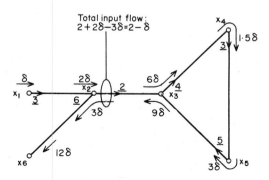

(d) Chain S with additional flow introduced

Fig. 11.17

vertex t if:
  (i)   Its gain is greater than unity,
  (ii) Its incremental capacity is non-zero,
and (iii) There is some vertex x_i on Φ so that an augmenting chain from x_i to t
        exists.

An active cycle with respect to the initial vertex s can be similarly defined.

From the above definition it should be clear that by circulating flow round an active cycle, extra flow can be created—by property (i)—and this extra flow can be transmitted to t— by property (iii). In the example of Fig. 11.16(c) given earlier, flow was created by circulation around active cycle (x_4, x_2, x_3, x_4).

An active cycle with respect to t is said to be *de-activated* if additional flow is imposed in G so that, either its incremental capacity is reduced to zero, or the capacities of all augmenting chains from any vertex on the cycle to t are reduced to zero (i.e. they are no longer augmenting in the new $G(\Xi)$).

In References [13] and [22] it is shown that:

THEOREM 7. *A flow pattern $\tilde{\Xi}$ is optimum if and only if $G(\tilde{\Xi})$ does not contain any active cycles with respect to s or t.*

THEOREM 8. *If $\tilde{\Xi}$ is an optimum flow pattern, then if flow is increased along the $(s$ to $t)$ augmenting chain of highest gain the resulting flow pattern is also optimum.*

THEOREM 9. *A flow pattern $\hat{\tilde{\Xi}}$ is optimum maximum if some cut-set $(X_o \to X - X_o)$, separating s from t is saturated and there is no active cycle with respect to t and with all its vertices in $X - X_o$.*

6.3 Optimum flow algorithm for graphs with gains

From the above three theorems the optimality of the algorithm described below follows immediately.

Step 1. Start with any feasible flow Ξ in G. (Zero flow in all arcs is acceptable.)

Step 2. Find an active cycle Φ in $G(\Xi)$, with respect to t.

Step 3. Let x_i be the vertex of Φ so that the augmenting chain is from x_i to t. (Note that x_i may be t itself.) Starting from x_i circulate flow δ round the active cycle and then transmit the excess flow at x_i along the augmenting chain to t. Choose δ so that (together with the existing flow Ξ, either the incremental capacity of Φ or that of the augmenting chain is reduced to zero.

Step 4. Update Ξ and return to step 2 until all active cycles are de-activated and no more active cycles are found, in which case go to step 5. (At this stage Ξ is the optimum flow $\tilde{\Xi}$ with $v_s = 0$.)

Step 5. Find the augmenting chain from s to t with the largest gain. Send flow along this chain until its incremental capacity is reduced to zero.

Step 6. Update Ξ and repeat step 5 until either the net output flow v_t is the required value of the optimum flow, or until no further augmenting chain can be found in which case the flow is the optimum maximum flow $\hat{\Xi}$.

The steps in the above algorithm are very simple and the method is efficient. Steps 2 and 5 can, once more, be executed by means of a shortest path algorithm. Thus, corresponding to the graph $G(\Xi)$ let us define the incremental graph $G^\mu(\Xi) = (X^\mu, A^\mu)$ with vertices $X^\mu = X$ and arcs A^μ so that:

$$(x_i^\mu, x_j^\mu) \in A^\mu \qquad \text{if} \qquad 0 \leqslant \xi_{ij}^e < q_{ij};$$

with incremental "cost"

$$c_{ij}^\mu = -\log(g_{ij})$$

and capacity

$$q_{ij}^\mu = q_{ij} - \xi_{ij}^e$$

and

$$(x_j^\mu, x_i^\mu) \in A^\mu \qquad \text{if} \qquad 0 < \xi_{ij}^e \leqslant q_{ij};$$

with incremental "cost"

$$c_{ji}^\mu = -\log(1/g_{ij})$$

and capacity

$$q_{ji}^\mu = \xi_{ij}^e \cdot g_{ij}.$$

A cycle Φ in $G(\Xi)$ with gain greater than unity then corresponds to a circuit of $G^\mu(\Xi)$ with negative "cost". This can be seen by taking logarithms of both sides of eqn (11.15) corresponding to a cycle Φ. Thus,

$$\log[g(\Phi)] = \sum_{(x_i, x_j) \in F} \log(g_{ij}) + \sum_{(x_j, x_i) \in B} \log(1/g_{ji})$$

$$= -\left[\sum_F c_{ij}^\mu + \sum_B c_{ij}^\mu\right]$$

where F and B are once more the set of forward and backward arcs of cycle Φ.

If $g(\Phi) > 1$ then $\log[g(\Phi)] > 0$ which implies that the cost of cycle Φ in $G^\mu(\Xi)$, i.e. $\sum_F c_{ij}^\mu + \sum_B c_{ij}^\mu$ must then be negative. Conversely, if $g(\Phi) \leqslant 1$ the cost of Φ in $G(\Xi)$ is then non-negative.

Thus, active cycles in $G(\Xi)$—at step 2 of the above algorithm—can be computed by attempting to find all shortest (least "cost") paths from the vertices of $G''(\varXi)$ to the terminal vertex t using the method of Section 2.2.1, Chapter 8 (as explained in Section 4 of that Chapter), and testing for negative cost circuits. However, if the flow δ circulated round the active cycle Φ at step 3 and then transmitted along an augmenting chain to t, does not reduce the capacity of Φ itself to zero but instead reduces to zero the capacity of the augmenting chain, then at the next execution of step 2 it is only necessary to find another augmenting chain leading from a vertex of Φ to t. Such a chain would render Φ active again and one can continue to step 3, etc., up to the time when Φ is de-activated and some different active cycle then has to be calculated at step 2.

Step 5 of the above algorithm can also be executed by means of the method of Section 2.2.1 of Chapter 8 since it is simply a shortest path calculation from s to t in the graph $G''(\varXi)$. This graph is free of negative cost circuits once the previous four steps of the algorithm have been executed.

6.4 Example

We wish to find the optimum maximum flow from vertex x_1 to vertex x_8 in the graph of Fig. 11.18 where the first label attached to an arc is its capacity and the second is its gain.

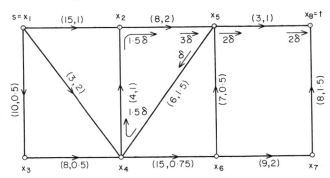

FIG. 11.18. Graph for example 6.4. First label is arc capacity, second is arc gain

Starting with $\varXi = (0, 0)$, the first active cycle found is $\Phi = (x_5, x_4, x_2, x_5)$ with the augmenting chain (x_5, x_8). Circulating a flow δ (where δ is the input flow of arc (x_5, x_4)) around this circuit as shown in Fig. 11.18, results in the maximum feasible value of δ of $1\frac{1}{2}$ units and for this value the resulting flow pattern is shown (with underlined numbers), in Fig. 11.19(a)—where arc (x_5, x_8) is saturated. The cycle Φ itself is not reduced to zero capacity, so at the next iteration when the shortest path algorithm (proceeding from all other vertices of $G''(\varXi)$ to vertex t), labels vertex x_4 of Φ; Φ becomes active

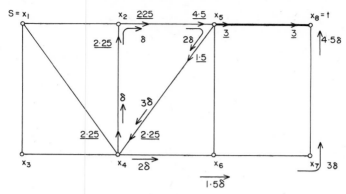

FIG. 11.19 (a). Flow pattern after first stage
————— Saturated arc

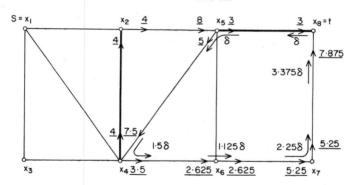

FIG. 11.19 (b). Flow pattern after 2nd stage
————— Saturated arcs

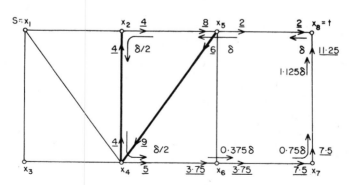

FIG. 11.19 (c). Flow pattern after 3rd stage
————— Saturated arcs

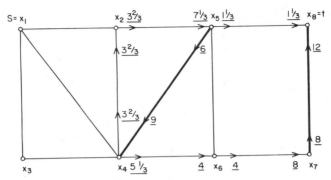

FIG. 11.19 (d). Optimum flow pattern
———— Saturated arcs

again with the new flow-augmenting path (x_4, x_6, x_7, x_8). Sending flow δ round this cycle and transmitting it along the augmenting path to t as shown in Fig. 11.19(a) produces the flow pattern shown underlined in Fig. 11.19(b).

A new active cycle $(x_8, x_5, x_4, x_6, x_7, x_8)$ is now determined at step 2 and the circulation of flow δ around this cycle as shown in Fig. 11.19(b) produces the flow pattern shown underlined in Fig. 11.19(c). Finally the active cycle $(x_8, x_5, x_2, x_4, x_6, x_7, x_8)$ is determined and its deactivation by flow circulation leads to the flow pattern of Fig. 11.19(d) which is the optimal flow pattern since no more active cycles can be found.

The (s to t) augmenting chain of largest gain is now $(x_1, x_4, x_2, x_5, x_8)$ of capacity 1/6. Once this chain is saturated (arc (x_4, x_2) is saturated), the augmenting chain with the next largest gain is (x_1, x_2, x_5, x_8) which when saturated produces the flow pattern of Fig. 11.19(e) which is the required optimum maximum flow pattern since no more augmenting (s to t) paths can be found.

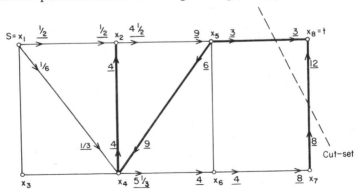

FIG. 11.19 (e). Optimum-maximum flow pattern
———— Saturated arcs

6.5 Graphs with gains at vertices

In Section 3.2 the maximum (s to t) flow problem with capacitated vertices was solved by simply replacing each vertex x_j by a pair of vertices x_j^+ and x_j^-. Similarly, if the vertices of a graph G have associated gains so that the flow continuity conditions (11.14) are not satisfied, we can form another graph G_o by replacing each vertex x_j of G by a vertex pair x_j^+ and x_j^- with an arc (x_j^+, x_j^-) between them; the gain of this arc being equal to the gain of vertex x_j. For an arc (x_i, x_j) of G we would then have an arc (x_i^-, x_j^+) in G_o and for an arc (x_j, x_k) of G we would have an arc (x_j^-, x_k^+) in G_o. The problem now becomes one of finding optimum flows in a graph G_o with arc gains only.

7. Problems P11

1. Find the maximum flow ξ^* from x_1 to x_7 in the graph G shown in Fig. 11.20, where arc capacities are shown next to be the arcs.

2. Draw the incremental graph $G^\mu(\xi)$ for the example of problem 1.

3. Decompose the flow pattern found in problem 1 into elementary (s to t) paths and circuit flows.

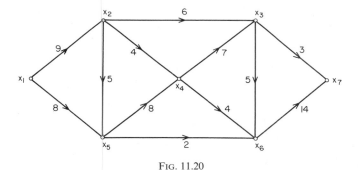

FIG. 11.20

4. In the graph shown in Fig. 11.20 assume that lower limits r_{ij} are placed on the flows in arcs (x_2, x_3), (x_5, x_4) and (x_2, x_5) of value $r_{2,3} = 4$, $r_{5,4} = 7$ and $r_{2,5} = 3$ respectively. Find the maximum flow from x_1 to x_7 if a feasible flow exists or indicate that no such flow is possible.

5. Prove Theorem 3 of Section 3.3.

6. Find the maximum flow between all pairs of vertices of the graph G shown in Fig. 11.21 and draw a flow-equivalent tree. Identify the cut-sets of G corresponding to this flow-equivalent tree, and find two other flow-equivalent trees to the graph.

7. Find the minimum cost maximum flow from x_1 to x_8 in the graph of Fig. 11.22 where the first number next to an arc is its capacity and the second its cost.

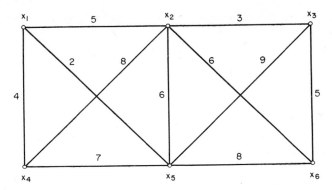

Fig. 11.21

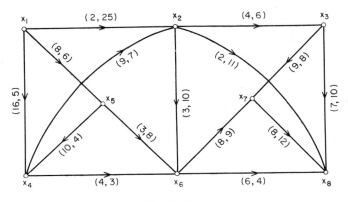

Fig. 11.22

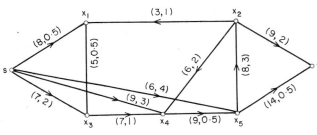

Fig. 11.23

8. What simplifications could be introduced into the minimum cost flow algorithms in the case where the required flow value is the maximum possible?

9. In the (s to t) maximum flow algorithm of Section 2.1, two of the many possible ways for choosing flow-augmenting chains are:

    (i) Choose the flow-augmenting chain of lowest cardinality,

and   (ii) Choose that chain which allows an augmenting flow of maximum value δ to be sent from s to t.

Which one of these two choices would one expect to be computationally superior and how does this choice depend on the magnitudes of the arc capacities? (See Ref. [9]).

10. Prove theorem 6 of Section 5.2.

11. Prove theorem 7 of Section 6.2.

12. Find the optimum flow, the maximum flow and the optimum maximum flow in the graph shown in Fig. 11.23, where the first number next to an arc is its capacity and the second is its gain.

8. References

1. Beale, E. M. L. (1959). An algorithm for solving the transportation problem when the shipping cost over each route is convex, *Nav. Res. Log. Quart.*, **6**, p. 43.
2. Bennington, G. E. (1972). An efficient minimal cost flow algorithm, Report No. 75, Department of Industrial Engineering, North Carolina State University at Raleigh.
3. Boldyreff, A. W. (1955). Determination of the maximal steady flow of traffic through a railroad network, *Ops. Res.*, **3**, p. 443.
4. Bray, J. A. and Witzgall, C. (1968). Algorithm 336 NETFLOW, *Comm. of ACM*, **11**, p. 631.
5. Busacker, R. G. and Saaty, T. L. (1965). "Finite graphs and networks", McGraw-Hill, New York.
6. Busacker, R. G. and Gowen, P. J. (1961). A procedure for determining a family of minimal-cost network flow patterns, Operations Research Office, Technical paper 15.
7. Clasen, R. J. (1967). RS-OKF3, SHARE IBM Users group SDA 3536.
8. Dantzig, G. B. and Fulkerson, D. R. (1956). On the max-flow min-cut theorem of networks, *In*: "Linear inequalities and related systems", *Ann. Math. Studies*, **38**, p. 215.
9. Edmonds, J. and Karp, R. M. (1972). Theoretical improvements in algorithmic efficiency for network flow problems, *Jl. of ACM*, **19**, p. 248.
10. Elias, P., Feinstein, A. and Shannon, C. E. (1956). A note on the maximum flow through a network, *IRE Trans.*, **IT-2**, p. 117.
11. Ford, L. R. and Fulkerson, D. R. (1956). Maximal flow through a network, *Can. Jl. of Math.*, **18**, 399.
12. Ford, L. R. and Fulkerson, D. R. (1962). "Flows in networks", Princeton University Press, Princeton.
13. Frank, H. and Frisch, I. T. (1971). "Communication, transmission and transportation networks", Addison Wesley, Reading, Massachusetts.
14. Gomory, R. E. and Hu, T. C. (1964). Synthesis of a communication network, *Jl. of SIAM (Appl. Math.)*, **12**, p. 348.

M

15. Grigoriadis, M. D. and White, W. W. (1972). A partitioning algorithm for the multicommodity network flow problem, *Math. Prog.*, **3**, p. 157.
16. Hu, T. C. (1969). "Integer programming and network flows", Addison Wesley, Reading, Massachusetts.
17. Hu, T. C. (1966). Minimum cost flow in convex cost networks, *Nav. Res. Log. Quart*, **13**, p. 1.
18. Jewell, W. S. (1968). Multicommodity network solutions, *In*: "Theorie de graphes", Dunod, Paris, p. 183.
19. Johnson, E. (1966). Networks and basic solutions, *Ops. Res.*, **14**, p. 619.
20. Klein, M. (1967). A primal method for minimal cost flows with applications to the assignment and transportation problems, *Man. Sci.*, **14**, p. 205.
21. Maurras, J. F. (1972). Optimization of the flow through networks with gains, *Math. Prog.*, **3**, p. 135.
22. Onaga, K. (1967). Optimum flows in general communication networks, *Jl. of the Franklin Institute*, **283**, p. 308.
23. Potts, R. B. and Oliver, R. M. (1972). "Flows in transportation networks", Academic Press, New York.
24. Sakarovitch, M. (1966). The multicommodity maximum flow problem, Report ORC-66-25, Operations Research Centre, Univ. of California, Berkeley.
25. Siagal, R. (1967). Multicommodity flows in directed networks, Report ORC-67-38, Operations Research Centre, Univ. of California, Berkeley.
26. Tomlin, J. A. (1966). Minimum cost multicommodity network flows, *Ops. Res.*, **14**, p. 45.
27. Wollmer, R. D. (1970). Multicommodity supply and transportation networks with resource constraints, The Rand Corporation, Report R.M.-6143-PR.

Chapter 12

Matchings, Transportation and Assignment Problems

1. Introduction

Consider the following problem of constructing a partial graph G_p of a general nondirected graph G, when the degree of every vertex with respect to G_p is prespecified.

DEGREE-CONSTRAINED PARTIAL GRAPH PROBLEM: Let $G = (X, A)$ be a nondirected graph with costs c_j associated with its links $a_j \in A$. Further, let nonnegative integers δ_i, $i = 1, \ldots, n$ be given. It is required to find a partial graph G_p^* of G so that the degrees of the vertices x_i with respect to G_p^* are the given numbers δ_i (i.e. $d_i^{G_p^*} = \delta_i$), and the sum of the costs of the links of G_p^* is a maximum (or minimum).

The above problem will be referred to in the present Chapter as the *general problem*. Obviously, given a graph G and the number δ_i it is quite possible that no partial graph G_p of G satisfying these degree constraints exists. Two necessary (but not sufficient—see Section 5) conditions that the δ_i must satisfy in order for some feasible partial graph to exist are:

$$\delta_i \leqslant d_i^G \quad \forall\, i = 1, \ldots, n$$

and

$$\sum_{i=1}^{n} \delta_i \text{ is even,}$$

(12.1)

the last condition being a direct result of the fact that for any nondirected graph the sum of its vertex degrees is twice the number of its links.

A *matching* of a nondirected general graph $G = (X, A)$ is a subset M of the set A of links of G. chosen so that no two links of M are adjacent (i.e. have a common terminal vertex). The following problem is then referred to as the *maximum matching problem* (MP).

MP: Find that matching M^* with the maximum cost; the cost of a matching

339

M being:

$$C_M = \sum_{a_j \in M} c_j.$$

M^* will be called the *maximum matching* of G.

The MP can also be expressed as a zero-one linear program:

$$\text{Maximize} \quad z = \sum_{j=1}^{m} c_j \xi_j \tag{12.2}$$

$$\text{subject to} \quad \sum_{j=1}^{m} b_{ij} \xi_j \leqslant 1 \quad \forall i = 1, \ldots, n \tag{12.3}$$

$$\xi_j = 0 \text{ or } 1, \tag{12.4}$$

where $[b_{ij}]$ is the incidence matrix of G and where $\xi_j = 1$ (or 0) depending on whether a_j is (or is not) in the matching.

It is quite obvious that the MP for a graph $\hat{G}$ is a special case of the general problem. If the number of vertices of $\hat{G}$ is even, all that is required is to add links of cost $-\infty$ to $\hat{G}$ until a new complete graph G is obtained. The MP then becomes the general problem for graph G with all δ_i set to 1. The solution to the MP is the solution (partial graph) G_p^* of the general problem, if one ignores those links of cost $-\infty$ appearing in G_p^*. If the number of vertices of $\hat{G}$ is odd, then an extra isolated vertex is added to $\hat{G}$ before the graph G is formed in the same way as explained above.

In the nomenclature of Chapter 3 a matching could reasonably be called an "independent link set" since no two links of a matching are adjacent. A "dominating link set" in the sense of Chapter 3 (i.e. a set of links "dominating" the vertices of G) would then be defined as a subset E of the links of G chosen so that every vertex has at least one link of E incident to it. Such a set E is generally referred to in the literature as a *covering* and this term will be used in the present Chapter.

Corresponding to the MP is then the *minimum covering problem* (CP) i.e.

CP: Find that covering E^* of G with the minimum cost $\sum_{a_j \in E^*} c_j$.

E^* will be called the *minimum covering* of G.

The zero-one linear programming formulation of the CP is:

$$\text{Minimize} \quad z = \sum_{j=1}^{m} c_j \xi_j \tag{12.5}$$

$$\text{subject to} \quad \sum_{j=1}^{m} b_{ij} \xi_j \geqslant 1 \quad \forall i = 1, \ldots, n \tag{12.6}$$

$$\xi_j = 0 \text{ or } 1 \tag{12.7}$$

where $\xi_j = 1$ (or 0) depending on whether a_j is (or is not) in the covering.

Figure 12.1(a) shows a graph with a matching shown by the heavy lines
and Figure 12.1(b) shows the same graph with a covering shown by the heavy
lines.

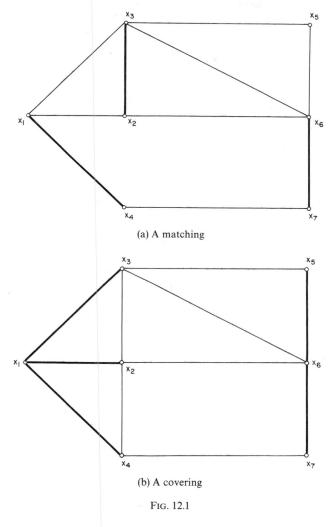

(a) A matching

(b) A covering

FIG. 12.1

In the special case where all link costs c_j are unity, the MP and CP are
reduced to the maximum *cardinality matching problem* (CMP) and the minimum *cardinality covering problem* (CCP). If a graph G has n vertices, the cardinality of a matching of G cannot, obviously, exceed $[n/2]$. This number,

however, is not always achieved; for example the "star" graph of Fig. 12.2 has a maximum cardinality matching of value 1.

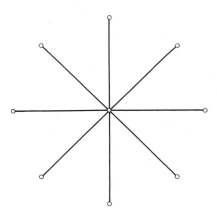

FIG. 12.2. Star graph

In the special case where the costs c_j are arbitrary but the graph itself is bipartite, the maximum matching problem reduces to the *assignment problem* (AP), a problem very well known in Operations Research. With this special graph structure, the general problem itself becomes another well known problem, the *transportation problem* (TP).

In the present Chapter we will give efficient methods of solution of the MP, CP, AP and TP. The MP appears as a subproblem in the graph traversal and Chinese postman's problem which occur in distribution problems such as milk and post delivery, spraying roads with salt grit in winter and refuse collection (see Chapter 9). The AP and TP are two basic problems with numerous direct and indirect applications in very many diverse areas other than the obvious ones from which they derived their names.

2. Maximum Cardinality Matchings

Given a matching M, a vertex x_i which is not the terminal vertex of any link in M is called an *exposed* vertex. In Fig. 12.3 where the matching is shown by thick lines, vertices x_6 and x_9 are exposed.

2.1 Alternating paths and trees

An *alternating path* is an elementary path whose links are alternately in M and not in M. $(x_8, x_7, x_5, x_3, x_4, x_{10}, x_1)$ is an example of such a path in

Fig. 12.3. An *augmenting path* is an alternating path whose initial and final vertices are exposed, $(x_9, x_{10}, x_4, x_1, x_2, x_3, x_5, x_6)$ is an example of such a path in the graph of Fig. 12.3.

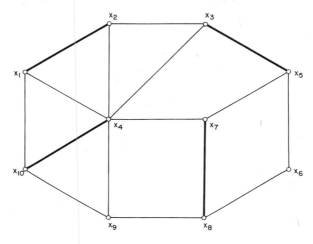

FIG. 12.3. Matching M

The basic theorem in connection with matchings can now be stated as follows:

THEOREM 1 [5,27]. *A matching M is a maximum cardinality matching if and only if there exists no augmenting path in G relative to M.*

PROOF

Necessity. Suppose an augmenting path P exists and let P also denote the set of links in the path. No link in the set $M - P \cap M$ can be adjacent to any link in P since the two terminal vertices of P are exposed (by definition), and all other vertices on P are already terminal vertices to some link in $P \cap M$. Hence the set of links $M' = (M - P \cap M) \cup (P - P \cap M) = M \cup P - P \cap M$, formed by interchanging those links in P not in M (i.e. $P - P \cap M$), with those links in P which are in M (i.e. $P \cap M$), is a matching. However, since both end links of P are not in in M, $|P - P \cap M|$ contains one link more than $|P \cap M|$ and therefore $|M'| = |M| + 1$, so that M is not a maximum cardinality matching.

Sufficiency. Let M be a matching which admits no augmenting path and let M^* be a maximum cardinality matching. Form the partial graph G_p

composed of links $M \cup M^* - M \cap M^*$, together with their associated vertices. No vertex of G_p can have degree greater than 2 since this would imply two or more links of M (or M^*) are adjacent which violates the definition of a matching. Thus G_p is composed of one or more connected components each one of which is either an isolated vertex, a simple path or a simple circuit, as shown in Fig. 12.4.

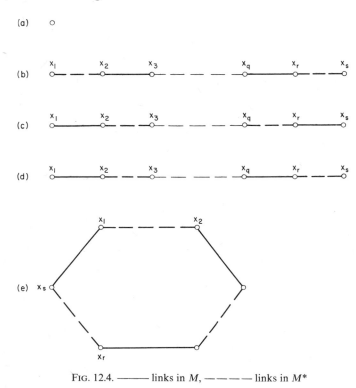

Fig. 12.4. ——— links in M, — — — — links in M^*

A path of type (b) cannot exist since it is an augmenting path relative to M contrary to the initial assumption. A path of type (c) cannot exist since it is an augmenting path relative to M^* and is contrary to the assumption that M^* is maximal. A circuit of type (e) with an odd number of links cannot exist since then either two links of M or two links of M^* would be adjacent to each other. What are then left are graphs of the form (a), (d), and (e) with an even number of links. For each one of these graphs the number of links $n(M)$ in M is equal to the number of links $n(M^*)$ in M^*. Since this applies to each connected component k of G_p we have:

$$\sum_k n_k(M) = \sum_k n_k(M^*)$$

where $n_k(M)$ and $n_k(M^*)$ is the number of links of component k belonging to M and M^* respectively. Therefore:

$$|M| \equiv |M \cap M^*| + \sum_k n_k(M) = |M \cap M^*| + \sum_k n_k(M^*) \equiv |M^*|$$

and hence M is a maximum cardinality matching.

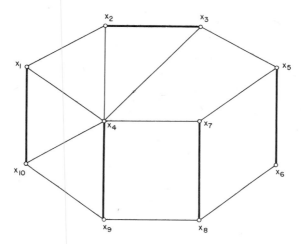

FIG. 12.5. Maximum cardinality matching M'

Considering the graph of Fig. 12.3 as an example, the interchange of the links in the augmenting path $P = (x_9, x_4, x_{10}, x_1, x_2, x_3, x_5, x_6)$ which are not in M, with those links in the path which are in M, produces the new matching M' shown in Fig. (12.5). In this figure there are no exposed vertices, hence no augmenting path exists and therefore M' is now a maximum cardinality matching according to Theorem 1 above.

An *alternating tree* relative to a given matching M is a tree T for which:

    (a) One vertex of T is exposed and is called the *root* of T.

    (b) All paths starting at the root are alternating paths.

and (c) All maximal paths from the root of T are of even cardinality, i.e. contain an even number of links.

Starting from the root of the tree and labelling it *outer* the vertices along any path starting from the root are labelled alternately *inner* and *outer*. Thus, if a vertex is at the end of an odd cardinality path from the root then the vertex is labelled inner and if it is at the end of an even cardinality path from the root it is labelled outer. Figure 12.6 shows an alternating tree. In this way, all vertices at the ends of the maximal paths from the root receive an "outer" label according to (c) above. Relative to the alternating tree the degree of all inner vertices is exactly 2 whereas the degree of an outer vertex can be any integer greater than or equal to 1.

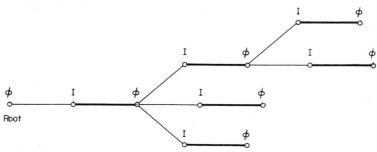

Fig. 12.6. Alternating tree. I: inner, $\varnothing$: outer

An *augmenting tree* is an alternating tree relative to a given matching whenever a link exists from an outer vertex x_o of the tree to an exposed vertex x_e not in the tree. The unique path from the root of the tree to x_o plus link (x_o, x_e) is then an augmenting path.

2.2 Blossoms

A *blossom* with respect to a matching M is an augmenting path for which the initial and final exposed vertices are identical—i.e. the path forms a circuit—and the number of links (or vertices) of the circuit is odd. Figure 12.7 shows a blossom consisting of a circuit of cardinality 5.

In the algorithms for the MP and CMP to be described later, blossoms are *shrunk* to derive a new simpler graph. The shrinking of a blossom B implies the replacement of all vertices of B (say X_B) by a single new *pseudo-vertex* x_b. Links (x_b, x_k) are added whenever a link exists from some vertex $x_i \in X_B$ to another vertex $x_k \notin X_B$. In the simpler graph resulting from such a shrinking, vertex x_b and other pseudo-vertices corresponding to previously shrunk blossoms may, at some later stage, form a new blossom which is shrunk again and so on. The final blossom B_o which is not contained in any other blossom is called an *outermost blossom*. The justification for the shrinking of blossoms is based on Theorem 2 given below.

A *blossomed tree* is an alternating tree relative to a given matching when-ever a link (x'_o, x''_o) exists between two outer vertices x'_o and x''_o of the tree. If P' is the set of links on the path from the root of the alternating tree to x'_o

and P'' the corresponding set for x_o'' then the set of links $[P' \cup P'' - P' \cap P'']$ together with link (x_o', x_o'') form a blossom. Figure 12.7 shows a blossomed tree.

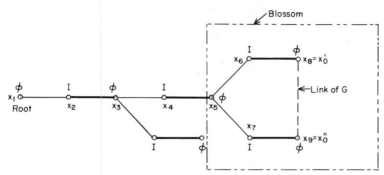

FIG. 12.7. Blossomed tree

Whenever a blossom B is shrunk, the resulting pseudo-vertex x_b is labelled an outer vertex. Any vertex of B could be labelled inner or outer as desired since these vertices are ends of one odd and one even cardinality path from the root of the tree, depending on the route that these paths take after they meet the first vertex of the blossom. Thus in Fig. 12.7 vertex x_6 can be labelled inner along the path $(x_1, x_2, x_3, x_4, x_5, x_6)$ or labelled outer along the path $(x_1, x_2, x_3, x_4, x_5, x_7, x_9, x_8, x_6)$. However, when a pseudo-vertex resulting from the shrinking of a blossom is labelled "outer", then the structure of the alternating tree remaining after such a shrinking is still correct—according to the definition of an alternating tree. Thus, after a blossom is shrunk an alternating tree is available in the resulting new graph.

An example of complex blossom formation and shrinking is shown in Figs. 12.8(a) to 12.8(d). In Fig. 12.8(a) a graph is shown with a matching M given by the heavy lines. An alternating tree T rooted at the exposed vertex x_1 is shown by continuous lines and other links of the graph not in T are shown dotted. Let us say that the last link added to the tree is (x_{11}, x_{12}). When this link is added the tree is seen to blossom because of the existence of a link between the two outer vertices x_{12} and x_{13}. The blossom B_1, shown ringed in Fig. 12.8(a), is shrunk into the pseudo-vertex x_{b_1} as shown in Fig. 12.8(b). The tree of Fig. 12.8(b) now blossoms once more into blossom B_2 because of the existence of link (x_{10}, x_{b_1}). This blossom is shrunk into x_{b_2} producing the graph of Fig. 12.8(c). The tree of Fig. 12.8(c) blossoms yet again because of the existence of link (x_3, x_{b_2}) to form blossom B_3. The shrinking of B_3 into vertex x_{b_3} produces the graph of Fig. 12.8(d). No further blossoming occurs, and B_3 is the outermost blossom.

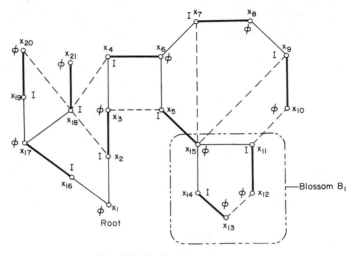

FIG. 12.8 (a). Shrinking of B_1

————: Links in alternating tree and also in matching
————: Links in alternating tree but not in matching
- - - - -: Other links of the graph

THEOREM 2. *If B is a blossom based on the odd vertex set X_B, and if x is any vertex $\in X_B$, then there exists a maximum cardinality matching in the subgraph $\langle X_B \rangle$ which leaves x exposed.*

Proof. Let $x_1, x_2, \ldots, x_\alpha$ be the odd cardinality circuit forming B. Then, either $x = x_i$ for some $i = 1, 2, \ldots, \alpha$, or else $x \in x_i$ for some pseudo-vertex x_i corresponding to a previously shrunk blossom B_i.

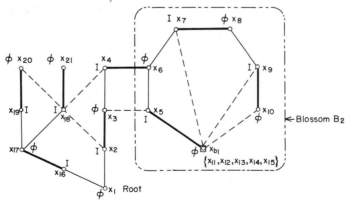

FIG. 12.8 (b). Shrinking of B_2

In the former case x can be left exposed and the other $\alpha - 1$ vertices of B matched completely. This is always possible since $\alpha - 1$ is an even number of vertices remaining on the circuit. If any one of these vertices, say x_p, corresponds to a previously shrunk blossom B_p, the matching of x_p would imply the matching of some vertex of B_p leaving once more an even number of vertices of B_p to be matched etc. until all vertices (other than x) are matched.

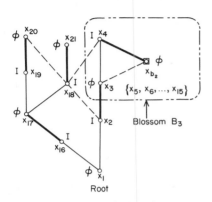

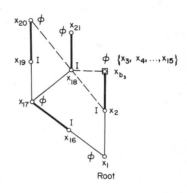

FIG. 12.8 (c). Shrinking of B_3

FIG. 12.8 (d). Alternating tree after the shrinking of blossoms

In the latter case vertex x_i can be left exposed and the other $\alpha - 1$ vertices of B matched as above. B_i can then be expanded into an odd cardinality circuit, say $x'_1, x'_2, \ldots, x'_\beta$, and one has the same situation relative to B_i as discussed above relative to B, so finally only vertex x is left exposed.

The importance of blossoms and circuits of odd cardinality in matching problems can be appreciated from the fact [8, 16, 17] that in the absence of such circuits, the linear programming formulation of the MP given by eqns (12.2) and (12.3) has inherently integer solutions so that the integrality constraints (12.4) are superfluous. Graphs without circuits of odd cardinality are known as bipartite graphs and are considered in later sections. The general properties of the polyhedron defined by constraints (12.3) were given by Balinski [1] as:

THEOREM 3. *All vertices of the convex polyhedron defined by:*

$$\sum_{j=1}^{m} b_{ij}\xi_j \leqslant 1, \qquad \forall i = 1, \ldots, n$$

where $[b_{ij}]$ is the incidence matrix of a graph, have coordinate values 0, $\frac{1}{2}$, or 1.

The causes of the noninteger ($\frac{1}{2}$) values are the odd circuits. For example, if the graph is a single circuit of odd cardinality as shown in Fig. 12.9, the solution to the linear program given by eqns (12.2) and (12.3) for the CMP is: $\xi_j = \frac{1}{2}$, for $j = 1$ to 5. The value of this solution is $2\frac{1}{2}$, whereas obviously the largest cardinality matching in the graph of Fig. 12.9 has the value 2.

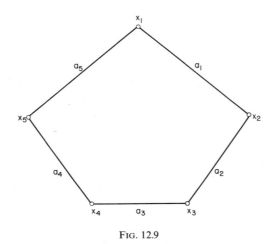

FIG. 12.9

2.3 Hungarian trees

A *hungarian tree* is an alternating tree in a graph in which all links having an outer vertex of the tree as one terminal, have an inner vertex also in the tree as the other terminal. Figure 12.10 shows a hungarian tree with its links drawn in continuous lines; heavy for those in the matching and light otherwise. Links of the graph not in the alternating tree are shown dotted.

The importance of hungarian trees in the MP is due to the following property:

THEOREM 4. [10]. *Let H be a hungarian tree in a graph $G = (X, A)$ and $G_0 = \langle X - X_H \rangle$ be the subgraph of G excluding the set X_H of vertices of H. Then, if M_H is the matching in the tree H and $M_{G_0}^*$ is any maximum cardinality matching in G_0, the set of links $M_H \cup M_{G_0}^*$ is a maximum cardinality matching in G.*

Proof. Let the set A of links of G be partitioned into three sets:

$$A_H = \{(x_i, x_j) | (x_i, x_j) \in A \quad \text{and} \quad x_i, x_j \in X_H\}$$
$$A_{HG_0} = \{(x_i, x_j) | (x_i, x_j) \in A, x_i \in X_H \quad \text{and} \quad x_j \in X - X_H\}$$
$$A_{G_0} = \{(x_i, x_j) | (x_i, x_j) \in A \quad \text{and} \quad x_i, x_j \in X - X_H\}.$$

If S is any arbitrary matching in G, then S can be partitioned in a way similar to that used for the partitioning of A so that

$$S = S_H \cup S_{HG_o} \cup S_{G_o}$$

where $S_H = S \cap A_H$, $S_{HG_o} = S \cap A_{HG_o}$ and $S_{G_o} = S \cap A_{G_o}$. By the definition of $M_{G_o}^*$ we have:

$$|M_{G_o}^*| \geq |S_{G_o}| \tag{12.8}$$

Consider now the graph G' consisting of the links $A_H \cup A_{HG_o}$ and their terminal vertices. Relative to M_H all vertices $x_k \in X - X_H$ and which are

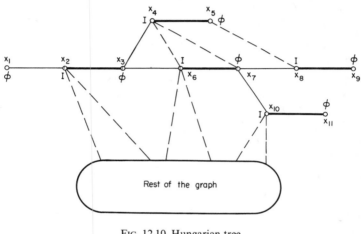

FIG. 12.10. Hungarian tree
——— Links both in the tree and in the matching
——— Links in the tree but not in the matching
----- Links not in the tree

terminals of links in A_{HG_o} are exposed. An alternating path starting from an exposed vertex $x_k \in X - X_H$ (or starting from the root vertex of H which is also exposed) can only be augmenting if it terminates at another vertex $x_k \in X - X_H$. However, an augmenting path must be of odd cardinality, which implies that if the first link of this path is from an exposed to an inner vertex, the last link must be from an outer to an exposed vertex. Thus, since by the definition of H all links in A_{HG_o} are between inner vertices of H and vertices in $X - X_H$, no augmenting path in graph G' can exist, and hence M_H is a maximum cardinality matching in G', i.e.

$$|M_H| \geq |S_H| + |S_{HG_o}| \tag{12.9}$$

Combining eqns (12.8) and (12.9) we get:

$$|M_H \cup M_{G_o}^*| \geqslant |S_H \cup S_{HG_o} \cup S_{G_o}|$$
$$\geqslant |S| \qquad (12.10)$$

i.e. $M_H \cup M_{G_o}^*$ is a maximum cardinality matching in G.

2.4 An algorithm for the CMP

Let an initial matching be given. (The empty matching can be used.) The algorithm proceeds by systematically searching for augmenting paths in an attempt to improve on the given matching according to Theorem 1. The algorithm was developed by Edmonds [10, 11].

An alternating tree is rooted at an exposed vertex and grown by alternately adding links which are in and not in the matching until:
either (i) The tree becomes augmenting,
   or (ii) The tree blossoms,
   or (iii) The tree becomes hungarian.

In case (i) the cardinality of the matching can be increased by one simply by tracing the augmenting path back to the root of the tree and then interchanging those links of the path that belong to the matching with the ones that do not. After augmentation the tree is discarded, and a new tree is rooted at some remaining exposed vertex, if one exists.

In case (ii), the resulting blossom is identified (as explained in Section 2.2), shrunk, and the growing of the tree continued in search for an augmenting path. As far as the computing is concerned, the shrinking of a vertex need not be done explicitly. All that is required is to mark all the vertices of the blossom as outer and set up labels on the vertices to indicate that they all belong to this blossom. The order in which these blossoms have been "shrunk" is important since at the end of the procedure the blossoms must be "expanded" in reverse order.

In case (iii) the vertices of the hungarian tree and their incident links are removed from the graph permanently (in accordance with Theorem 4) and the algorithm is reapplied to the remaining subgraph.

2.4.1 DESCRIPTION OF THE ALGORITHM

Initialization
Step 1. Choose an initial matching M in G.

Choose root
Step 2. If at least two exposed vertices exist in G, choose one as the root of the tree, mark it outer and go to step 3, else go to step 8.

Step 3. Choose an outer vertex x_o of the tree. For each link (x_o, x_i): if x_i is an exposed vertex goto step 4; if x_i is an inner vertex goto step 7; if x_i is not in the tree but is not exposed goto step 5.

Step 4. An augmenting path from the root to x_i has been found. Form a new improved matching, discard the current tree and the vertex labels and goto step 2.

Step 5. Add link (x_o, x_i) to the alternating tree and mark x_i inner. Find the link (x_i, x_k) which is in the current matching and add it to the tree marking x_k outer. If there exists a link between x_k and another outer vertex goto Step 6 else goto step 3.

Step 6. Identify and shrink the blossom which is formed. Mark the resulting pseudo-vertex outer. Update the blossom partial ordering and return to step 3.

Step 7. Return to step 3 until case 7 is the only case that can occur. In that event remove the vertices of the tree and all links of G incident on these vertices. Call the remaining subgraph G and return to step 2.

Step 8. Individually induce into the last graph G, and into each hungarian graph that has been removed, the optimum matching as follows. Expand the outermost blossom first and induce a matching which leaves exposed relative to it that vertex x which is matched in the unexpanded blossom. (See theorem 2.) Continue the expansion in the reverse ordering recorded at step 6 until the entire graph has been reexpanded and the maximum cardinality matching has been obtained.

2.5 Example

We wish to find the maximum cardinality matching in the graph shown in Fig. 12.11(a) where an initial matching is represented by the heavy lines. We will apply the algorithm of the previous section with the rules that at step 3 the outer vertices are scanned in ascending order of their index, and given an outer vertex, the links leading from it are considered in ascending order of the index of their second terminal vertex.

Step 2 of the algorithm chooses vertex x_1 as the root of the tree, and after repeated application of steps 3 and 5 the alternating tree of Fig. 12.11(b) results. At this stage, just after links (x_3, x_8) and (x_8, x_9) have been added to the alternating tree, step 6 identifies the blossom $(x_3, x_8, x_9, x_5, x_4)$ which is shrunk into the pseudo-vertex b_1 to produce the tree of Fig. 12.11(c).

The next application of step 3 adds links (b_1, x_{10})—actually (x_4, x_{10})—and (x_{10}, x_{11}) as shown in Fig. 12.11(d). A second blossom $(b_1, x_{10}, x_{11}, x_7, x_6)$ is

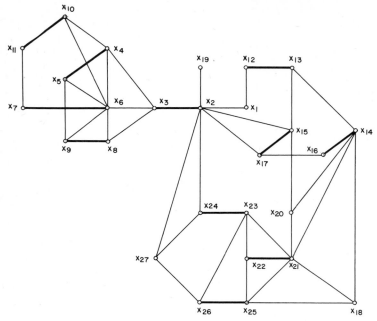

FIG. 12.11 (a). Graph for example 2.5
——— Links in matching

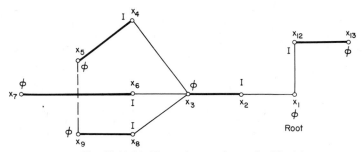

FIG. 12.11 (b). Alternating tree in graph of fig. (a)

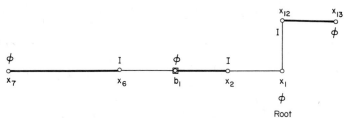

FIG. 12.11 (c). Tree after shrinking blossom b_1

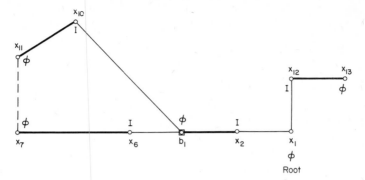

FIG. 12.11 (d). Blossom b_2 formed

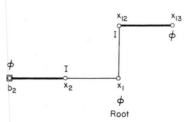

FIG. 12.11 (e). Tree after shrinking blossom b_2

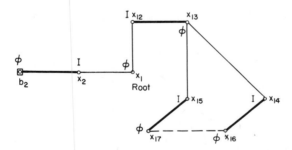

FIG. 12.11 (f). Formation of blossom b_3

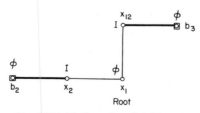

FIG. 12.11 (g). Tree after shrinking of b_3

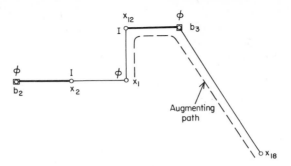

FIG. 12.11 (h). Augmenting path found

identified by step 6, which is again shrunk into the pseudo-vertex b_2 to produce the tree of Fig. 12.11(e). Expansion of this tree continues from vertex x_{13} and after the links $(x_{13}, x_{14}), (x_{14}, x_{16}), (x_{13}, x_{15})$ and (x_{15}, x_{17}) are added a third blossom $(x_{13}, x_{14}, x_{16}, x_{17}, x_{15})$ is identified as shown in Fig. 12.11(f) and is shrunk into vertex b_3 to produce the tree of Fig. 12.11(g).

The next application of step 3 adds link (b_3, x_{18}) and an augmenting path $(x_1, x_{12}, b_3, x_{18})$ is recognized at step 4. (See Fig. 12.11(h).) When b_3 is expanded (see Fig. 12.11(i)), two possible paths exist from x_1 to x_{18}; and the augmenting path is the one of odd cardinality, i.e. $P_2 = (x_1, x_{12}, x_{13}, x_{15}, x_{17}, x_{16}, x_{14}, x_{18})$. When links of this path which are in the matching are interchanged with those that are not, a new improved matching is obtained as shown in Fig. 12.12(a).

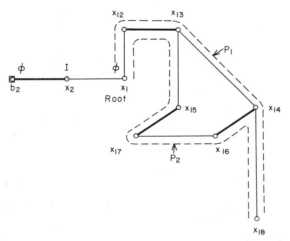

FIG. 12.11 (i). Expansion of blossom b_3. P_2 is augmenting path

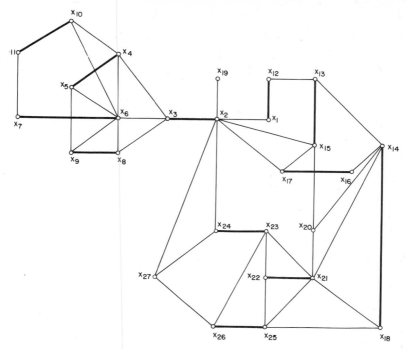

Fig. 12.12 (a). Improved matching

Discarding the tree and starting again, step 2 chooses vertex x_{19} to be the new root. After several applications of steps 3, 5 and 6 the tree becomes as shown in Fig. 12.12(b) where b_2 has the same meaning as before. This tree is recognized as a hungarian tree by step 7 and after its removal the remaining subgraph is as shown in Fig. 12.13(a).

Step 2 now chooses x_{20} to be the new root and after several applications of steps 3 and 5, an augmentation is detected just after links (x_{18}, x_{25}) and (x_{25}, x_{26}) are added by step 5. The final tree is shown in Fig. 12.13(b) and the augmenting path is $(x_{20}, x_{14}, x_{18}, x_{25}, x_{26}, x_{27})$. Exchanging links of this path which are in the matching with those that are not, produces a matching with

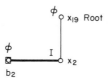

Fig. 12.12 (b). Hungarian tree in the graph of fig. 12.12 (a)

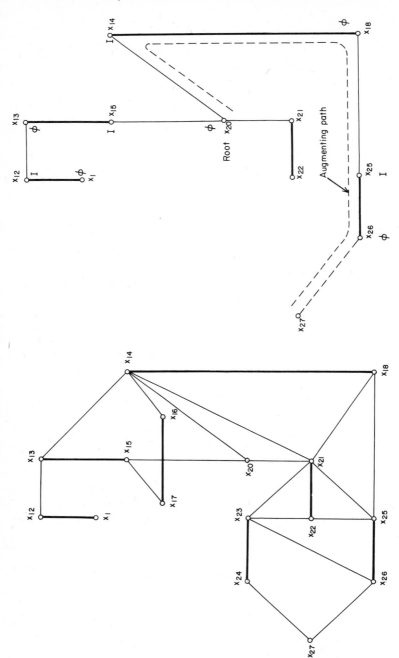

Fig. 12.13 (a). Graph after removal of hungarian tree   Fig. 12.13 (b). Alternating tree and augmenting path in the graph of fig. 12.13 (a)

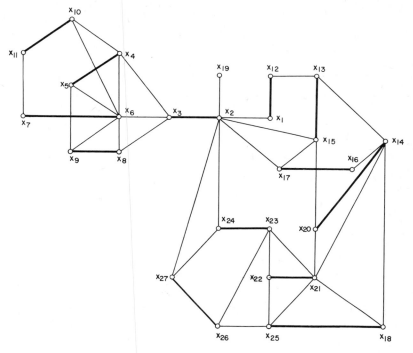

FIG. 12.14. Maximum cardinality matching for example 2.5

no exposed vertices left. (Note that vertex x_{19} has been removed with the hungarian tree earlier.) The resulting matching shown in Fig. 12.14 is, therefore, the maximum cardinality matching.

3. Maximum Matchings with Costs

Consider now the matching problem for the general graph $G = (X, A)$, with costs c_j associated with its links $a_j \in A$. In this section we examine the problem of finding a *perfect* matching (i.e. a matching where every vertex is matched to some other vertex), and which has maximum cost. This is the problem defined by eqns (12.2) to (12.4) except that the inequality in constraint (12.3) is replaced by equality. It is explained in the Introduction how the problems with equality or inequality constraints can be made equivalent by adding links of cost $-\infty$ and adding an extra artifical vertex in the case where the number of vertices is odd. Thus, the *MP* we consider is:

Maximize

$$z = \sum_{j=1}^{m} c_j \xi_j \qquad (12.11)$$

subject to

$$\sum_{a_j \in T_i} \xi_j = 1 \qquad \forall i = 1, \ldots, n \qquad (12.12)$$

where

$$T_i = \{(x_i, x_k) \,|\, x_k \in \Gamma(x_i)\},$$

i.e. the set of all links incident at vertex x_i; and

$$\xi_j \in \{0, 1\} \qquad (12.13)$$

As mentioned earlier, the integrality constraints (12,13) are not superfluous when circuits of odd cardinality exist in the graph. Edmonds [10, 11, 12], however, has shown that these constraints can be replaced by a system of linear constraints and proved the following theorem.

THEOREM 5 [11]. *For any graph, the convex hull of the solutions to* (12.12) *and* (12.13) *is the polyhedron defined by* (12.12) *and in addition:*

(i) *For every subset* $S_r \subset X$ *containing an odd number of vertices (say,* $|S_r| = 2q_r + 1$ *for some positive integer* q_r*), add the constraints*

$$\sum_{a_j \in A_r} \xi_j \leqslant q_r \qquad (12.14)$$

where A_r *is the set of links of the subgraph* $\langle S_r \rangle$,
and (ii) *Add the constraints* $\xi_j \geqslant 0$ *for all* $j = 1, \ldots, m,$ $\qquad (12.15)$

Obviously, any matching of G satisfies constraints (12.14); what is not so obvious is that these constraints are also sufficient. We will show this constructively by providing a perfect matching of G which is a solution to the linear program defined by (12.11), (12.12), (12.14) and (12.15) for any given link cost vector $[c_j]$.

Let us call the above linear program P. The linear program dual to P is [8]:
Minimize

$$u = \sum_{i=1}^{n} \pi_i + \sum_{S_r} q_r \lambda_r \qquad (12.16)$$

subject to

$$\pi_i + \pi_k + \sum_{S_r \in F_j} \lambda_r \geqslant c_j \qquad \forall j = 1, \ldots, m \qquad (12.17)$$

and

$$\lambda_r \geqslant 0 \qquad \text{for all } S_r \qquad (12.18)$$

Thus, a variable π_i is now associated with each vertex x_i of G, and a variable λ_r associated with every odd set, S_r, of vertices. In constraints (12.17) c_j is the cost of link a_j assumed to be (x_i, x_k) and $F_j = \{S_r \mid a_j \in A_r\}$, i.e. F_j is the family of subsets S_r which contain both x_i and x_k.

According to the complementary slackness theorem of linear programming a vector $[\xi_j^*]$ maximizes z subject to constraints (12.12), (12.14) and (12.15) and a vector $[\pi_i^*, \lambda_r^*]$ minimizes u subject to constraints (12.17) and (12.18) if and only if:

(i) For every λ_r^* either $\lambda_r^* = 0$ or else equality holds for the corresponding constraint (12.14).

and (ii) For every ξ_j^* either $\xi_j^* = 0$ or else equality holds in the corresponding constraint (12.17).

We will proceed to prove theorem 5 by describing a procedure which finds a perfect matching M (and hence a vector $[\xi_j]$), and also a vector $[\pi_i^*, \lambda_r^*]$ which satisfies constraints (12.17), (12.18) and the complementary conditions (i) and (ii); thus providing a proof that M is optimal.

3.1 An algorithm for the MP

Suppose we start with the graph G and place weights π_i on its vertices, choosing the π_i's large enough for constraints (12.17) to be satisfied when all λ_r's are taken to be zero. Now let G' be the partial graph of G containing only those links for which equality holds in constraint (12.17). We will call G' the *equality* partial graph of G. If in G' we can find a perfect matching (by applying the algorithm for the CMP described in the last section), then this matching is guaranteed to be optimal. This is so because $[\pi_i, \lambda_r]$ satisfy constraints (12.17) and (12.18), complementary condition (i) is satisfied automatically since $\lambda_r = 0$ for all r, and condition (ii) is satisfied by the fact that only links in G' are considered to be in the matching, so that ξ_j cannot be $\neq 0$ if a_j does not satisfy constraint (12.17).

In general, however, it will not be possible to find a perfect matching in G' and the algorithm for the CMP will detect a hungarian tree H. Since, as mentioned in Section 2.3, the maximum cardinality matching in G' is partly composed from the matching in H which leaves the root of H exposed, the existence of a hungarian tree implies that there is no perfect matching in G'. Thus, when such a tree is detected, the weight vector $[\pi_i, \lambda_r]$ has to be changed so that a new equality graph G' of G is generated and tested. The way that such weights are calculated is explained later on.

Let us for the moment assume that for any feasible weight vector $[\pi_i, \lambda_r]$ (i.e. one which satisfies eqns (12.17) and (12.18)), a given λ_r is non-zero only if S_r is a set of vertices contained in a pseudo-vertex of the current G', or in a pseudo-vertex contained in a pseudo-vertex of G' at any level of the shrinking process. Moreover, assume that for all the links which at some stage formed blossoms that were later shrunk into the pseudo-vertices of G', equality holds in expression (12.17). The above two assumptions are certainly true initially, since the first equality graph G' formed as mentioned earlier, has no pseudo-vertices at all, and has all λ_r's set to zero.

We will now introduce the following indexing notation in order to clarify the explanation of the algrithm. The initial graph G will be referred to as G_0. The equality partial graph of G_0 is G'_1 and this is the graph in which an alternating tree is grown as was done in the algorithm for the CMP. When the first blossom eventually forms and is shrunk, the graph itself will be denoted G_1 and the equality partial graph of G_1 will be G'_2. After f blossoms are shrunk the graph will be G_f and its equality partial graph G'_{f+1}.

The general step of the algorithm is then as follows.

Let the current graph (after a number of blossom shrinkings), be G_{f-1} and its equality partial graph be G'_f. In G'_f start growing an alternating tree T—as was done for the CMP—until the tree either:

    A. blossoms
    B. augments
    C. becomes hungarian.

Case A. In this case the blossom is shrunk thus obtaining a new graph G_f and its equality partial graph G'_{f+1}. As mentioned earlier, the shrinking of a blossom leaves the remaining T with the correct structure of an alternating tree so T could be retained and growing of the tree can continue.

Case B. In this case a better matching with larger cardinality is obtained as explained for the CMP. T must now be discarded and a new tree grown in the same G'_f by choosing some remaining exposed vertex of G'_f as the new root. One should note here, that once T is discarded and a new tree is started to be grown in G'_f some of the pseudo-vertices of G'_f (formed by the shrinking of earlier blossoms which produced $G_0, G_1, \ldots, G_{f-1}$) can now appear labelled inner in the new tree.

Case C. In this case the weight vector $[\pi_i, \lambda_r]$ is changed for the current graph G_{f-1} so that a new equality partial graph, different from the current G'_f is obtained. The changes in $[\pi_i, \lambda_r]$ are chosen so that the new G'_f will:

    (a) Continue to satisfy the assumptions made earlier; i.e. that λ_r is nonzero only for vertex sets corresponding to pseudo-vertices of G_{f-1}, and that equality holds for all links of blossoms that were shrunk into the pseudo-vertices of G_{f-1}.
and
    (b) Either: the current alternating tree in the old G'_f can be grown further,
           or it blossoms, or augments—using new links that enter the new G'_f;

Or: some pseudo-vertex of the current alternating tree which was labelled inner is disposed of;

Or: we prove that no perfect matching exists in G_o.

The changes in the weight vector $[\pi_i, \lambda_r]$ are calculated as follows:

For the links $a_j = (x_i, x_k)$ in G_{f-1} but not in G'_f, with one terminal vertex in the current alternating tree T and labelled outer, and the other terminal vertex not in T calculate:

$$\Delta_1 = \min_{a_j} [\pi_i + \pi_k - c_j] \qquad (12.19)$$

For the links $a_j = (x_i, x_k)$ in G_{f-1} but not in G'_f, with both terminal vertices in T and both labelled outer calculate:

$$\Delta_2 = \tfrac{1}{2} \min_{a_j} [\pi_i + \pi_k - c_j] \qquad (12.20)$$

For the sets S_r of vertices of G forming an outermost pseudo-vertex of T labelled inner calculate:

$$\Delta_3 = \tfrac{1}{2} \min_{S_r} [\lambda_r] \qquad (12.21)$$

Take

$$\Delta = \min [\Delta_1, \Delta_2, \Delta_3] \qquad (12.22)$$

Now change the weight vector $[\pi_i, \lambda_r]$ as follows:

For every vertex x_i of G which is an outer vertex of T or is contained in a pseudo-vertex of T labelled outer, decrease π_i to $\pi_i - \Delta$.

For every vertex x_i of G which is an inner vertex of T or is contained in a pseudo-vertex of T labelled inner, increase π_i to $\pi_i + \Delta$.

For every set S_r of vertices of G which forms an outermost pseudo-vertex of T labelled outer, increase λ_r to $\lambda_r + 2\Delta$.

For every set S_r of vertices of G which forms an outermost pseudo-vertex of T labelled inner decrease λ_r to $\lambda_r - 2\Delta$.

Regardless of the value of Δ, all the links of the old G'_f which form the current T remain in the new G'_f since the weights π_i on all inner and outer vertices of T are increased and decreased by the same amount so that equality in expression (12.17) still holds. Equality in expression (12.17) also continues to hold for all links of blossoms that have been shrunk to produce pseudo-vertices of G_{f-1}. Thus, let a pseudo-vertex of G_{f-1} be labelled outer in the alternating tree T in the old G'_f. Then for some link $a_j = (x_i, x_k)$ which was in the blossom whose shrinking produced the pseudo-vertex, we will have both π_i and π_k reduced by Δ. However, λ_r for the vertex set S_r corresponding to this pseudo-vertex is increased by 2Δ so that equality in expression (12.17) still holds for link a_j. Similarly for links of blossoms which produced pseudo-vertices of G_{f-1} that are labelled inner in the tree T in the old G'_f.

Suppose now that $\Delta = \Delta_1$ in eqn (12.22) and that the link which produced this value of Δ_1 in eqn (12.19) is a^*_j. After the Δ-change of the weight vector $[\pi_i, \lambda_r]$, link a^*_j will satisfy expression (12.17) with equality so it will enter the new G'_f. Thus, the tree T (which as mentioned above is still an alternating tree in the new G'_f), can make use of a^*_j to either expand the tree or augment it depending on whether the terminal vertex of a^*_j not in T is exposed or not.

If now $\Delta = \Delta_2$ in eqn (12.22) and a^*_j is the link which produced the value Δ_2 in eqn (12.20), then the addition of a^*_j to the tree T will make the tree blossom.

Finally, if $\Delta = \Delta_3$ in eqn (12.22), then the subtraction of 2Δ from λ_r during the change of the weight vector $[\pi_i, \lambda_r]$ will cause some λ_r (say λ^*_r) to become zero according to eqn (12.21). Let x^* be the inner pseudo-vertex of the current T corresponding to λ^*_r and B^* be the outermost blossom whose shrinking produced x^*. Pseudo-vertex x^* of graph G_{f-1} can now be expanded to B^* and this expansion produces another graph G_{f-2}. (Note, however, that this G_{f-2} is not the same graph from which G_{f-1} was obtained after the shrinking of a blossom, since the blossom that is now expanded to obtain G_{f-2} from G_{f-1} is not the same outermost blossom that was shrunk to obtain G_{f-1}.) Let G'_{f-1} be the equality partial graph of the new G_{f-2}. Obviously, since equality holds in expression (12.17) for all links of B^*, these links are also in G'_{f-1}. If B^* consists of say $2q + 1$ links, a matching of q links can easily be induced on B^* as shown in Fig. 12.15. The expansion of B^* separates the current alternating tree into two components, one component touching the blossom at a vertex x_1 and the other at a vertex x_2. Using only links of B^*, there are two paths (one of even and one of odd cardinality), between x_1 and x_2 and the addition of the even cardinality path to the links of the current T will reconnect T into a single tree (see Fig. 12.15(c)). This new alternating tree can then be expanded until cases A, B or C above occur again.

The algorithm terminates when a perfect matching is obtained in some G'_f at a stage f. The corresponding graph G_{f-1} is then expanded backwards into the initial graph G_o by expanding blossoms in reverse order than the one in which they were shrunk, and inducing a perfect matching into each blossom as it is expanded—just as was done at step 8 of the CMP algorithm of Section 2.4.1.

When a perfect matching is obtained it is a maximum perfect matching since we then have a solution (i.e. a vector $[\xi_j]$), and a dual weighting vector

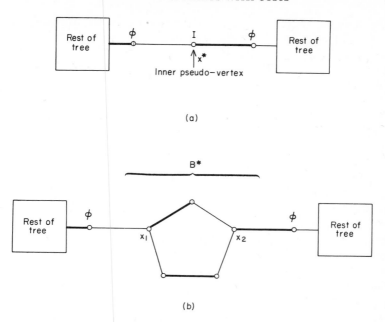

(a)

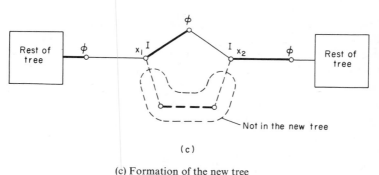

(b)

(b) Expansion of the inner pseudo-vertex into the outmost blossom B^*

(c)

(c) Formation of the new tree

Fig. 12.15

$[\pi_i, \lambda_r]$ which satisfy constraints (12.17) and (12.18) and also the complementarity conditions (i) and (ii) mentioned earlier.

The alternative termination, is when case C occurs but no constraints (12.19), (12.20) and (12.21) exist. In this case Δ can be taken as large as desired, and it is then quite apparent that no perfect matching in G exists.

The above method of obtaining a solution to the MP was developed by Edmonds [11]. It is an efficient algorithm (see Section 3.3) and is also a constructive proof of theorem 5.

3.1.1 DESCRIPTION OF THE ALGORITHM

Initialization

Step 1. On the vertices of G_o place weights π_i so that for all links in G_o, expression (12.17) is satisfied with all λ_r set to 0. Set $G = G_o$.

Form G'

Step 2. Form the equality partial graph G' of the current G.

Grow tree

Step 3. If no alternating tree T exists in G' root one at some exposed vertex of G' and grow it. If no exposed vertices remain goto step 8. If an alternating tree already exists in T continue growing it. If T blossoms goto step 4, if T augments goto step 5, if T becomes hungarian goto step 6.

Tree blossoms

Step 4. Shrink the blossoms to a pesudo-vertex. Call the resulting graph G, its equality partial graph G', and the remaining tree T. Goto step 3.

Tree augments

Step 5. Improve current matching by interchanging links that are in, and not in the matching along the augmenting chain. Discard T and goto step 3.

Tree is hungarian

Step 6. Calculate Δ_1, Δ_2, Δ_3 and Δ from eqns (12.19) to (12.22). If no constraints (12.19) to (12.21) exist Stop; the graph has no perfect matching. Otherwise change the weight vector $[\pi_i, \lambda_r]$ as mentioned earlier. If $\Delta = \Delta_1$ or $\Delta = \Delta_2$, retain the current T and goto step 2. If $\Delta = \Delta_3$ goto step 7.

Step 7. Expand the pseudo-vertex which produced Δ_3 in eqn (12.21) into a blossom B. Call the resulting graph G and its equality partial graph G'. Induce a perfect matching in B. Reconstruct the alternating tree by adding to the links of T the necessary path from B and call this tree T. Goto step 3.

End

Step 8. Re-expand all blossoms in reverse order inducing the perfect maximum matching after each expansion.

3.2 Example

We want to find the maximum perfect matching of the graph G_o shown in Fig. 12.16. We start by placing arbitrary weights π_i on the vertices x_i, $i = 1, \ldots, 12$ given by the vector $[\pi_i]^1 = (9, 2, 11, 7, 3, 4, 8, 6, 6, 5, 3, 4)$, and taking all $\lambda_r = 0$.

The equality partial graph G'_1 of graph G_o is shown in Fig. 12.17. Let us choose vertex x_1 to be the root and grow an alternating tree in G'_1 at step 3 of the algorithm. Step 5 immediately picks link (x_1, x_2) to be in the matching.

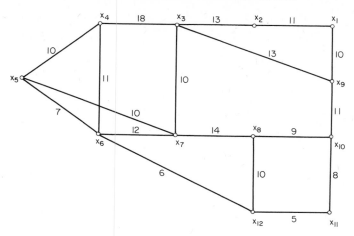

FIG. 12.16. Graph G_0 for example 3.2

Choosing vertices x_3, x_5 and x_7 (in that order) as roots for new trees, augmentations follow instantly so that a stage is reached where the links in the matching are those shown by the heavy lines in Fig. 12.17. Choosing vertex x_{12} as the root of a new tree, the tree is grown at step 3 of the algorithm until it reaches the stage shown in Fig. 12.18 when it blossoms. Step 4 shrinks this blossom into pseudo-vertex x_{b_1} and the growth of the tree continued at step 3 until it reaches the state shown in Fig. 12.19 and becomes hungarian. The graph

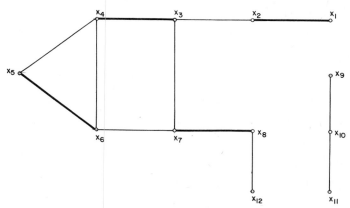

FIG. 12.17. Equality partial graph G_1'

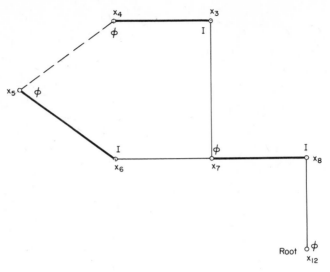

FIG. 12.18. Alternating tree in G_1'

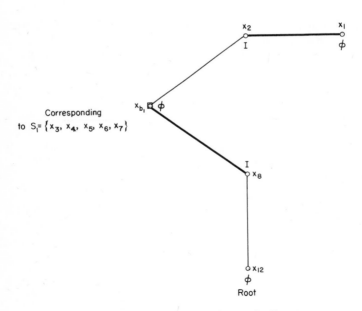

Corresponding
to $S_1 = \{x_3, x_4, x_5, x_6, x_7\}$

FIG. 12.19. Hungarian tree in G_1'

itself (after the shrinking), is G_1 and its equality partial graph is G_2' as shown in Figs 12.20(a) and (b) respectively.

Step 6 of the algorithm calculates:

$$\Delta_1 = \min [\underbrace{9 + 6 - 10}_{\text{link } (x_1, x_9)}, \quad \underbrace{11 + 6 - 13}_{\text{link } (x_3, x_9)}, \quad \underbrace{4 + 3 - 5}_{\text{link } (x_{12}, x_{11})}]$$

$$= 2 \quad \text{Corresponding to link } (x_{12}, x_{11})$$

$$\Delta_2 = \tfrac{1}{2}[4 + 4 - 6] = 1 \quad \text{Corresponding to link } (x_6, x_{12})$$

$$\Delta_3 = \infty \quad \text{(No constraint exists)}$$

Thus $\quad \Delta = \min [2, 1, \infty] = 1.$

The weights are now changed so that the new π vector becomes $[\pi_i]^2 = (8, 3, 10, 6, 2, 3, 7, 7, 6, 5, 3, 3)$ and the weight λ_1 corresponding to the odd vertex set $S_1 = \{x_3, x_4, x_5, x_6, x_7\}$ is $\lambda_1 = 2$.

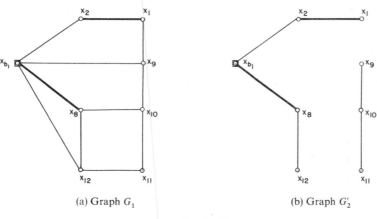

(a) Graph G_1 (b) Graph G_2'

FIG. 12.20

With the new weight vector $[\pi_i, \lambda_r]$, the new equality partial graph G_2' of the graph G_1 becomes as shown in Fig. 12.21, i.e. link (x_{12}, x_{b_1}) enters the old G_2'. Since the tree T of Fig. 12.19 is kept, the new entering link causes a blossom to form—consisting of odd circuit (x_{12}, x_8, x_{b_1}). This is detected immediately at step 3 and step 4 shrinks this blossom to the pseudo-vertex x_{b_2} to obtain the new tree of Fig. 12.22. The graph itself is now G_2 and its equality partial graph is G_3' as shown in Figs 12.23(a) and (b) respectively.

The tree shown in Fig. 12.22 is a hungarian tree in G_3', so step 6 calculates

N

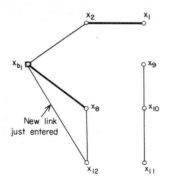

FIG. 12.21. New equality partial graph G_2' for the graph G_1 of Fig. 12.20 (a)

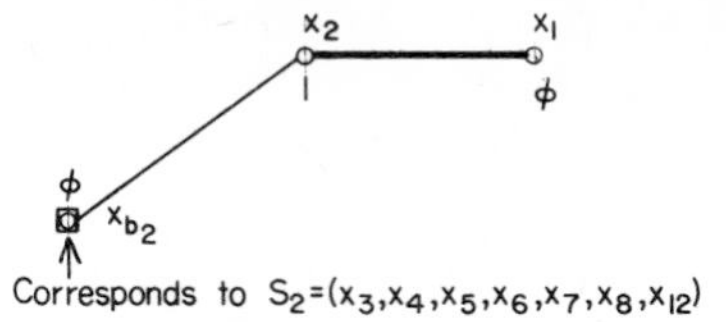

FIG. 12.22. Alternating tree in G_2' of Fig. 12.21.

a new change in the weight vector as follows:

$$\Delta_1 = \min \left[\underbrace{8 + 6 - 10}_{\text{link } (x_1, x_9)}, \quad \underbrace{10 + 6 - 13}_{\text{link } (x_3, x_9)}, \quad \underbrace{7 + 5 - 9}_{\text{link } (x_8, x_{10})}, \quad \underbrace{3 + 3 - 5}_{\text{link } (x_{12}, x_{11})} \right]$$

$= 1$ Corresponding to link (x_{12}, x_{11})

Δ_2 and $\Delta_3 = \infty$

Therefore $\Delta = 1$. Using this value of Δ to change the weight vector we get $[\pi_i]^3 = (7, 4, 9, 5, 1, 2, 6, 6, 6, 5, 3, 2)$ and the weight λ_2 corresponding to the odd vertex set $S_o = \{x_3, x_4, x_5, x_6, x_7, x_8, x_{12}\}$ is $\lambda_2 = 2$.

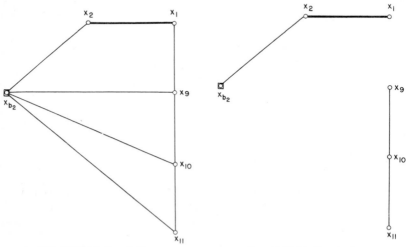

FIG. 12.23 (a). Graph G_2

FIG. 12.23 (b). Graph G_3'

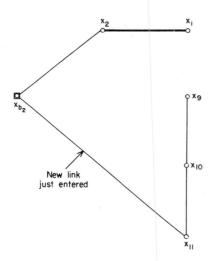

FIG. 12.24. New equality partial graph G'_3 for the graph G_2 of fig. 12.23 (a)

With these new weights $[\pi_i, \lambda_r]$, the new equality partial graph of graph G_2 is as shown in Fig. 12.24. With the introduction of the new link (x_{12}, x_{11}) the alternating tree of Fig. 12.22 augments so that the new link enters the matching. The tree is now discarded and a new tree is rooted at one of the remaining exposed vertices of G'_3, i.e. either at x_9 or at x_{10}. In either case link (x_9, x_{10}) immediately enters the matching and a perfect matching results. This perfect matching of G_2 is shown in Fig. 12.25(a). To find the corresponding matching in G_o we first expand blossom b_2 and induce in it a perfect matching as shown in Fig. 12.25(b), and then expand blossom b_1 and again induce in it a perfect matching as shown in Fig. 12.25(c). This last figure then shows the maximum perfect matching of the graph G_o and the cost of the matching is 66. Since, from linear programming theory, the maximum of z given by eqn (12.11) is equal to the minimum of u given by eqn (12.16), we have a check. Thus, since $|S_1| = 2q_1 + 1 = 5$ and $|S_2| = 2q_2 + 1 = 7$ we have $q_1 = 2$ and $q_2 = 3$ so that:

$$u = (7 + 4 + 9 + 5 + 1 + 2 + 6 + 6 + 6 + 5 + 3 + 2) + (2 \times 2 + 3 \times 2)$$

$$= 66.$$

3.3 Some computational results

The algorithm for the CMP, and hence also that for the MP, is a good algorithm in the sense that the number of operations involved is a polynomial function of the number of vertices n of the graph. For the CMP the growth in computation time is theoretically $O[n^4]$ although in practice the growth is appreciably slower [30]. A computer code given in [30] finds a dependence of $O[n^{2.7}]$ and another program [29] finds a dependence of $O[n^{2.3}]$. Edmonds

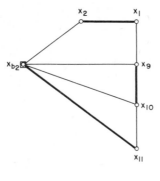

(a) Perfect matching in G_2

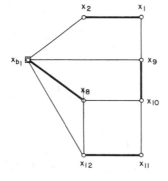

(b) Matching with b_2 expanded

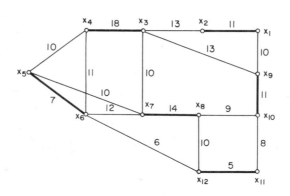

(c) Maximum matching

FIG. 12.25

and Johnston [29] report a typical result for a matching problem (in fact for the general problem discussed in Section 5 with $\delta_i = 1$ or 2 and $1 \leqslant c_j \leqslant 10$), with 300 vertices, 1500 links and which took about 30 seconds to solve on an IBM 360/91.

4. The Assignment Problem (AP)

In this section we consider the special case of matchings in bipartite graphs [9, 18, 25]. Given a bipartite graph $G = (X^a \cup X^b, A)$, where X^a and X^b are independent sets of vertices and links $a_j = (x_i, x_k) \in A$ have $x_i \in X^a$, $x_k \in X^b$ and cost c_j, we want to find a perfect matching of G with maximum (or minimum) cost. The minimization version of this problem is well known in the literature as the *Assignment Problem* (AP), and—without loss of generality —is often discussed in connection with complete bipartite graphs.

Let $|X^a| = |X^b| = n$. Considering the complete $K_{n,n}$ graph, it is easy to see how the matching problem for this special case can be solved using the network flow methods of Chapter 11 instead of the general matching algorithm developed in Section 3 of this chapter. Thus, let an artificial source vertex s be added with arcs (s, x_i), $\forall x_i \in X^a$ having unit capacity and zero cost, and an artifical sink vertex t added with arcs (x_k, t), $\forall x_k \in X^b$, also with unit capacity and zero cost. Then, if the links A are considered as directed arcs (x_i, x_k) with unit capacity, the maximum-flow minimum-cost calculation from s to t would have value n, and have nonzero flow (of value 1)† in those arcs of A with minimum summed cost. These arcs of A would then form a minimum cost perfect matching of G. Obviously, with this special structure many of the steps of the general minimum cost network flow algorithm can be particularized to produce a more efficient procedure for the AP.

Alternatively, the AP can be viewed as a special case of the MP and could be solved by a specialized version of the algorithm of Section 3.1.1. The main complication associated with this general algorithm—i.e. the formation and shrinking of blossoms—does not appear for the AP since in a bipartite graph no odd cardinality circuits can exist. Since it is customary to consider the AP as a minimization problem, we will follow this custom and describe a *minimizing* version of the algorithm of Section 3.1.1 which is applicable to bipartite graphs only. The change from a maximizing to a minimizing problem is quite trivial.

4.1 An algorithm for the AP (minimization case)

The algorithm described in this section was developed by Konig [22], Egervary [15], and Kuhn [23, 24] and is referred to as the *hungarian* method.

† Note from Chapter 11, that with all capacities integer, the arc flows are also integers, i.e. either 0 or 1 in this case.

4.1.1 Description of the algorithm

Initialization

Step 1. On the vertices X^a and X^b of G place weights π_i^a and π_k^b respectively, so that for any link $a_j = (x_i, x_k)$ we have $\pi_i^a + \pi_k^b \leqslant c_j$.

Form G'

Step 2. Form the equality partial graph G' of G given the current $[\pi]$ weights.

Grow tree

Step 3. If no alternating tree T exists in G' root one at some exposed vertex $x_i \in X^a$ of G' and grow it. If no exposed vertex remains goto step 7. If an alternating tree already exists in T continue growing it. If T augments goto step 4; if the tree becomes hungarian goto step 5.

Tree augments

Step 4. Improve current matching by interchanging links that are in, and not in the matching along the augmenting chain. Discard T and goto step 3.

Tree is hungarian

Step 5. For those links $a_j = (x_i, x_k)$ not in G', with one terminal vertex in the current tree T and labelled outer and the other terminal vertex not in T, calculate:

$$\Delta = \min_{a_j} \left[c_j - \pi_i^a - \pi_i^b \right]$$

Step 6. For every vertex $x_i \in X^a$ of G which is an outer vertex of T change π_i^a to $\pi_i^a + \Delta$, and for every $x_k \in X^b$ of G which is an inner vertex of T change π_k^b to $\pi_k^b - \Delta$. Retain the current T and goto step 2.

End

Step 7. The current perfect matching is minimal.

4.1.2 The algorithm in matrix form

Often, when the graph is complete. the operations of the above algorithm are performed on an $n \times n$ matrix $\mathbf{C}$ whose rows correspond to the vertices of X^a and whose columns correspond to the vertices of X^b. The initial entries c_{ik} are the costs of links (x_i, x_k).

At step 1 of the algorithm the initial weight π_i^a for vertex $x_i \in X^a$ is taken to be the smallest entry in row i. The weight π_i^a is subtracted from all entries in row i and this is done for all rows. The weight π_k^b of vertex $x_k \in X^b$ is then taken to be the smallest entry in column k of the resulting matrix. Again π_k^b is subtracted from all entries of column k and this is done for all columns to obtain a new matrix $\mathbf{C}'$ with $c_{ik}' = c_{ik} - \pi_i^a - \pi_k^b \geqslant 0$. The matrix $\mathbf{C}'$ is referred to as the *reduced* matrix. The partial graph G' would then be the $n \times n$ bipartite graph with arcs (x_i, x_k) only in those positions corresponding to $c_{ik}' = 0$. A perfect matching of G' would correspond to a choice of 0 entries in matrix $\mathbf{C}'$ so that there is exactly one chosen 0 entry in each row and each column of $\mathbf{C}'$.

The growing of an alternating tree in this implied partial graph G' is done by labelling the rows and columns of $\mathbf{C}'$ as follows. Start by forming an arbitrary matching in G', let us say by marking 0 entries in $\mathbf{C}'$ so that no two marked 0's are in the same row or column. (A marked zero in position (i, k) is taken to imply that link (x_i, x_k) is in the matching, and vice versa.)

Find a row s with no marked entry and set a label $p_s = 0$. If no such row is found, the current matching is the minimum perfect matching. (Vertex x_s —if one exists—now corresponds to an exposed vertex chosen as the root of the alternating tree, and the p_i are predecessor labels used to store this tree. For storage purposes the tree is considered as a rooted arborescence so that if arc (x_α, x_β) is in the arborescence then $p_\beta = x_\alpha$. For the root of the tree we take $p_s = 0$. (See Section 2.2.1, Chapter 7.) Starting with the label $p_s = 0$ and with all other rows and columns unlabelled:

(i) Label every unlabelled column k containing an unmarked 0 in a labelled row i with $p_k = i$. If more than one choice exists pick any one.

(ii) Label every unlabelled row i containing a marked 0 in a labelled column k with $p_i = k$.

Repeat labellings (i) and (ii). These two operations implicitly grow a tree in G' the tree being given by the predecessor labels. It is worthwhile to note here that all "outer" vertices of this tree correspond to rows of $\mathbf{C}'$ and all "inner" vertices to columns. Operations (i) and (ii) are repeated until a column t with no marked entries is labelled (by p_t say), or until no further labelling is possible. In the former case an augmenting path $t, p_t, p_{t'}, p_{t''}, \ldots, s$ has been found, where $t' = p_t$, $t'' = p_{t'}$ etc. The 0's in locations $(t, p_t), (p_t, p_{t'}), (p_{t'}, p_{t''}), \ldots, (., s)$ are then alternately marked and unmarked to produce a better matching. The labels p_i are erased and the procedure repeated. In the latter case a hungarian tree has been detected. If I^+ and K^+ are the sets of all labelled rows and columns and I^- and K^- the set of all unlabelled rows and columns respectively, calculate:

$$\Delta = \min_{\substack{i \in I^+ \\ k \in K^-}} [c'_{ik}]$$

(Note that $c'_{ik} = c_{ik} - \pi^a_i - \pi^b_k$.)
Update. $\pi_i = \pi_i + \Delta$ for $i \in I^+$, $\pi_k = \pi_k - \Delta$ for $k \in K^+$, $c'_{ik} = c'_{ik} - \Delta$ for $i \in I^+$ and $k \in K^-$, $c'_{ik} = c'_{ik} + \Delta$ for $i \in I^-$ and $k \in K^+$, and c'_{ik} unchanged otherwise. Operations (i) and (ii) can now be re-applied to the new $\mathbf{C}'$ matrix, etc.

4.2 Example

Solve the AP for a completely connected bipartite graph G where the cost matrix $\mathbf{C}$ is given by:

	1	2	3	4	5	6	7	8
1	13	21	20	12	8	26	22	11
2	12	36	25	41	40	11	4	8
3	35	32	13	36	26	21	13	37
4	34	54	7	8	12	22	11	40
5	21	6	45	18	24	34	12	48
6	42	19	39	15	14	16	28	46
7	16	34	38	3	34	40	22	24
8	26	20	5	17	45	31	37	43

$\mathbf{C} =$ (to the left of the matrix)

The reduced matrix $\mathbf{C}'$ is then the 8×8 matrix shown below, where the vectors to the left and top of the matrix are the weight vectors $[\pi_i]$ and $[\pi_k]$ respectively.

$\pi_k \rightarrow$	5	0	0	0	0	2	0	3

$\pi_i \downarrow$		1	2	3	4	5	6	7	8	
8	1	0*	13	12	4	0	16	14	0	
4	2	3	32	21	37	36	5	0*	1	7
13	3	17	19	0*	23	13	6	0	21	3
7	4	22	47	0	1	5	13	4	30	0
6	5	10	0*	39	12	18	26	6	39	
14	6	23	5	25	1	0*	0	14	29	
3	7	8	31	35	0*	31	35	19	18	
5	8	16	15	0	12	40	24	32	35	

$\uparrow \ p_i$

		4				3		$\leftarrow p_k$

We start by randomly choosing a matching i.e. choosing any set of 0's no two of which lie on the same row or column. The chosen 0's are shown marked with a star. Row 4 does not have a marked entry so it receives the label $p_{\text{row }4} = 0$. The following labels are then placed in order: $p_{\text{col. }3} = 4$, $p_{\text{row }3} = 3$, $p_{\text{col. }7} = 3$, $p_{\text{row }2} = 7$. The final label vectors $[p_i]$ and $[p_k]$ are shown to the right and bottom of the $\mathbf{C}'$ matrix above.

The labelling cannot continue further. We now have $I^+ = \{2, 3, 4\}$, $K^+ = \{3, 7\}$, $I^- = \{1, 5, 6, 7, 8\}$ and $K^- = \{1, 2, 4, 5, 6, 8\}$. Δ is calculated to be 1, and after the change to the weights $[\pi_i]$ and $[\pi_k]$, the new matrix $\mathbf{C}' = [c_{ik} - \pi_i - \pi_k]$ is as shown below.

$\pi_k \to$		5	0	-1	0	0	2	-1	3	
$\pi_i \downarrow$		1	2	3	4	5	6	7	8	
8	1	0*	13	13	4	0	16	15	0	
5	2	2	31	21	36	35	4	0*	0	7
14	3	16	18	0*	22	12	5	0	20	3
8	4	21	46	0	0	4	12	4	29	0
6	5	10	0*	40	12	18	26	7	39	
14	6	23	5	26	1	0*	0	15	29	
3	7	8	31	36	0*	31	35	20	18	
5	8	16	15	1	12	40	24	33	35	
				4	4			3	2	$\leftarrow p_k$

$\uparrow p_i$

The labelling continues so that $p_{\text{col. }4} = 4$ and $p_{\text{col. }8} = 2$. Since column 8 has no marked 0 an augmenting path has now been found and is shown pictorially in the matrix above. Exchanging marked and unmarked 0's along this path gives the new improved matching.

At the second iteration the labels are erased, row 8 is chosen as a row with no marked entries, and marked with $p_{\text{row }8} = 0$. The labelling cannot continue further, Δ is calculated to be 1, $\pi_{\text{row }8}$ is changed to 6 and the resulting new C' is shown below. (Note that at this second iteration the hungarian tree is, in fact, the single isolated—in terms of the current G'—vertex corresponding to row 8.) The labelling then continues to produce the vectors $[p_i]$ and $[p_k]$ as shown below.

$\pi_k \rightarrow$		5	0	−1	0	0	2	−1	3	
π_i ↓		1	2	3	4	5	6	7	8	
8	1	0*	13	13	4	0	16	15	0	
5	2	2	31	21	36	35	4	0	0*	
14	3	16	18	0	22	12	5	0*	20	
8	4	21	46	0*	0	4	12	4	29	3
6	5	10	0*	40	12	18	26	7	39	
14	6	23	5	26	1	0*	0	15	29	
3	7	8	31	36	0*	31	35	20	18	4
6	8	15	14	0	11	39	23	32	34	0

↑ p_i

			8	4					← p_k

Again labelling cannot continue further, Δ is now calculated to be 4, the weights $[\pi_i]$ and $[\pi_k]$ are changed, and the resulting new matrix C' is shown below.

$\pi_k \rightarrow$		5	0	−5	−4	0	2	−1	3		
$\pi_i \downarrow$		1	2	3	4	5	6	7	8		
8	1	0*	13	17	8	0	16	15	0		
5	2	2	31	25	40	35	4	0	0*		
14	3	16	18	4	26	12	5	0*	20	7	
12	4	17	42	0*	0	0	8	0	25	3	
6	5	10	0*	44	16	18	26	7	39		
14	6	23	5	30	5	0*	0	15	29	5	
7	7	4	27	36	0*	27	31	16	14	4	
10	8	11	10	0	11	35	19	28	30	0	

$\uparrow$ p_i

				8	4	4	6	4		$\leftarrow p_k$

The labelling now continues with the following labels placed in order: $P_{\text{col. }5} = P_{\text{col. }7} = 4$, $P_{\text{row }6} = 5$, $P_{\text{row }3} = 7$, $P_{\text{col. }6} = 6$. Since column 6 has no marked entry and it is labelled, an augmenting path has been found and is shown in the above matrix. Changing the marked and unmarked 0's along this chain results in a perfect matching which is therefore the minimum cost one. The links in the minimum cost matching are then: (1, 1), (2, 8), (3, 7), (4, 5), (5, 2), (6, 6), (7, 4) and (8, 3) with total cost 76. (As a check note that $\sum_i \pi_i + \sum_k \pi_k = 76$.)

5. The General Degree-Constrained Partial Graph Problem

For a given graph $G = (X, A)$ with associated arc costs, the general problem was defined in the Introduction as the problem of finding a partial graph G_p^* with respect to which the degrees of the vertices x_i are equal to the given integers δ_i, and so that the sum of the costs of the links in G_p^* is a maximum (or minimum). It was mentioned in the Introduction that the MP is a special case of this general problem. We will now show that the general problem can itself be solved as a MP using the algorithm of Section 3.1.1.

From the graph G let us form a graph $\hat{G}$ as follows. If δ_i is required to be the degree of vertex x_i then for every vertex x_i of G place δ_i vertices $x_i^1, x_i^2, \ldots, x_i^{\delta_i}$ in $\hat{G}$. For each link $a_j = (x_i, x_k)$ in G place two additional vertices r_j and s_j in $\hat{G}$ and add the links (x_i^α, r_j), $\alpha = 1, \ldots, \delta_i$ all with cost $\frac{1}{2}c_{ik}$, the links (s_j, x_k^β), $\beta = 1, \ldots, \delta_k$ with cost $\frac{1}{2}c_{ik}$ and link (r_j, s_j) of cost 0.

It is now quite apparent, that the general problem for graph G corresponds to the MP for graph $\hat{G}$. If link (r_j, s_j) is in the matching of $\hat{G}$, this corresponds to link (x_i, x_k) not being in G_p^*. If two links (x_i^α, r_j) and (s_j, x_k^β) are in the matching of $\hat{G}$ this corresponds to link (x_i, x_k) being in G_p^*. There is a one-to-one equivalence between matchings in $\hat{G}$ and degree-feasible partial graphs of G. However, since the number of vertices of $\hat{G}$ is much larger than n, the above transformation from G to $\hat{G}$ is computationally inefficient. In fact, it is unnecessary to construct $\hat{G}$ explicitly, and it is quite possible [13] to modify the algorithm of Section 3.1.1 so that it operates on the original graph G directly, with $\hat{G}$ being implied.

5.1 The transportation problem

The general problem for a bipartite graph $G = (X^a \cup X^b, A)$ is well known as the transporation problem (TP) [8, 16, 18, 25, 4, 21], and is a generalization of the AP discussed in Section 4. The TP receives its name from the following interpretation. If there are N sources (supply points) and M sinks (demand points), with source i capable of supplying amount α_i and sink k having demand β_k, with

$$\sum_{i=1}^{N} \alpha_i = \sum_{k=1}^{M} \beta_k,$$

find the least cost transportation pattern from the sources to the sinks, when the cost of transporting a unit amount from i to k is c_{ik}.

When the α_i and β_k are integers, the TP can also be considered as a general matching problem in a bipartite graph, by first forming the graph $\hat{G}$ as explained earlier and then solving the AP for this graph. Alternatively, the TP could be considered as a network flow problem in a way exactly analogous to that explained for the AP in Section 4.

Because of the practical importance of the TP, we will state the matrix form of the hungarian algorithm for this problem. The algorithm is very similar to that given in Section 4.1.2 for the AP and no further explanations are given here.

5.1.1 THE HUNGARIAN ALGORITHM FOR THE TP
Initialization
Step 1. Starting with the cost matrix **C**, reduce it to the matrix **C′**.
Find augmenting path
Step 2. Label a row s for which $\alpha_s \neq 0$ with $p_s = 0$. If no such row exists goto step 9.
Step 3. Label every unlabelled column k containing a 0 in a labelled row i (i.e. having $c'_{ik} = 0$), with $p_k = i$. If more than one choice exists choose any one.
Step 4. Label every unlabelled row i containing a marked entry in a labelled column k with $p_i = k$.
Step 5. Repeat steps 3 and 4 until either a column t with $\beta_t \neq 0$ is labelled in which case go to step 6, or until no further labelling is possible in which case goto step 8.
Augment
Step 6. Identify the augmenting path P between s and t and calculate ε as the minimum between the numbers:

$$\min\left[\alpha_s, \beta_t\right] \qquad \text{and} \qquad \min_{\substack{(i, k) \text{ on } P \\ \text{and marked}}} \left[c'_{ik}\right].$$

Step 7. Alternatingly add and subtract (in that order) ε from all (i, k) entries on the alternating path P. Mark those entries where ε was added and remove the marks from any entry whose value was reduced to zero after subtracting ε.

Reduce α_s and β_t by ε. Remove all labels and return to step 2.
Change matrix **C′**
Step 8. Calculate $\Delta = \min_{\substack{i \in I^+ \\ k \in K^-}} \left[c'_{ik}\right]$

and update $c'_{ik} = c'_{ik} - \Delta$ if $i \in I^+$ and $k \in K^-$, $c'_{ik} = c'_{ik} + \Delta$ if $i \in I^-$ and $k \in K^+$ and c'_{ik} unchanged otherwise. Retain the labels and return to step 3.
Step 9. The optimal solution has been found. The amounts are to be transported only along the links that are marked, and if entry (i, k) is marked, c'_{ik} represents the amount to be transported along link (x_i, x_k).

It should be noted that in the above algorithm the matrix **C′** serves two purposes. The entries which are unmarked represent the costs as previously. (Some of these entries may be 0.) All the entries which are marked, however, have zero costs, but the location itself is instead used to store the amount to be transported along the corresponding link. This removes the necessity of using a separate matrix in order to store the amounts to be transported.

5.2 Example
Solve the TP with the cost matrix shown below, where the vectors to the left and top represent the supplies α_i and demands β_k of the sources and sinks

respectively. (The weight vectors $[\pi_i]$ and $[\pi_k]$ are not kept here.)

β_k	2	3	3	5	2

α_i		1	2	3	4	5
5	1	5	3	4	5	6
4	2	2	6	5	3	2
6	3	6	4	3	4	4

The above matrix is reduced to obtain:

β_k	2	3	3	5	2

α_i		1	2	3	4	5
5	1	2	0	1	1	3
4	2	0	4	3	0	0
6	3	3	1	0	0	1

After several applications of steps 2, 3, 6 and 7 with immediate augmentations occurring, the matrix C', the vectors α_i and β_k and the labels become:

β_k	0	0	0	0	2

α_i		1	2	3	4	5		p_i
2	1	2	3*	1	1	3		0
0	2	2*	4	3	2*	0		
0	3	3	1	3*	3*	1		

		p_k
1		

Δ is calculated to be 1 and the new matrix $\mathbf{C}'$ is shown below:

β_k	0	0	0	0	2

α_i		1	2	3	4	5	
2	1	1	3*	0	$\boxed{0}$	2	0
0	2	2*	5	3	$\boxed{2*} \leftarrow \boxed{0}$		4
0	3	3	2	3*	3*	1	3

p_i

		1	1	1	2	p_k

The labels are placed in the following order $p_{\text{col. 3}} = p_{\text{col. 4}} = 1$, $p_{\text{row 3}} = 3$ (or 4), $p_{\text{row 2}} = 4$, $p_{\text{col. 5}} = 2$, so that column 5 with $\beta_5 = 2$ is labelled and an augmenting path (shown above) is found.

ε is calculated to be 2, and after step 7 the matrix is as shown below:

β_k	0	0	0	0	0

α_i		1	2	3	4	5
0	1	1	3*	0	2*	2
0	2	2*	5	3	0	2*
0	3	3	2	3*	3*	1

Step 2 terminates the algorithm since all $\alpha_i = 0$, and the solution is then shown by the marked entries in the above matrix. The minimum cost for this TP is $3 \times 3 + 2 \times 5 + 2 \times 2 + 2 \times 2 + 3 \times 3 + 3 \times 4 = 48$.

6. The Covering Problem

The minimum cardinality covering problem (CCP), has been defined in the Introduction as that of finding a set E of links of the graph $G = (X, A)$, so that the cardinality of E is a minimum and every vertex of G is the terminal of at least one link in E. We will now show that the CCP and the CMP are equivalent by showing that:

THEOREM 6. *If* E^* *is a minimum cardinality covering, and for every vertex* x_i *having degree* $d_i^{E^*} > 1$, *all but one of the links incident at* x_i *are removed, the resulting set of links* M^* *is a maximum cardinality matching.*

Proof. Since with the above construction we have $d_i^{M^*} \leqslant 1$, no two links in M^* are adjacent and M^* is a matching. What we therefore have to show is that M^* is of maximum cardinality. Let us suppose that $|M^*|$ is not maximum. Then, by theorem 1, an augmenting path exists from some exposed (i.e. with $d_i^{M^*} = 0$) vertex x_i to another exposed vertex x_k. Since $d_i^{M^*} = d_k^{M^*} = 0$, it implies that at least one link from x_i to another vertex ($\neq x_k$) has been removed, and conversely for x_k. Let one of the removed links incident at x_i be a_1 and one of the removed links incident at x_k be a_2. Let P be the augmenting path (of cardinality $2p + 1$) from x_i to x_k. If now the links of M^* lying on path P and the two links a_1 and a_2 are removed from E^* and instead the links of path P not in M^* are entered into E^*, the resulting set ($\bar{E}$ say) would still be a covering since every vertex of P would be the terminal of some link in the set E^*. However, $\bar{E}$ has been derived by removing $p + 2$ links from E^* and adding $p + 1$ links, hence $|\bar{E}| < |E^*|$ which contradicts the assumption that E^* is a minimum cardinality covering. Hence M^* constructed according to the theorem is a maximum cardinality matching.

Corresponding to theorem 6 and having an exactly analogous proof [30] is:

THEOREM 7. *If* M^* *is a maximum cardinality matching of a graph* G, *and for every exposed vertex* x_i *a link incident at* x_i *is added, the resulting set of links* E^* *is a minimum cardinality covering of* G.

The general CP with costs associated with the links of G, can be solved by an algorithm very similar to that for the MP and will not be discussed here further. A detailed algorithm for the CP is given by White [30].

7. Problems P12

1. Find the maximum cardinality matching for the graph shown in Fig. 12.26. Hence derive the minimum cardinality covering for this graph.

2. Find the maximum matching of the graph given in Fig. 12.27 where the numbers next to the links are the link costs.

3. Show that in a bipartite graph $G = (X^a \cup X^b, A)$ a perfect matching exists if and only if for every $S \subset X$, $|S| \leqslant |\Gamma(S)|$. (This is known as the theorem of König and Hall [5]).

4. For the graph shown in Fig. 12.28 find any partial graph G_p with vertex degrees given by the vector $[\delta_i] = [1, 2, 1, 1, 2, 3, 1, 3, 2]$.

5. Show that in a matrix having some 0 and some non-zero elements, the minimal

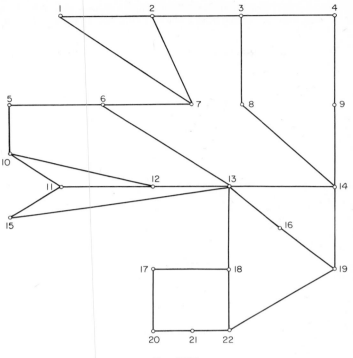

Fig. 12.26

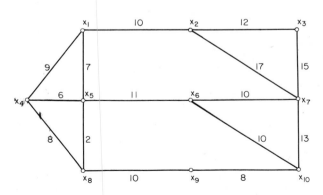

Fig. 12.27

O

number of lines (rows and/or columns) containing all the zeros is equal to the maximum number of zeros situated in separate rows and columns. (This is known as König's theorem [5]).

6. Find the minimum cost solution to the AP whose cost matrix **C** is given by:

$$
\mathbf{C} = \begin{array}{c c} & \begin{array}{cccccc} 1 & 2 & 3 & 4 & 5 & 6 \end{array} \\ \begin{array}{c} 1 \\ 2 \\ 3 \\ 4 \\ 5 \\ 6 \end{array} & \begin{array}{|c|c|c|c|c|c|} \hline 17 & 8 & 12 & 9 & 14 & 6 \\ 15 & 13 & 18 & 15 & 10 & 4 \\ 14 & 16 & 8 & 12 & 5 & 9 \\ 18 & 7 & 14 & 9 & 11 & 13 \\ 7 & 16 & 11 & 14 & 6 & 10 \\ 18 & 9 & 13 & 5 & 15 & 7 \\ \hline \end{array} \end{array}
$$

7. Define the minimax AP to be the problem of finding an assignment in a matrix **C** so that the maximum cost of an entry in this assignment is as small as possible. Solve the minimax AP for the matrix **C** of problem 6.

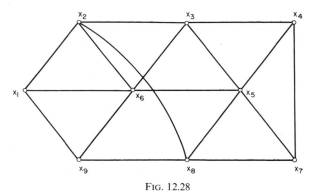

FIG. 12.28

8. Solve the transportation problem given below, where the 4 × 5 matrix is the cost matrix **C**, the column vector is the vector of supplies and the row vector is the vector of demands.

	3	4	2	7	6
9	17	18	12	9	14
6	15	13	18	15	10
5	14	16	8	12	5
7	18	7	14	9	11

(Note: Convert the problem into an equivalent problem in which the sum of demands is equal to the sum of supplies).

9. Consider the transportation problem with cost matrix **C** as shown and with the supply and demands given by the vectors

$$[\alpha_i]^1 = [10, 15, 20]$$
$$[\beta_k]^1 = [5, 12, 13, 15]$$

	1	2	3	4
C = 1	25	10	2	30
2	5	15	20	10
3	100	65	0	2

The optimal solution to this problem is given by the shipments shown in the top left hand corners of the following matrix.

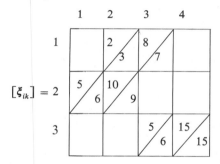

The cost of this solution is 241 and the total amount transported is 45 units.

Consider now the problem with the same cost matrix but with supply and demands given by the vectors:

$$[\alpha_i]^2 = [10, 15, 21]$$
$$[\beta_k]^2 = [6, 12, 13, 15]$$

The optimal solution is now given by the shipments shown in the bottom right hand corners of matrix $[\xi_{ik}]$ above.

The cost of this solution is 239 and the total amount transported is 46 units.

Explain how it is possible to transport a larger amount for less total cost for a given cost matrix and particularly note that the arcs along which the shipments are made remain the same. (See Ref. [28]).

8. References

1. Balinski, M. L. (1965), Integer programming; Methods, uses, computations, *Man. Sci.*, **12**, p. 253.
2. Balinski, M. L. (1969), Labelling to obtain a maximum matching, *In*: "Combina-

torial Math. and its Appl.", Bose and Dowling, Eds., Univ. of North Carolina Press, p. 585.

3. Balinski, M. L. (1970), On maximum matching, minimum covering and their connections, *In*: "Proc. Princeton Symp. on Math. Programming", Kahn, Ed., Princeton Univ. Press, p. 303.

4. Balinski, M. L. and Gomory, R. E. (1964). A primal method for the assignment and transportation problems, *Man. Sci.*, **10**, p. 578.

5. Berge, C. (1962), "The theory of graphs", Methuen, London.

6. Bourgeois, F. and Lassalle, J. C. (1971). Algorithm 415—Algorithm for the assignment problem, *Comm. of ACM*, **14**, p. 805.

7. Bourgeois, F. and Lassalle, J. C. (1971), An extension of the Munkres algorithm for the assignment problem to rectangular matrices, *Comm. of ACM*, **14**, p. 802.

8. Dantzig, G. B. (1963), "Liner programming and extensions", Princeton Univ. Press, Princeton, New Jersey.

9. Desler, J. F. and Hakimi, S. L. (1969), A graph-theoretic approach to a class of integer programming problems, *Ops. Res.*, **17**, p. 1017.

10. Edmonds, J. (1965), Paths, trees and flowers, *Canadian Math. Jl.*, **17**, p. 449.

11. Edmonds, J. (1965), Maximum matchings and a polyhedron with 0–1 vertices, *Jl. of Res., Nat. Bur. of Stand.*, **69B**, p. 125.

12. Edmonds, J. (1967), An introduction to matchings, *In*: "Summer seminar on mathematics of the decision sciences", Stanford University.

13. Edmonds, J. and Johnson, E. L. (1970), Matching: A well solved class of integer linear programs, *In*: "Proc. Calgary Intl. Conf. on Combinatorial Structures and their Appl.", Gordon and Breach, New York, p. 89.

14. Edmonds, J. (1962). Covers and packings in a family of sets, *Bull. Amer. Math. Soc.*, **68**, p. 494.

15. Egervary, E. (1953), On combinatorial properties of matrices, translated by H. W. Kuhn, Office of Naval Research Logistics Project Report, Dept. Math., Princeton University.

16. Ford, L. R. and Fulherson, D. R. (1962), "Flows in Networks", Princeton Univ. Press, Princeton, New Jersey.

17. Garfinkel, R. S. and Nemhauser, G. L. (1972), "Integer Programming", Wiley, New York.

18. Glover, F. (1967), Maximum matching in a convex bipartite graph, *Nav. Res. Log. Quart.*, **14**, p. 313.

19. Gordon, B. E. (1971), The maximum matching problem—A comparison of the Edmonds and Balinski algorithms, Graduate School of Management, University of Rochester.

20. Ivanescu, P. L. and Rudeanu, S. (1968). A pseudo-boolean approach to matching problems in graphs with applications to assignment and transportation problems, *In*: Théorie des Graphes", Dunod, Paris.

21. Klein, M. (1967). A primal method for minimal cost flows with applications to the assignment and transportation problems, *Man. Sci.*, **14**, p. 205.

22. Konig, D. (1950), "Theorie der endlichen und unendlichen Graphen", Chelsea, New York.

23. Kuhn, H. W. (1955), The Hungarian method for the assignment problem, *Nav. Res. Log. Quart.*, **2**, p. 83.

24. Kuhn, H. W. (1956). Variants of the Hungarian method for the assignment problem, *Nav. Res. Log. Quart.,* **3**, p. 253.

25. Morrison, D. R. (1969). Matching algorithms, *Jl. of Combinatorial Theory*, **6**, p. 20.
26. Munkres, J. (1957). Algorithms for the assignment and transportation problems, *Jl. of SIAM (Appl. Math.)*, **5**, p. 32.
27. Norman, R. Z. and Rabin, M. O. (1959). An algorithm for a minimum cover of a graph, *Proc. Amer. Math. Soc.*, **10**, p. 315.
28. Szwarc, W. (1970). The transportation paradox, Report 199, Carnegie—Mellon University, Pittsburg, Pennsylvania.
29. Thacker, B. G. (1972). Matchings in weighted graphs, M.Sc. Thesis, Imperial College, London University.
30. White, L. J. (1967), A parametric study of matchings and covering in weighted graphs, Ph.D. Thesis, University of Michigan.
31. Witzgall, C. and Zahn, C. T. Jr. (1965). Modification of Edmonds' maximum matching algorithm, *Jl. of Res., Nat. Bur. of Stand.*, **69B**, p. 91.
32. Yaspan, A. (1966). On finding a maximal assignment, *Ops. Res.*, **14**, p. 646.

Appendix I

Decision-tree Search Methods

1. The Decision-tree Search Principle

The basic principle involved in decision-tree search methods is the partition of an initial problem P_o into a number of subproblems P_1, P_2, ..., P_k, (whose totality represent problem P_o), followed by an attempt to *resolve* each one of these subproblems. By resolve we mean:

either (i) Find an optimal solution,

or (ii) Show that the value of the optimal solution is worse than the best solution obtained so far,

or (iii) Show that the subproblem is infeasible.

This partition is represented by the *tree* of Fig. A.1, where a vertex represents a subproblem.

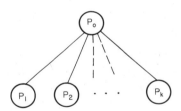

FIG. A.1. Partition of problem P_0 into subproblems

The reason for partitioning a problem P_o into a number of subproblems is that these subproblems are easier to resolve, either because of their smaller size, or because of their structure which may not be shared by the initial problem P_o. However, in general, it may still be impossible to resolve a subproblem P_i and this subproblem is then partitioned further into yet smaller subproblems P_{i_1}, P_{i_2}, ..., P_{i_r} as shown in Fig. A.2. This partitioning, (also called *branching*), is repeated for every subproblem which cannot be resolved.

At any one stage the complete set of subproblems requiring resolution is represented by all the extreme vertices (i.e. vertices of degree 1), on all paths emanating from the root of the decision-tree. (The root of this tree represents the initial problem P_o). These vertices are called *pendant* vertices and in Fig. A.2 they are: $P_1, \ldots, P_{i-1}, P_{i_1}, \ldots, P_{i_r}, P_{i+1}, \ldots, P_k$.

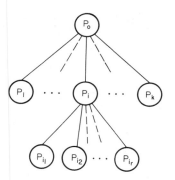

FIG. A.2. Tree after branching from P_i

Obviously, if the search is to be exhaustive, the set of sub-problems into which a problem is partitioned must represent the whole problem. Thus, if problem P_i is partitioned into r subproblems $P_{i_1}, \ldots, P_{i_r}$ we must have:

$$\{P_{i_1}\} \cup \{P_{i_2}\} \cup \ldots \cup \{P_{i_r}\} = \{P_i\} \qquad \text{(A.1)}$$

where $\{P\}$ is used to represent the set of all feasible solutions to problem P.

Since eqn (A.1) must apply for every partition we have:

$$\{P_o\} = \bigcup \{P(j)| P(j): \text{pendant vertex of tree}\} \qquad \text{(A.2)}$$

In situations where it is required to enumerate all solutions to a problem P_o (rather than just find the optimal solution), it is desirable to be able to enumerate these by partitioning the problem into subproblems as above and then enumerate the solutions to each one of the subproblems. What is then required is to have no duplications in the generated solutions, i.e. a problem P_i must be partitioned into subproblems $P_{i_1}, \ldots, P_{i_r}$ in such a way so that

$$\{P_{i_s}\} \cap \{P_{i_q}\} = \varnothing \qquad \text{(A.3)}$$

for any two subproblems P_{i_s} and P_{i_q} with $s \neq q$.

Equation (A.3) then defines a proper partition of problem P_i. Although condition (A.3) is not necessary for a valid decision-tree search it is nevertheless computationally very advantageous since then:

(a) For an optimization problem P_o: the optimal solution is a solution to one and only one subproblem represented by a pendant vertex,

(b) For a complete enumeration problem: the union of the sets of solutions to the subproblems represented by the pendant vertices is the complete set of solutions to P_o without duplications.

2. Some Examples of Valid Branching

Consider a problem P_i involving n variables in which some variable ξ can have one of four possible values say a, b, c or d.

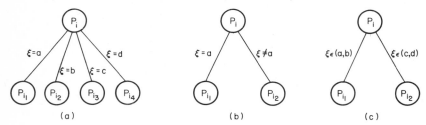

FIG. A.3. Three possible ways of branching from P_i

(a) One possible partition of P_i is into four subproblems P_{i_1}, P_{i_2}, P_{i_3} and P_{i_4}, where for subproblem P_{i_1} we set $\xi = a$, for P_{i_2} we set $\xi = b$, for P_{i_3} we set $\xi = c$ and for P_{i_4} we set $\xi = d$. The subproblems $P_{i_1}, \ldots, P_{i_4}$ each have $n - 1$ variables and hence are easier to solve than problem P_i. (Fig. A.3(a)).

(b) Another possible partition of P_i is into two subproblems P_{i_1} and P_{i_2}, where for P_{i_1} we set $\xi = a$ and for P_{i_2} we set $\xi \neq a$, i.e. ξ is either b, c or d. (Fig. A.3(b)).

(c) Yet another possible partition of P_i is into two subproblems P_{i_1} and P_{i_2}, where for P_{i_1} we set $\xi = a$ or b and for P_{i_2} we set $\xi = c$ or d. (Fig. A.3(c)).

All the above branching possibilities are feasible and satisfy condition A.3. Which one is preferable depends on the nature of the problem being solved, with possibilities of types (a) or (b) being the ones which are most often found to be useful.

3. Types of Decision-tree Search

From what has been said above, any subproblem represented by a pendant vertex and which cannot be resolved, can be branched from at any time by partitioning it into smaller subproblems. There are, however, only two basic types of search depending on the way of choosing the next pendant vertex to be branched from:

3.1 Depth-first search

In this type of search branching is continued from the latest generated subproblem until finally a subproblem is generated which can be resolved. At that point a *backtracking* step is taken, i.e. the last-but-one subproblem generated is chosen and branching continues again from the corresponding vertex. In this type of search problems are stored in a *pushdown stack* with the topmost problem chosen for investigation (resolution or partitioning) at every stage. The newly generated subproblems are put on top of this stack, and when a subproblem is resolved it is removed from the stack. The

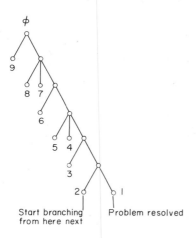

Fig. A.4 (a). Depth-first tree

FIG. A.4 (b). Breadth-first tree
●: Problem resolved

shape of the decision tree when the first subproblem is resolved with this type of search is shown in Fig. A.4(a), where the order of priority for investigation amongst the existing subproblems at this stage is indicated by the numbering.

3.2 Breadth-first search

In breadth-first search the branching proceeds from level to level so that if the initial problem P_o is partitioned into subproblems $P_1, P_2, \ldots, P_k$ at level 1, each one of these subproblems is investigated before the investigation of any subproblem at level 2. The subproblems at level 1 which could not be resolved, are partitioned into subproblems at level 2 and, once more, all of these are investigated before investigating any subproblems that may be produced at level 3 etc. The shape of the decision-tree resulting from this type of search is shown in Fig. A.4(b).

4. Use of Bounds

If the problem P_o to be solved is an optimization problem, then regardless of the type of search, the search is complete only when all the subproblems which are represented by pendant vertices are resolved. In order to speed up this resolution process, lower or upper bounds (for the cases of minimization or maximization problems respectively), are calculated for each of the pendant vertices. The bound represents the lowest (or highest) possible value of an optimal solution to the subproblem corresponding to the pendant vertex. Thus, (for a minimization problem), if the lower bound for a vertex corresponding to P_i happens to be greater than the value of the best answer obtained in the search so far, then it is unnecessary to consider further branching from P_i since no solution in $\{P_i\}$ can be better than the currently best answer. Subproblem P_i is then automatically resolved according to (ii) of section 1.

5. Branching Functions

In both the depth-first and breadth-first searches the choice of which vertex to branch from next has not been completely specified.

For the depth-first search where a problem P_i has just been branched from to produce subproblems $P_{i_1}, \ldots, P_{i_r}$, then we have already said that the next branching must be from one of these last-generated subproblems; we did not, however, specify which one, and any of these could be considered as "last-generated". For the breadth-first search we have already said that all subproblems at a given level must be investigated before proceeding to the next level but did not specify in which order they are to be investigated.

The *branching function* is a function that calculates which one of the allowable vertices should be branched from next. For a vertex corresponding to subproblem P_j this function is some measure of the probability that the optimal solution to the whole problem P_o is a solution to P_j. Quite obviously, that vertex corresponding to the subproblem with the greatest chance of containing the optimal solution may be the best choice for the next branching. Several heuristic measures of this probability can be used, a useful one being simply related to the lower or upper bounds calculated for the vertices. With such a measure, a vertex with a low lower bound (for the minimization case) is considered as having a high probability.

By using a branching function another type of decision-tree search immediately suggests itself in addition to the depth and breadth-first searches described earlier. Thus, one could use the branching function to completely choose which vertex to branch from next. For example, if the vertex bounds are also used as the branching function as mentioned earlier, then one could always branch from that pendant vertex whose lower bound is smallest. This type of search is, in general, a hybrid between the depth and breadth-first searches, although in the literature it is often referred to as breadth-first.

Index

396